Undergraduate Texts in Mathematics

Springer
New York
Berlin
Heidelberg
Barcelona
Hong Kong
London
Milan
Paris
Singapore
Tokyo

Undergraduate Texts in Mathematics

Abbott: Understanding Analysis.
Anglin: Mathematics: A Concise History and Philosophy.
Readings in Mathematics.
Anglin/Lambek: The Heritage of Thales.
Readings in Mathematics.
Apostol: Introduction to Analytic Number Theory. Second edition.
Armstrong: Basic Topology.
Armstrong: Groups and Symmetry.
Axler: Linear Algebra Done Right. Second edition.
Beardon: Limits: A New Approach to Real Analysis.
Bak/Newman: Complex Analysis. Second edition.
Banchoff/Wermer: Linear Algebra Through Geometry. Second edition.
Berberian: A First Course in Real Analysis.
Bix: Conics and Cubics: A Concrete Introduction to Algebraic Curves.
Brémaud: An Introduction to Probabilistic Modeling.
Bressoud: Factorization and Primality Testing.
Bressoud: Second Year Calculus.
Readings in Mathematics.
Brickman: Mathematical Introduction to Linear Programming and Game Theory.
Browder: Mathematical Analysis: An Introduction.
Buchmann: Introduction to Cryptography.
Buskes/van Rooij: Topological Spaces: From Distance to Neighborhood.
Callahan: The Geometry of Spacetime: An Introduction to Special and General Relavitity.
Carter/van Brunt: The Lebesgue–Stieltjes Integral: A Practical Introduction.
Cederberg: A Course in Modern Geometries. Second edition.
Childs: A Concrete Introduction to Higher Algebra. Second edition.
Chung: Elementary Probability Theory with Stochastic Processes. Third edition.
Cox/Little/O'Shea: Ideals, Varieties, and Algorithms. Second edition.
Croom: Basic Concepts of Algebraic Topology.
Curtis: Linear Algebra: An Introductory Approach. Fourth edition.
Devlin: The Joy of Sets: Fundamentals of Contemporary Set Theory. Second edition.
Dixmier: General Topology.
Driver: Why Math?
Ebbinghaus/Flum/Thomas: Mathematical Logic. Second edition.
Edgar: Measure, Topology, and Fractal Geometry.
Elaydi: An Introduction to Difference Equations. Second edition.
Exner: An Accompaniment to Higher Mathematics.
Exner: Inside Calculus.
Fine/Rosenberger: The Fundamental Theory of Algebra.
Fischer: Intermediate Real Analysis.
Flanigan/Kazdan: Calculus Two: Linear and Nonlinear Functions. Second edition.
Fleming: Functions of Several Variables. Second edition.
Foulds: Combinatorial Optimization for Undergraduates.
Foulds: Optimization Techniques: An Introduction.
Franklin: Methods of Mathematical Economics.
Frazier: An Introduction to Wavelets Through Linear Algebra.
Gamelin: Complex Analysis.
Gordon: Discrete Probability.
Hairer/Wanner: Analysis by Its History.
Readings in Mathematics.
Halmos: Finite-Dimensional Vector Spaces. Second edition.

(continued after index)

Anthony L. Peressini
Francis E. Sullivan
J. J. Uhl, Jr.

The Mathematics of Nonlinear Programming

With 66 Illustrations

Springer

Anthony L. Peressini
J. J. Uhl, Jr.
Department of Mathematics
University of Illinois
Urbana, IL 61801
USA

Francis E. Sullivan
Supercomputing Research Center
17100 Science Drive
Bowie, MD 20715-4300
USA

Mathematics Subject Classification (2000): 49-01, 90C25

Library of Congress Cataloging-in-Publication Data
Peressini, Anthony L.
The mathematics of nonlinear programming
(Undergraduate texts in mathematics)
Includes index.
1. Mathematical optimizations. 2. Nonlinear programming.
I. Sullivan, Francis E. I. Uhl, J. Jerry. III. Title. IV. Series.
QA402.5.P42 1988 519.7′6 87-23411

Printed on acid-free paper.

Typeset by Asco Trade Typesetting Ltd., Hong Kong.
Printed and bound by Braun-Brumfield, Inc., Ann Arbor, MI.
Printed in the United States of America.

9 8 7 6

ISBN 0-387-96614-5
ISBN 3-540-96614-5 SPIN 10797285

Springer-Verlag New York Berlin Heidelberg
A member of BertelsmannSpringer Science+Business Media GmbH

Preface

Existing texts on the mathematics of nonlinear programming seem, for the most part, to be written for advanced students having substantial sophistication and experience in mathematics, numerical analysis, and scientific computing. This is unfortunate because nonlinear programming provides an excellent opportunity to explore an interesting variety of mathematics that is quite accessible to students with some background in advanced calculus and linear algebra. Linear algebra becomes alive only when it is applied outside of its own sphere. Linear algebra is very much alive in this book.

We have endeavored to write an undergraduate mathematics textbook which develops some of the ideas and techniques that have proved to be useful in the optimization of nonlinear functions. This book is written for students with a working knowledge of matrix algebra and advanced calculus, but with no previous assumed acquaintance with modern optimization theory. We attempt to provide such students with a careful, clear development of the mathematical underpinnings of optimization theory and a flavor of some of the basic methods. This background will prepare them to read the more advanced literature in the field and to understand, at least in rough outline, the inner workings of some of the professionally written software that is used by practitioners to solve problems arising in such fields as engineering, statistics, and commerce. Although we have included detailed proofs we have been careful to include plenty of informal discussion of the meaning of these theorems, and we have illustrated their application with numerous examples and exercises. The level of sophistication gradually increases as the text proceeds, but we pay careful attention to the development of the reader's intuition for the content throughout. We have done this so that readers with very modest backgrounds in formal mathematics can gain an understanding of the essential content without studying all of the proofs. Indeed, when we teach this material in our own courses, we sometimes omit detailed proofs and devote the time instead to the discussion of examples or to outlines of the main ideas in these proofs. In this way, and through associated homework exercises, we try to increase gradually the student's appreciation of and facility with the mathematical development.

All of the material in this book has been tested in our introductory mathematics course in nonlinear programming at the University of Illinois and in similar courses at other universities over the last ten years. The content and mode of presentation that evolved worked well with mathematics majors as well as with other students who represent a wide variety of majors from economics to electrical engineering.

We begin with a study of the classical optimization methods of calculus. This leads, in a natural way, to the study of convex sets and convex functions in Chapter 2. Included in this chapter is a treatment of unconstrained geometric programming. In Chapter 3 we consolidate our position with a discussion of basic numerical methods for unconstrained minimization. There is no attempt to be encyclopedic, but we do consider the classical techniques of Newton's Method and the Method of Steepest Descent, and then we proceed to study the more modern approaches of Broyden's Method, the Davidon–Fletcher–Powell Method and the Broyden–Fletcher–Goldfarb–Shanno Method in some detail. In each case, we try to convey to the student heuristic reasons why each of these methods has advantages and, in some cases, disadvantages. Issues from numerical analysis are identified and addressed to some extent, but we do not pursue this important aspect of nonlinear programming in detail. We do think we have gone far enough to make the student aware of some of the numerical issues involved and to lay the groundwork for a more extensive study later on.

Chapter 4 is devoted to least squares approximation. The basic problems here are best least squares solutions to inconsistent (that is, overdetermined) linear systems and minimum norm solutions to underdetermined linear systems. Next, in Chapter 5, the book turns to a study of the Separation Theorem for Convex Sets and a development of the Karush–Kuhn–Tucker Theorem. Our development makes use of the physical meaning of the Karush–Kuhn–Tucker multipliers as a motivation for the proof of the Karush–Kuhn–Tucker Theorem. This theorem is then used to establish duality in convex programming and the general theory of constrained geometric programming.

Penalty functions for constrained problems are the topic of Chapter 6. The general theory is established and then we follow Duffin's amazingly elementary penalty function approach to the Karush–Kuhn–Tucker Theorem.

The last chapter deals with classical Lagrange multipliers and problems with both equality and inequality constraints. Wolfe's Algorithm for quadratic programs is studied at the conclusion of the chapter.

The content and organization of this text allow the instructor considerable flexibility in the presentation of the material. Our one-semester introductory course in nonlinear programming usually covers the material in the unstarred sections of the first six chapters. However, we have also taught the content of Chapters 1, 2, 3, 7, and 6 in that order in the same course with good results.

None of us has done research in nonlinear programming, but nevertheless we have enjoyed the blend of mathematics found in this book and we hope that you do too.

A number of friends and colleagues have helped us directly, or indirectly, or inadvertently. Some of them are Tom Morley, Dennis Karney, Lawrence Riddle, Robert Bartle, Tenney Peck, Donald Sherbert, Shih-Ping Han, Paul Boggs, Horacio Porta, James Burke, and James Crenshaw. We are also grateful for the skillful assistance of Cherri Davison, who typed the final manuscript, as well as Mabel Jones and Sandy McGary, who typed earlier drafts.

Urbana, Illinois A. L. P.
and Gaithersburg, Maryland F. E. S.
June 1987 J. J. U.

Table of Contents

CHAPTER 1

Unconstrained Optimization via Calculus

1.1. Functions of One Variable

The object of this section is to review the fundamental results of calculus related to the optimization of real-valued functions defined on the real line, R, or some interval, I, of the real line. These results are based on the following theorem from calculus, known as Taylor's Formula or the Extended Law of the Mean:

(1.1.1) Theorem. *Suppose that $f(x)$, $f'(x)$, $f''(x)$ exist on the closed interval $[a, b] = \{x \in R: a \leq x \leq b\}$. If x^*, x are any two different points of $[a, b]$, then there exists a point z strictly between x^* and x such that*

$$f(x) = f(x^*) + f'(x^*)(x - x^*) + \frac{f''(z)}{2}(x - x^*)^2.$$

Here is an indication why this formula is useful for optimization: If $f(x)$ is a function such that $f''(x)$ is positive for all real x, and if x^* is a point such that $f'(x^*) = 0$, then Taylor's Formula tells us that

$$f(x) = f(x^*) + 0 + \text{a positive number}$$

for all real numbers $x \neq x^*$. Hence $f(x) > f(x^*)$ for all $x \neq x^*$, that is, x^* is the point that minimizes the value of $f(x)$. Essentially the same reasoning shows that if $f''(x)$ is always negative and $f'(x^*) = 0$, then x^* is the point that maximizes the value of $f(x)$. This simple observation, which is essentially the Second Derivative Test, forms the basis for the entire development of this chapter. Here is an example of an application.

(1.1.2) Example. If $f(x) = e^{x^2}$, then $f'(x) = 2xe^{x^2}$ and $f''(x) = 4x^2e^{x^2} + 2e^{x^2} = (4x^2 + 2)e^{x^2}$. Since $f''(x) > 0$ for all real x and since $f'(0) = 0$, we learn that $f(0) = 1$ is smaller than any other value of $f(x)$.

Let us fix some terminology.

(1.1.3) Definitions. Suppose $f(x)$ is a real-valued function defined on some interval I. (The interval I may be finite or infinite, open or closed, or half-open.) A point x^* in I is:

(a) a *global minimizer* for $f(x)$ on I if $f(x^*) \leq f(x)$ for all x in I;
(b) a *strict global minimizer* for $f(x)$ on I if $f(x^*) < f(x)$ for all x in I such that $x \neq x^*$;
(c) a *local minimizer* for $f(x)$ if there is a positive number δ such that $f(x^*) \leq f(x)$ for all x in I for which $x^* - \delta < x < x^* + \delta$;
(d) a *strict local minimizer* for $f(x)$ if there is a positive number δ such that $f(x^*) < f(x)$ for all x in I for which $x^* - \delta < x < x^* + \delta$ and $x \neq x^*$;
(e) a *critical point* of $f(x)$ if $f'(x^*)$ exists and is equal to zero.

Obvious modifications of the preceding definitions yield definitions of *global maximizer*, *strict global maximizer*, *local maximizer*, and *strict local maximizer* of $f(x)$. Because the maximizers of $f(x)$ are simply the minimizers of $-f(x)$, we will concentrate most of our attention on minimizers.

The following theorems summarize the basic facts about minimization of functions of one variable.

(1.1.4) Theorem. *Suppose that $f(x)$ is a differentiable function on an interval I. If x^* is a local minimizer or maximizer of $f(x)$, then either x^* is an endpoint of I or $f'(x^*) = 0$.*

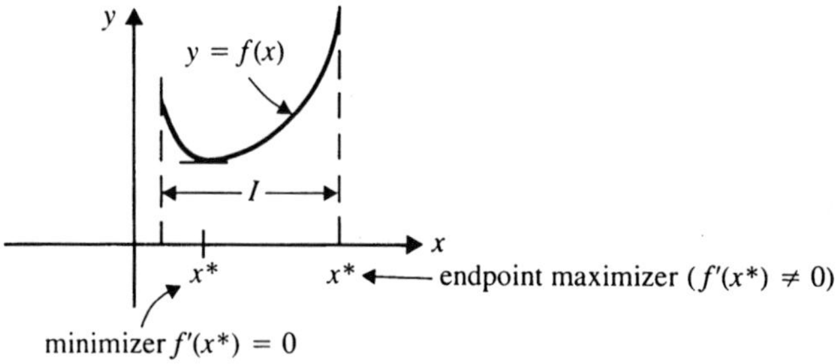

PROOF. Suppose x^* is a local minimizer of $f(x)$ and that x^* is not an endpoint of I. By hypothesis, $f'(x^*)$ exists; we must show that $f'(x^*) = 0$. Recall that

$$f'(x^*) = \lim_{x \to x^*} \frac{f(x) - f(x^*)}{x - x^*}. \tag{1}$$

Since $f(x^*) \leq f(x)$ for x sufficiently close to x^*, the expression $f(x) - f(x^*)$ is nonnegative for all x sufficiently close to x^*. Hence, since $x - x^* > 0$ for $x^* < x$, and $x - x^* < 0$ for $x^* > x$, we see that

$$\frac{f(x) - f(x^*)}{x - x^*} \geq 0 \quad \text{for } x^* < x,$$

$$\frac{f(x) - f(x^*)}{x - x^*} \leq 0 \quad \text{for } x < x^*,$$

provided x is sufficiently close to x^*. These observations, together with (1), show that $f'(x^*) \geq 0$ and $f'(x^*) \leq 0$, that is, $f'(x^*) = 0$.

Once the critical points of a function have been identified, the following result can be used to determine whether these points are minimizers.

(1.1.5) Theorem. *Suppose that $f(x), f'(x), f''(x)$ are all continuous on an interval I and that $x^* \in I$ is a critical point of $f(x)$.*

(a) *If $f''(x) \geq 0$ for all $x \in I$, then x^* is a global minimizer of $f(x)$ on I.*
(b) *If $f''(x) > 0$ for all $x \in I$ such that $x \neq x^*$, then x^* is a strict global minimizer of $f(x)$ on I.*
(c) *If $f''(x^*) > 0$, then x^* is a strict local minimizer of $f(x)$.*

Proof. If $x \in I$ and $x \neq x^*$, then Taylor's Formula (Theorem (1.1.1)) and the hypothesis that $f'(x^*) = 0$ yield

$$f(x) - f(x^*) = \frac{f''(z)}{2}(x - x^*)^2, \tag{2}$$

where z is a point strictly between x^* and x. Consequently, if $f''(x) \geq 0$ for all $x \in I$, then $f(x) \geq f(x^*)$ for all $x \in I$ since $(x - x^*)^2/2 \geq 0$ for all $x \in I$. This proves (a), and an obvious modification of this argument establishes (b). Finally, if $f''(x^*) > 0$, the continuity of $f''(x)$ implies that there is a $\delta > 0$ such that $f''(x) > 0$ for all $x \in I$ such that $x^* - \delta < x < x^* + \delta$. But then (2) shows that $f(x) > f(x^*)$ for all $x \in I$ such that $x \neq x^*$, $x^* - \delta < x < x^* + \delta$, that is, x^* is a strict local minimizer of $f(x)$.

Of course, the test for maximizers corresponding to (1.1.5) can be obtained by replacing the conditions $f''(x) \geq 0$, $f''(x) > 0$, $f''(x^*) > 0$ in (a), (b), (c) by $f''(x) \leq 0$, $f''(x) < 0$, $f''(x^*) < 0$, respectively. Note that in the statement of (1.1.5) and the corresponding result for maximizers, global information about $f''(x)$ yields information about global minimizers and maximizers while local information about $f''(x)$ provides information about local minimizers and maximizers.

(1.1.6) Examples

(a) Consider $f(x) = 3x^4 - 4x^3 + 1$. Since

$$f'(x) = 12x^3 - 12x^2 = 12x^2(x - 1),$$

the only critical points of $f(x)$ are $x = 0$ and $x = 1$. Also, since

$$f''(x) = 36x^2 - 24x = 12x(3x - 2),$$

we see that $f''(0) = 0$ and $f''(1) = 12$. Therefore, $x = 1$ is a strict local minimizer of $f(x)$ by (1.1.5)(c)), but (1.1.5) provides no information about the critical point $x = 0$. To analyze the behavior of $f(x)$ near $x = 0$, we observe that $x^4 < x^3$ for $0 < x < 1$ so that $f(x) < 1$ just to the right of the origin, while $f(x) > 1$ to the left of the origin. Consequently, the critical point $x = 0$ is neither a maximizer nor a minimizer of $f(x)$; rather, it is a "horizontal point of inflection" for $f(x)$. Note that

$$\lim_{x \to +\infty} f(x) = +\infty, \qquad \lim_{x \to -\infty} f(x) = +\infty,$$

so $f(x)$ has no global maximizer on R. The strict local minimizer $x = 1$ is also a strict global minimizer.

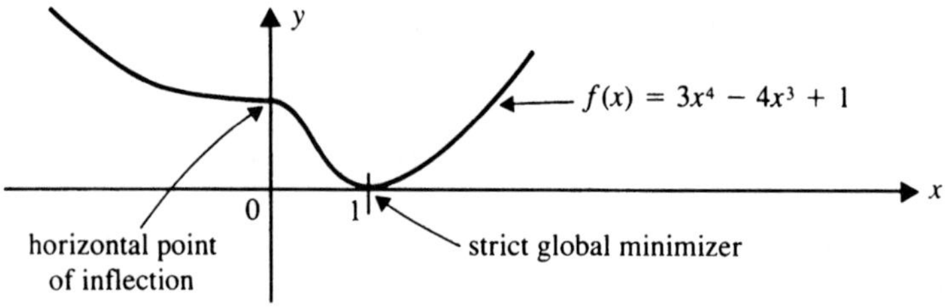

(b) The function $f(x) = \ln(1 - x^2)$ is defined on the interval $I = (-1, +1)$. Since $f'(x) = -2x/(1 - x^2)$, the function $f(x)$ has only one critical point $x = 0$ on I, and this point is a strict global maximizer of $f(x)$ on I since

$$f''(x) = \frac{(1 - x^2)(-2) - (-2x)(-2x)}{(1 - x^2)^2} = \frac{-2(1 + x^2)}{(1 - x^2)^2} < 0$$

for all $x \in I$.

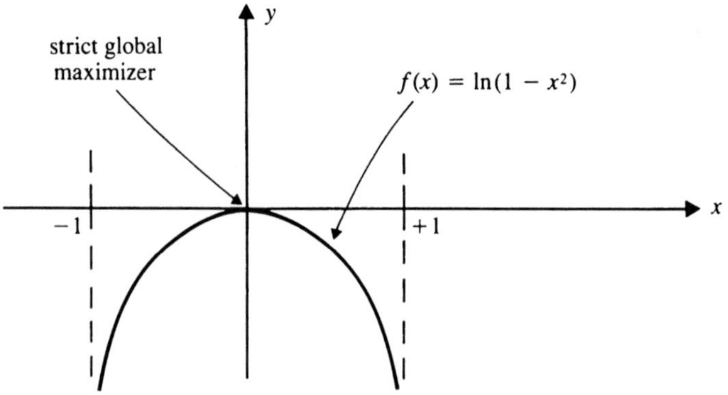

1.2. Functions of Several Variables

The next objective is to extend the results of the preceding section to functions of more than one variable by blending some calculus and linear algebra. We will set the stage for this by reviewing some terminology and notation.

An *n-vector* or *vector in* R^n is an ordered n-tuple $\mathbf{x} = (x_1, x_2, \ldots, x_n)$ of real numbers x_i called the *components* of $\mathbf{x}$. It is very convenient to think of a function $f(x_1, x_2, \ldots, x_n)$ of n variables as a function $f(\mathbf{x})$ of a single vector variable $\mathbf{x} = (x_1, x_2, \ldots, x_n)$.

Although we have described an n-vector as a "row" vector, it is often convenient to think of an n-vector as a "column" vector:

$$\mathbf{x} = \begin{pmatrix} x_1 \\ x_2 \\ \vdots \\ x_n \end{pmatrix}.$$

We will make use of both these interpretations without special comment. Often, it does not matter which interpretation is used, and when it does, the correct interpretation is usually clear from the context.

We define addition of two vectors $\mathbf{x} = (x_1, x_2, \ldots, x_n)$ and $\mathbf{y} = (y_1, y_2, \ldots, y_n)$ in R^n by

$$\mathbf{x} + \mathbf{y} = (x_1 + y_1, x_2 + y_2, \ldots, x_n + y_n),$$

and multiplication of $\mathbf{x}$ and a real number λ by

$$\lambda\mathbf{x} = (\lambda x_1, \lambda x_2, \ldots, \lambda x_n).$$

The set R^n of all n-vectors is a real vector space for these definitions of addition and scalar multiplication. We will assume some familiarity with the basic concepts concerning the vector space R^n such as linear independence and dependence, bases, dimension, subspaces, etc.

If $\mathbf{x} = (x_1, x_2, \ldots, x_n)$ and $\mathbf{y} = (y_1, y_2, \ldots, y_n)$ are vectors in R^n, their *dot product* or *inner product* $\mathbf{x}\cdot\mathbf{y}$ is defined by

$$\mathbf{x}\cdot\mathbf{y} = x_1y_1 + x_2y_2 + \cdots + x_ny_n = \sum_{k=1}^{n} x_ky_k.$$

The dot product is linear in both variables; that is,

$$(\alpha\mathbf{x} + \beta\mathbf{y})\cdot\mathbf{z} = \alpha(\mathbf{x}\cdot\mathbf{z}) + \beta(\mathbf{y}\cdot\mathbf{z}),$$

$$\mathbf{x}\cdot(\alpha\mathbf{y} + \beta\mathbf{z}) = \alpha(\mathbf{x}\cdot\mathbf{y}) + \beta(\mathbf{x}\cdot\mathbf{z}),$$

for all vectors $\mathbf{x}$, $\mathbf{y}$, $\mathbf{z}$ in R^n and all real numbers α, β. Two vectors $\mathbf{x}$ and $\mathbf{y}$ are *orthogonal* if $\mathbf{x}\cdot\mathbf{y} = 0$.

The *norm* or *length* $\|\mathbf{x}\|$ of a vector $\mathbf{x} = (x_1, x_2, \ldots, x_n)$ in R^n is defined by

$$\|\mathbf{x}\| = (x_1^2 + x_2^2 + \cdots + x_n^2)^{1/2} = (\mathbf{x}\cdot\mathbf{x})^{1/2}.$$

The norm is a real-valued function on R^n with the following properties:

(1) $\|\mathbf{x}\| \geq 0$ for all vectors $\mathbf{x}$ in R^n.
(2) $\|\mathbf{x}\| = 0$ if and only if $\mathbf{x}$ is the zero vector $\mathbf{0}$.
(3) $\|\alpha\mathbf{x}\| = |\alpha|\,\|\mathbf{x}\|$ for all vectors $\mathbf{x}$ in R^n and all real numbers α.
(4) $\|\mathbf{x}+\mathbf{y}\| \leq \|\mathbf{x}\| + \|\mathbf{y}\|$ for all vectors $\mathbf{x}$, $\mathbf{y}$ in R^n (the Triangle Inequality).
(5) $|\mathbf{x}\cdot\mathbf{y}| \leq \|\mathbf{x}\|\,\|\mathbf{y}\|$ for all vectors $\mathbf{x}$, $\mathbf{y}$ in R^n with equality holding in this inequality if and only if one vector is a multiple of the other (the Cauchy–Schwarz Inequality).

For nonzero vectors $\mathbf{x}$ and $\mathbf{y}$ in R^2 or R^3, the dot product $\mathbf{x}\cdot\mathbf{y}$ is usually defined by

$$\mathbf{x}\cdot\mathbf{y} = \|\mathbf{x}\|\,\|\mathbf{y}\| \cos\theta, \tag{6}$$

where θ is the angle in the range $[0, \pi]$ between $\mathbf{x}$ and $\mathbf{y}$.

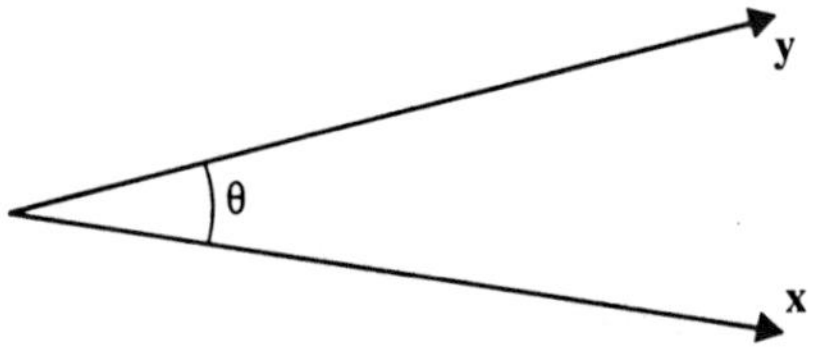

For vectors $\mathbf{x}$ and $\mathbf{y}$ in R^n with $n > 3$, formula (6) for the dot product is still correct if we define $\cos\theta$ properly. For $\mathbf{x}, \mathbf{y} \in R^n$, *define*

$$\cos\theta = \frac{\mathbf{x}\cdot\mathbf{y}}{\|\mathbf{x}\|\,\|\mathbf{y}\|}.$$

By the Cauchy–Schwarz inequality, $-1 \leq \cos\theta \leq 1$ and $\cos\theta = 1$ if and only if one vector is a positive multiple of the other. The Cauchy–Schwarz Inequality is usually proved in most linear algebra courses. A proof of this result can be found immediately after the proof of Corollary 2.6.2.

If $\mathbf{x}$ and $\mathbf{y}$ are vectors in R^n, the *distance* $d(\mathbf{x}, \mathbf{y})$ between $\mathbf{x}$ and $\mathbf{y}$ is defined by

$$d(\mathbf{x}, \mathbf{y}) = \|\mathbf{x}-\mathbf{y}\| = \left(\sum_{i=1}^{n} (x_i - y_i)^2\right)^{1/2}.$$

The *ball* $B(\mathbf{x}, r)$ *centered at* $\mathbf{x}$ *of radius* r is the set of all vectors $\mathbf{y}$ in R^n whose distance from $\mathbf{x}$ is less than r, that is,

$$B(\mathbf{x}, r) = \{\mathbf{y} \in R^n\colon \|\mathbf{y}-\mathbf{x}\| < r\}.$$

Note that in R^1, the ball $B(\mathbf{x}, r)$ is just the open interval $(\mathbf{x} - r, \mathbf{x} + r)$ centered at $\mathbf{x}$ of length $2r$; in R^2, $B(\mathbf{x}, r)$ is the interior of the circle centered at $\mathbf{x}$ of radius r; in R^3, $B(\mathbf{x}, r)$ is the interior of the sphere centered at $\mathbf{x}$ of radius r.

A point $\mathbf{x}$ in a subset D of R^n *is an interior point of* D if there is an $r > 0$ such that the ball $B(\mathbf{x}, r)$ is contained in D. The *interior* D^0 of D is the set of all interior points of D. A set G in R^n is *open* if $G^0 = G$, that is, if all of its points are interior points. A set F in R^n is *closed* if F contains every point $\mathbf{x}$ for which there is a sequence $\{\mathbf{x}^{(k)}\}$ of points in F with

$$\lim_k \|\mathbf{x}^{(k)} - \mathbf{x}\| = 0.$$

It is not difficult to verify that a set F in R^n is closed if and only if its complement $G = F^c$ in R^n is open.

A set D in R^n is *bounded* if there is a constant $M > 0$ such that $\|\mathbf{x}\| < M$ for all $\mathbf{x} \in D$, that is, D is bounded if and only if D is contained in some suitably large ball $B(\mathbf{0}, M)$ centered at $\mathbf{0}$.

Subsets of R^n that are both closed and bounded are called *compact* sets. Any finite set $F = \{\mathbf{x}^{(1)}, \mathbf{x}^{(2)}, \ldots, \mathbf{x}^{(k)}\}$ in R^n is compact. Also, if F is the set consisting of a convergent sequence $\{\mathbf{x}^{(k)}\}$ and its limit $\mathbf{x}^{(0)}$ (that is, $\lim \|\mathbf{x}^{(k)} - \mathbf{x}^{(0)}\| = 0$), then F is compact. On the other hand, one of the most important features of compact subsets of R^n is embodied in the following *Bolzano–Weierstrass Property*:

If D is a compact subset of R^n, then any sequence $\{\mathbf{x}^{(k)}\}$ in D contains a subsequence $\{\mathbf{x}^{(k_j)}\}$ that converges to a point $\mathbf{x}^{(0)}$ in D.

(1.2.1) Examples

(a) In R^2, the set G of points inside the first quadrant, that is,

$$G = \{\mathbf{x} = (x_1, x_2) \in R^2 \colon x_1 > 0, x_2 > 0\},$$

is open but not bounded.

(b) In R^2, the set F of points with nonnegative components, that is,

$$F = \{\mathbf{x} = (x_1, x_2) \in R^2 \colon x_1 \geq 0, x_2 \geq 0\},$$

is closed but not bounded. A point $\mathbf{x} = (x_1, x_2)$ of F is an interior point of F if and only if $x_1 > 0$, $x_2 > 0$. (Note that if $\mathbf{x} = (x_1, x_2)$ and $x_1 > 0$, $x_2 > 0$, then the ball $B(\mathbf{x}, r)$ is contained in F whenever r is a positive number smaller than both x_1 and x_2.)

(c) In R^3, any line or plane D is a closed set that is not bounded and does not contain any interior points.

(d) In R^3, the set $D = \{\mathbf{x} = (x_1, x_2, x_3) \in R^3 \colon 0 \leq x_i \leq 1 \text{ for } i = 1, 2, 3\}$ is a compact set. The interior points of D are precisely those $\mathbf{x} = (x_1, x_2, x_3)$ with $0 < x_i < 1$ for all $i = 1, 2, 3$. The following definitions are completely analogous to those in (1.1.3).

(1.2.2) Definitions. Suppose that $f(\mathbf{x})$ is a real-valued function defined on a subset D of R^n. A point $\mathbf{x}^*$ in D is:

(a) a *global minimizer* for $f(\mathbf{x})$ on D if $f(\mathbf{x}^*) \leq f(\mathbf{x})$ for all $\mathbf{x} \in D$;
(b) a *strict global minimizer* for $f(\mathbf{x})$ on D if $f(\mathbf{x}^*) < f(\mathbf{x})$ for all $\mathbf{x} \in D$ such that $\mathbf{x} \neq \mathbf{x}^*$;
(c) a *local minimizer* for $f(\mathbf{x})$ if there is a positive number δ such that $f(\mathbf{x}^*) \leq f(\mathbf{x})$ for all $\mathbf{x} \in D$ for which $\mathbf{x} \in B(\mathbf{x}^*, \delta)$;

(d) a *strict local minimizer* for $f(\mathbf{x})$ if there is a positive number δ such that $f(\mathbf{x}^*) < f(\mathbf{x})$ for all $\mathbf{x} \in D$ for which $\mathbf{x} \in B(\mathbf{x}^*, \delta)$ and $\mathbf{x} \neq \mathbf{x}^*$;

(e) a *critical point* for $f(\mathbf{x})$ if the first partial derivatives of $f(\mathbf{x})$ exist at $\mathbf{x}^*$ and

$$\frac{\partial f}{\partial x_i}(\mathbf{x}^*) = 0, \qquad i = 1, 2, \ldots, n.$$

These definitions can be modified in an obvious way to yield definitions of *global maximizer*, *strict global maximizer*, *local maximizer*, and *strict local maximizer* for $f(\mathbf{x})$. For the sake of simplicity, we will often limit our discussion to minimizers and leave to the reader the minor task of interpreting the results for maximization problems by replacing $f(\mathbf{x})$ by $-f(\mathbf{x})$.

The following theorem is the analog for functions of several variables of (1.1.4). Note that the proof reduces the consideration of functions of several variables to the case of functions of one variable.

(1.2.3) Theorem. *Suppose that $f(\mathbf{x})$ is a real-valued function for which all first partial derivatives of $f(\mathbf{x})$ exist on a subset D of R^n. If $\mathbf{x}^*$ is an interior point of D that is a local minimizer of $f(\mathbf{x})$, then $\mathbf{x}^*$ is a critical point of $f(\mathbf{x})$, that is, $(\partial f/\partial x_i)(\mathbf{x}^*) = 0$ for $i = 1, 2, \ldots, n$.*

PROOF. Since $\mathbf{x}^*$ is a local minimizer for $f(\mathbf{x})$ and an interior point of D, there is a positive number r such that the ball $B(\mathbf{x}^*, r)$ is contained in D and $f(\mathbf{x}^*) \leq f(\mathbf{x})$ for all $\mathbf{x} \in B(\mathbf{x}^*, r)$. We will show that $(\partial f/\partial x_1)(\mathbf{x}^*) = 0$; the proof that $(\partial f/\partial x_i)(\mathbf{x}^*) = 0$ for $i = 2, \ldots, n$ is entirely similar.

To this end, note that the function $g(x)$ of one variable defined by

$$g(x) = f(x, x_2^*, x_3^*, \ldots, x_n^*)$$

is differentiable and satisfies $g(x_1^*) \leq g(x)$ for all x such that $x_1^* - r < x < x_1^* + r$. Hence, x_1^* is a local minimizer for $g(x)$ on $I = (x_1^* - r, x_1^* + r)$. Consequently, since x_1^* is not an endpoint of I, it follows from (1.1.4) that $g'(x_1^*) = 0$. But

$$g'(x_1^*) = \frac{\partial f}{\partial x_1}(x_1^*, x_2^*, \ldots, x_n^*) = \frac{\partial f}{\partial x_1}(\mathbf{x}^*)$$

so $(\partial f/\partial x_1)(\mathbf{x}^*) = 0$, which is the result we set out to prove.

The preceding proof illustrates an important idea. Often, seemingly difficult facts about functions of several variables can be easily derived by reducing them to corresponding facts about functions of one variable. We will make further use of this idea.

Our minimizer test (1.1.5) for functions of one variable was based on Taylor's Formula:

$$f(x) = f(x^*) + f'(x^*)(x - x^*) + \frac{f''(z)}{2}(x - x^*)^2,$$

where z is a point strictly between x^* and x. If we can determine the corresponding formula for functions of several variables, then it should be possible to use it to develop tests for minimizers of functions of several variables. We will now show that the appropriate version of Taylor's Formula for functions of several variables can be obtained by reduction to the single variable case. We begin by considering the case of a function of two variables.

Suppose that $f(\mathbf{x}) = f(x_1, x_2)$ is a function defined on R^2 and that $\mathbf{x}^* = (x_1^*, x_2^*)$ and $\mathbf{x} = (x_1, x_2)$ are fixed points. Define $\varphi(t)$ for $t \in R$ by

$$\varphi(t) = f(\mathbf{x}^* + t(\mathbf{x} - \mathbf{x}^*)) = f(x_1^* + t(x_1 - x_1^*), x_2^* + t(x_2 - x_2^*)).$$

Then $\varphi(t)$ is a function of a single variable t such that

$$\varphi(0) = f(\mathbf{x}^*) = f(x_1^*, x_2^*); \qquad \varphi(1) = f(\mathbf{x}) = f(x_1, x_2).$$

Consequently, if $\varphi'(t)$ and $\varphi''(t)$ are continuous, we can apply Taylor's Formula to $\varphi(t)$ at the points $t^* = 0$, $t = 1$ to obtain

$$f(\mathbf{x}) = f(\mathbf{x}^*) + \varphi'(0)(1 - 0) + \frac{\varphi''(s)}{2}(1 - 0)^2, \tag{7}$$

where s is a point between 0 and 1. Moreover, if $f(\mathbf{x})$ has continuous first and second partial derivatives, then $\varphi(t)$ has continuous first and second derivatives which can be computed by the Chain Rule as follows: If $t \in R$ and $\mathbf{w} = \mathbf{x}^* + t(\mathbf{x} - \mathbf{x}^*)$, then

$$\varphi(t) = f(\mathbf{w}) = f(x_1^* + t(x_1 - x_1^*), x_2^* + t(x_2 - x_2^*)).$$

According to the Chain Rule,

$$\begin{aligned} \varphi'(t) &= \frac{\partial f}{\partial x_1}(\mathbf{w})(x_1 - x_1^*) + \frac{\partial f}{\partial x_2}(\mathbf{w})(x_2 - x_2^*) \\ &= \nabla f(\mathbf{w}) \cdot (\mathbf{x} - \mathbf{x}^*), \end{aligned} \tag{8}$$

where $\nabla f(\mathbf{w}) = ((\partial f/\partial x_1)(\mathbf{w}), (\partial f/\partial x_2)(\mathbf{w}))$ is the gradient of $f(\mathbf{x})$ evaluated at $\mathbf{w}$. By making use of the Chain Rule again, we obtain

$$\begin{aligned} \varphi''(t) &= \frac{\partial}{\partial x_1}\left[\frac{\partial f}{\partial x_1}(\mathbf{w})(x_1 - x_1^*) + \frac{\partial f}{\partial x_2}(\mathbf{w})(x_2 - x_2^*)\right](x_1 - x_1^*) \\ &\quad + \frac{\partial}{\partial x_2}\left[\frac{\partial f}{\partial x_1}(\mathbf{w})(x_1 - x_1^*) + \frac{\partial f}{\partial x_2}(\mathbf{w})(x_2 - x_2^*)\right](x_2 - x_2^*) \\ &= \frac{\partial^2 f}{\partial x_1^2}(\mathbf{w})(x_1 - x_1^*)^2 + 2\frac{\partial^2 f}{\partial x_1\, \partial x_2}(\mathbf{w})(x_1 - x_1^*)(x_2 - x_2^*) \\ &\quad + \frac{\partial^2 f}{\partial x_2^2}(\mathbf{w})(x_2 - x_2^*)^2. \end{aligned}$$

(In obtaining the last equation, we made use of the fact that the "cross partials" $(\partial^2 f/\partial x_1\, \partial x_2)(\mathbf{w})$ and $(\partial^2 f/\partial x_2\, \partial x_1)(\mathbf{w})$ are equal since $f(\mathbf{x})$ has continuous second partial derivatives.) The preceding formula for $\varphi''(t)$ can be expressed in matrix form as

$$\varphi''(t) = (x_1 - x_1^*, x_2 - x_2^*)\cdot\begin{pmatrix} \dfrac{\partial^2 f}{\partial x_1^2}(\mathbf{w}) & \dfrac{\partial^2 f}{\partial x_1\,\partial x_2}(\mathbf{w}) \\ \dfrac{\partial^2 f}{\partial x_2\,\partial x_1}(\mathbf{w}) & \dfrac{\partial^2 f}{\partial x_2^2}(\mathbf{w}) \end{pmatrix}\begin{pmatrix} x_1 - x_1^* \\ x_2 - x_2^* \end{pmatrix}$$

$$= (\mathbf{x} - \mathbf{x}^*)\cdot(Hf(\mathbf{w})(\mathbf{x} - \mathbf{x}^*)), \tag{9}$$

where $Hf(\mathbf{w})$ is the 2×2-symmetric matrix

$$\left(\frac{\partial^2 f}{\partial x_i\,\partial x_j}(\mathbf{w}) \colon i, j = 1, 2\right)$$

of all second-order partial derivatives evaluated at $\mathbf{w}$. $Hf(\mathbf{w})$ is called the *Hessian* of $f(x)$ evaluated at $\mathbf{w}$.

We can use (8) and (9) to express (7) as follows:

$$f(\mathbf{x}) = f(\mathbf{x}^*) + \nabla f(\mathbf{x}^*)\cdot(\mathbf{x} - \mathbf{x}^*) + \tfrac{1}{2}(\mathbf{x} - \mathbf{x}^*)\cdot Hf(\mathbf{z})(\mathbf{x} - \mathbf{x}^*), \tag{10}$$

where $\mathbf{z} = \mathbf{x}^* + s(\mathbf{x} - \mathbf{x}^*)$ and $0 \le s \le 1$. This is Taylor's Formula for a function of two variables. It is valid for any choice of $\mathbf{x}$ and $\mathbf{x}^*$ in R^2 if $f(\mathbf{x})$ has continuous first and second partial derivatives on R^2. As you can see, the gradient $\nabla f(\mathbf{x}^*)$ plays the role of the first derivative and the Hessian $Hf(z)$ that of the second derivative in the single-variable version of Taylor's Theorem.

The version (10) of Taylor's Formula persists in all higher dimensions. More precisely, if $f(\mathbf{x}) = f(x_1, \ldots, x_n)$ is a function of n variables with continuous first and second partial derivatives on R^n and if the *gradient* ∇f of $f(\mathbf{x})$ is the n-vector

$$\nabla f = \left(\frac{\partial f}{\partial x_1}, \frac{\partial f}{\partial x_2}, \ldots, \frac{\partial f}{\partial x_n}\right),$$

while the *Hessian* Hf of $f(\mathbf{x})$ is the symmetric $n \times n$-matrix

$$Hf = \begin{pmatrix} \dfrac{\partial^2 f}{\partial x_1^2} & \dfrac{\partial^2 f}{\partial x_1\,\partial x_2} & \cdots & \dfrac{\partial^2 f}{\partial x_1 x_n} \\ \dfrac{\partial^2 f}{\partial x_2\,\partial x_1} & \dfrac{\partial^2 f}{\partial x_2^2} & & \vdots \\ \vdots & & & \\ \dfrac{\partial^2 f}{\partial x_n\,\partial x_1} & \dfrac{\partial^2 f}{\partial x_n\,\partial x_2} & \cdots & \dfrac{\partial^2 f}{\partial x_n^2} \end{pmatrix},$$

then Taylor's Formula (10) is valid for all choices of $\mathbf{x}$ and $\mathbf{x}^*$ in R^n. The proof of (10) for functions of n variables is essentially the same as that for functions of two variables.

If the function $f(\mathbf{x})$ is not defined on all of R^n, then Taylor's Formula (10) remains valid for given $\mathbf{x}$ and $\mathbf{x}^*$ in the domain of $f(\mathbf{x})$, provided that $f(\mathbf{x})$ has continuous first and second partial derivatives on some open set containing the "line segment $[\mathbf{x}^*, \mathbf{x}]$ joining $\mathbf{x}^*$ and $\mathbf{x}$," that is,

$$[\mathbf{x}^*, \mathbf{x}] = \{\mathbf{w}: \mathbf{w} = \mathbf{x}^* + t(\mathbf{x} - \mathbf{x}^*); 0 \le t \le 1\}.$$

In particular, if $f(\mathbf{x})$ has continuous second partial derivatives on some ball $B(\mathbf{x}^*, r)$ centered at $\mathbf{x}^*$, then Taylor's Formula (5) is valid for all $\mathbf{x} \in B(\mathbf{x}^*, r)$.

The following result summarizes these observations:

(1.2.4) Theorem. *Suppose that $\mathbf{x}^*$, $\mathbf{x}$ are points in R^n and that $f(\mathbf{x})$ is a function of n variables with continuous first and second partial derivatives on some open set containing the line segment $[\mathbf{x}^*, \mathbf{x}] = \{\mathbf{w} \in R^n: \mathbf{w} = \mathbf{x}^* + t(\mathbf{x} - \mathbf{x}^*); 0 \le t \le 1\}$ joining $\mathbf{x}^*$ and $\mathbf{x}$. Then there exists a $\mathbf{z} \in [\mathbf{x}^*, \mathbf{x}]$ such that*

$$f(\mathbf{x}) = f(\mathbf{x}^*) + \nabla f(\mathbf{x}^*)\cdot(\mathbf{x} - \mathbf{x}^*) + \tfrac{1}{2}(\mathbf{x} - \mathbf{x}^*)\cdot Hf(\mathbf{z})(\mathbf{x} - \mathbf{x}^*).$$

Now that we are armed with Taylor's Formula for functions of several variables we can return to our primary objective—to develop tests for maximizers and minimizers among the critical points of a function. We begin with a straightforward result concerning global maximizers and minimizers.

(1.2.5) Theorem. *Suppose that $\mathbf{x}^*$ is a critical point of a function $f(\mathbf{x})$ with continuous first and second partial derivatives on R^n. Then:*

(a) *$\mathbf{x}^*$ is a global minimizer for $f(\mathbf{x})$ if $(\mathbf{x} - \mathbf{x}^*)\cdot Hf(\mathbf{z})(\mathbf{x} - \mathbf{x}^*) \ge 0$ for all $\mathbf{x} \in R^n$ and all $\mathbf{z} \in [\mathbf{x}^*, \mathbf{x}]$;*
(b) *$\mathbf{x}^*$ is a strict global minimizer for $f(\mathbf{x})$ if $(\mathbf{x} - \mathbf{x}^*)\cdot Hf(\mathbf{z})(\mathbf{x} - \mathbf{x}^*) > 0$ for all $\mathbf{x} \in R^n$ such that $\mathbf{x} \ne \mathbf{x}^*$ and for all $\mathbf{z} \in [\mathbf{x}^*, \mathbf{x}]$;*
(c) *$\mathbf{x}^*$ is a global maximizer for $f(\mathbf{x})$ if $(\mathbf{x} - \mathbf{x}^*)\cdot Hf(\mathbf{z})(\mathbf{x} - \mathbf{x}^*) \le 0$ for all $\mathbf{x} \in R^n$ and all $\mathbf{z} \in [\mathbf{x}^*, \mathbf{x}]$;*
(d) *$\mathbf{x}^*$ is a strict global maximizer for $f(\mathbf{x})$ if $(\mathbf{x} - \mathbf{x}^*)\cdot Hf(\mathbf{z})(\mathbf{x} - \mathbf{x}^*) < 0$ for all $\mathbf{x} \in R^n$ such that $\mathbf{x} \ne \mathbf{x}^*$ and for all $\mathbf{z} \in [\mathbf{x}^*, \mathbf{x}]$.*

PROOF. Since $\mathbf{x}^*$ is a critical point of $f(\mathbf{x})$, the first partial derivatives of $f(\mathbf{x})$ are zero at $\mathbf{x}^*$ so $\nabla f(\mathbf{x}^*) = 0$. Therefore, if $\mathbf{x}$ is any point of R^n other than $\mathbf{x}^*$, (1.2.4) asserts that

$$f(\mathbf{x}) = f(\mathbf{x}^*) + \tfrac{1}{2}(\mathbf{x} - \mathbf{x}^*)\cdot Hf(\mathbf{z})(\mathbf{x} - \mathbf{x}^*),$$

where $\mathbf{z} \in [\mathbf{x}^*, \mathbf{x}]$. This equation yields each of the assertions in the theorem. For example, for (b) we note that

$$f(\mathbf{x}) - f(\mathbf{x}^*) = \tfrac{1}{2}(\mathbf{x} - \mathbf{x}^*)\cdot Hf(\mathbf{z})(\mathbf{x} - \mathbf{x}^*) > 0,$$

and so $f(\mathbf{x}) > f(\mathbf{x}^*)$ for all $\mathbf{x} \in R^n$ such that $\mathbf{x} \ne \mathbf{x}^*$. The remaining assertions of the theorem are verified in a similar way.

Theorem (1.2.5) cannot be regarded as a practical test for global maximizers and minimizers until we have some convenient criteria for determining the sign of

$$(\mathbf{x} - \mathbf{x}^*) \cdot Hf(\mathbf{z})(\mathbf{x} - \mathbf{x}^*).$$

Fortunately, such criteria are available in a somewhat more general context in linear algebra. We will now describe this context and then develop the corresponding criteria in the next section.

We have already observed that the Hessian $Hf(\mathbf{x})$ of a function $f(\mathbf{x})$ of n variables with continuous first and second partial derivatives is an $n \times n$-symmetric matrix. Any $n \times n$-symmetric matrix A determines a function $Q_A(\mathbf{y})$ on R^n called the *quadratic form associated with A*

$$Q_A(\mathbf{y}) = \mathbf{y} \cdot A\mathbf{y}, \qquad \mathbf{y} \in R^n.$$

(1.2.6) Example. If A is the 3×3-symmetric matrix

$$A = \begin{pmatrix} 2 & -1 & 2 \\ -1 & 3 & 0 \\ 2 & 0 & 5 \end{pmatrix},$$

then the quadratic form $Q_A(\mathbf{y})$ associated with A is

$$\begin{aligned} Q_A(\mathbf{y}) = \mathbf{y} \cdot A\mathbf{y} = \mathbf{y} \cdot \left(\begin{pmatrix} 2 & -1 & 2 \\ -1 & 3 & 0 \\ 2 & 0 & 5 \end{pmatrix} \begin{pmatrix} y_1 \\ y_2 \\ y_3 \end{pmatrix} \right) \\ = (y_1, y_2, y_3) \cdot (2y_1 - y_2 + 2y_3, -y_1 + 3y_2, 2y_1 + 5y_3) \\ = 2y_1^2 + 3y_2^2 + 5y_3^2 - 2y_1 y_2 + 4y_1 y_3. \end{aligned}$$

In general, $Q_A(\mathbf{y})$ is a sum of terms of the form $c_{ij} y_i y_j$ where $i, j = 1, \ldots, n$ and c_{ij} is a coefficient which may be zero, that is, every term in $Q_A(\mathbf{y})$ is of second degree in the variables $y_1, y_2, \ldots, y_n$. On the other hand, any function $q(y_1, \ldots, y_n)$ that is the sum of second-degree terms in $y_1, y_2, \ldots, y_n$ can be expressed as the quadratic form associated with an $n \times n$-symmetric matrix A by "splitting" the coefficient of $y_i y_j$ between the (i, j) and (j, i) entries of A.

(1.2.7) Example. The function

$$q(y_1, y_2, y_3) = y_1^2 - y_2^2 + 4y_3^2 - 2y_1 y_2 + 4y_2 y_3$$

is a sum of second-degree terms in the variables y_1, y_2, y_3. If

$$A = \begin{pmatrix} 1 & -1 & 0 \\ -1 & -1 & 2 \\ 0 & 2 & 4 \end{pmatrix},$$

then A is a 3×3-symmetric matrix and it is easy to check that

$$q(y_1, y_2, y_3) = \mathbf{y} \cdot A\mathbf{y} = Q_A(\mathbf{y}), \qquad \mathbf{y} \in R^3.$$

If $f(\mathbf{x})$ is a function of n variables with continuous first and second partial derivatives, and if $H = Hf(\mathbf{z})$ is the Hessian of $f(\mathbf{x})$ evaluated at a point $\mathbf{z}$, then H is an $n \times n$-symmetric matrix. For $\mathbf{x}$, $\mathbf{x}^*$ in R^n, the quadratic form Q_H associated with H evaluated at $\mathbf{x} - \mathbf{x}^*$ is

$$Q_H(\mathbf{x} - \mathbf{x}^*) = (\mathbf{x} - \mathbf{x}^*) \cdot Hf(\mathbf{z})(\mathbf{x} - \mathbf{x}^*).$$

This is precisely the expression that occurs in the statement of (1.2.5). The following definitions introduce the types of sign restrictions on Q_H that occur in that result.

(1.2.8) Definitions. Suppose that A is an $n \times n$-symmetric matrix and that $Q_A(\mathbf{y}) = \mathbf{y} \cdot A\mathbf{y}$ is the quadratic form associated with A. Then A and Q_A are called:

(a) *positive semidefinite* if $Q_A(\mathbf{y}) = \mathbf{y} \cdot A\mathbf{y} \geq 0$ for all $\mathbf{y} \in R^n$;
(b) *positive definite* if $Q_A(\mathbf{y}) = \mathbf{y} \cdot A\mathbf{y} > 0$ for all $\mathbf{y} \in R^n$, $\mathbf{y} \neq \mathbf{0}$;
(c) *negative semidefinite* if $Q_A(\mathbf{y}) = \mathbf{y} \cdot A\mathbf{y} \leq 0$ for all $\mathbf{y} \in R^n$;
(d) *negative definite* if $Q_A(\mathbf{y}) = \mathbf{y} \cdot A\mathbf{y} < 0$ for all $\mathbf{y} \in R^n$, $\mathbf{y} \neq \mathbf{0}$;
(e) *indefinite* if $Q_A(\mathbf{y}) = \mathbf{y} \cdot A\mathbf{y} > 0$ for some $\mathbf{y} \in R^n$ and $Q_A(\mathbf{y}) < 0$ for other $\mathbf{y} \in R^n$.

With this terminology established, we can now reformulate (1.2.5) as follows:

(1.2.9) Theorem. *Suppose that* $\mathbf{x}^*$ *is a critical point of a function* $f(\mathbf{x})$ *with continuous first and second partial derivatives on* R^n *and that* $Hf(\mathbf{x})$ *is the Hessian of* $f(\mathbf{x})$. *Then* $\mathbf{x}^*$ *is:*

(a) *a* global minimizer *for* $f(\mathbf{x})$ *if* $Hf(\mathbf{x})$ *is positive semidefinite on* R^n;
(b) *a* strict global minimizer *for* $f(\mathbf{x})$ *if* $Hf(\mathbf{x})$ *is positive definite on* R^n;
(c) *a* global maximizer *for* $f(\mathbf{x})$ *if* $Hf(\mathbf{x})$ *is negative semidefinite on* R^n;
(d) *a* strict global maximizer *for* $f(\mathbf{x})$ *if* $Hf(\mathbf{x})$ *is negative definite on* R^n.

1.3. Positive and Negative Definite Matrices and Optimization

Now we take up the search for convenient ways to recognize positive and negative definite, positive and negative semidefinite, and indefinite symmetric matrices. Here are some examples.

(1.3.1) Examples

(a) *A symmetric matrix whose entries are all positive need not be positive definite.* For example, the matrix

$$A = \begin{pmatrix} 1 & 4 \\ 4 & 1 \end{pmatrix}$$

is not positive definite. For if $\mathbf{x} = (1, -1)$, then

$$Q_A(\mathbf{x}) = (1, -1)\begin{pmatrix} 1 & 4 \\ 4 & 1 \end{pmatrix}\begin{pmatrix} 1 \\ -1 \end{pmatrix} = (1, -1)\begin{pmatrix} -3 \\ 3 \end{pmatrix} = -6 < 0.$$

(b) *A symmetric matrix with some negative entries may be positive definite.* For example, the matrix

$$A = \begin{pmatrix} 1 & -1 \\ -1 & 4 \end{pmatrix}$$

corresponds to the quadratic form

$$Q_A(\mathbf{x}) = \mathbf{x} \cdot A\mathbf{x} = x_1^2 - 2x_1x_2 + 4x_2^2.$$

Since $Q_A(\mathbf{x}) = (x_1 - x_2)^2 + 3x_2^2$, we see that if $\mathbf{x} = (x_1, x_2) \neq (0, 0)$, then $Q_A(\mathbf{x}) > 0$ since $(x_1 - x_2)^2 > 0$ if $x_1 \neq x_2$ and $3x_2^2 > 0$ if $x_1 = x_2$.

(c) The matrix

$$A = \begin{pmatrix} 1 & 0 & 0 \\ 0 & 3 & 0 \\ 0 & 0 & 2 \end{pmatrix}$$

is positive definite because the associated quadratic form $Q_A(\mathbf{x})$ is

$$Q_A(\mathbf{x}) = \mathbf{x} \cdot A\mathbf{x} = x_1^2 + 3x_2^2 + 2x_3^2,$$

and so $Q_A(\mathbf{x}) > 0$ unless $x_1 = x_2 = x_3 = 0$.

(d) A 3×3-diagonal matrix

$$A = \begin{pmatrix} d_1 & 0 & 0 \\ 0 & d_2 & 0 \\ 0 & 0 & d_3 \end{pmatrix}$$

is:

(1) positive definite if $d_i > 0$ for $i = 1, 2, 3$;
(2) positive semidefinite if $d_i \geq 0$ for $i = 1, 2, 3$;
(3) negative definite if $d_i < 0$ for $i = 1, 2, 3$;
(4) negative semidefinite if $d_i \leq 0$ for $i = 1, 2, 3$;
(5) indefinite if at least one d_i is positive and at least one d_i is negative for $i = 1, 2, 3$.

For example, in case (2), if $d_1 > 0$, $d_2 > 0$, $d_3 = 0$, then

$$Q_A(\mathbf{x}) = d_1x_1^2 + d_2x_2^2 \geq 0$$

for all $\mathbf{x} \neq \mathbf{0}$ since $d_1 > 0$, $d_2 > 0$, but if $\mathbf{x} = (0, 0, 1)$, then $Q_A(\mathbf{x}) = 0$ even though $\mathbf{x} \neq \mathbf{0}$.

(e) If a 2×2-symmetric matrix

$$A = \begin{pmatrix} a & b \\ b & c \end{pmatrix}$$

is positive definite, then $a > 0$ and $c > 0$. For if $\mathbf{x} = (1, 0)$, then $\mathbf{x} \neq \mathbf{0}$ and so

$$0 < Q_A(\mathbf{x}) = a \cdot 1^2 + 2b \cdot 1 \cdot 0 + c \cdot 0^2 = a.$$

Similarly, if $\mathbf{x} = (0, 1)$, then $0 < Q_A(\mathbf{x}) = c$. However, (a) shows that there are 2×2-symmetric matrices with $a > 0$, $c > 0$ that are not positive definite. We will see later that the size of b relative to the size of the product ac is the determining factor for positive definiteness.

These examples show that for general symmetric matrices there is little relationship between the signs of the matrix entries and the positive or negative definite features of the matrix. They also show that for diagonal matrices, these features are completely transparent. We will develop two basic tests for positive and negative definiteness—one in terms of determinants, and in Section 1.5 we will develop the other in terms of eigenvalues. We take up the determinant approach now and we begin by looking at functions of two variables.

If A is a 2×2-symmetric matrix

$$A = \begin{pmatrix} a_{11} & a_{12} \\ a_{12} & a_{22} \end{pmatrix}$$

then the associated quadratic form is

$$Q_A(\mathbf{x}) = \mathbf{x} \cdot A\mathbf{x} = a_{11}x_1^2 + 2a_{12}x_1x_2 + a_{22}x_2^2.$$

For any $\mathbf{x} \neq \mathbf{0}$ in R^2, either $\mathbf{x} = (x_1, 0)$ with $x_1 \neq 0$ or $\mathbf{x} = (x_1, x_2)$ with $x_2 \neq 0$. Let us analyze the sign of $Q_A(\mathbf{x})$ in terms of the entries of A in each of these two cases.

Case 1. $\mathbf{x} = (x_1, 0)$ with $x_1 \neq 0$.

In this case, $Q_A(\mathbf{x}) = a_{11}x_1^2$ so $Q_A(\mathbf{x}) > 0$ if and only if $a_{11} > 0$, while $Q_A(\mathbf{x}) < 0$ if and only if $a_{11} < 0$.

Case 2. $\mathbf{x} = (x_1, x_2)$ with $x_2 \neq 0$.

In this case, $x_1 = tx_2$ for some real number t and

$$Q_A(\mathbf{x}) = [a_{11}t^2 + 2a_{12}t + a_{22}]x_2^2 = \varphi(t)x_2^2,$$

where $\varphi(t) = a_{11}t^2 + 2a_{12}t + a_{22}$. Since $x_2 \neq 0$, we see that $Q_A(\mathbf{x}) > 0$ for all such $\mathbf{x}$ if and only if $\varphi(t) > 0$ for all $t \in R$.

Note that

$$\varphi'(t) = 2a_{11}t + 2a_{12},$$

$$\varphi''(t) = 2a_{11},$$

so that $t^* = -a_{12}/a_{11}$ is a critical point of $\varphi(t)$ and this critical point is a strict minimizer if $a_{11} > 0$ and a strict maximizer if $a_{11} < 0$. If $a_{11} > 0$ and if $t \in R$, then

$$\varphi(t) \geq \varphi(t^*) = \varphi\left(-\frac{a_{12}}{a_{11}}\right) = -\frac{a_{12}^2}{a_{11}} + a_{22} = \frac{1}{a_{11}} \det\begin{pmatrix} a_{11} & a_{12} \\ a_{12} & a_{22} \end{pmatrix}.$$

Thus, if $a_{11} > 0$ and $\det\begin{pmatrix} a_{11} & a_{12} \\ a_{12} & a_{22} \end{pmatrix} > 0$, then $\varphi(t) > 0$ for all $t \in R$ and so $Q_A(\mathbf{x}) > 0$ for all $\mathbf{x} = (x_1, x_2)$ with $x_2 \neq 0$. On the other hand, if $Q_A(\mathbf{x}) > 0$ for all such $\mathbf{x}$, then $\varphi(t) > 0$ for all $t \in R$ and so $a_{11} > 0$ and the discriminant of $\varphi(t)$

$$4a_{12}^2 - 4a_{11}a_{22} = -4\det\begin{pmatrix} a_{11} & a_{12} \\ a_{12} & a_{22} \end{pmatrix}$$

is negative, that is, $a_{11} > 0$ and $\det\begin{pmatrix} a_{11} & a_2 \\ a_{12} & a_{22} \end{pmatrix} > 0$. An entirely similar analysis shows that $Q_A(\mathbf{x}) < 0$ for all $\mathbf{x} = (x_1, x_2)$ with $x_2 \neq 0$ if and only if $a_{11} < 0$ and $\det\begin{pmatrix} a_{11} & a_{12} \\ a_{12} & a_{22} \end{pmatrix} > 0$. This proves the following result:

(1.3.2) Theorem. A 2×2-*symmetric matrix*

$$A = \begin{pmatrix} a_{11} & a_{12} \\ a_{12} & a_{22} \end{pmatrix}$$

is:

(a) *positive definite if and only if*

$$a_{11} > 0, \qquad \det\begin{pmatrix} a_{11} & a_{12} \\ a_{12} & a_{22} \end{pmatrix} > 0;$$

(b) *negative definite if and only if*

$$a_{11} < 0, \qquad \det\begin{pmatrix} a_{11} & a_{12} \\ a_{12} & a_{22} \end{pmatrix} > 0.$$

The 2×2 case and a little imagination suggest the correct formulation of the general case.

Suppose A is an $n \times n$-symmetric matrix. Define Δ_k to be the determinant of the upper left-hand corner $k \times k$-submatrix of A for $1 \leq k \leq n$. The determinant Δ_k is called the kth *principal minor* of A.

$$A = \begin{pmatrix} \Delta_1 & \Delta_2 & \Delta_3 & & \\ a_{11} & a_{12} & a_{13} & \cdots & a_{1n} \\ a_{12} & a_{22} & a_{23} & \cdots & a_{2n} \\ a_{13} & a_{23} & a_{33} & \cdots & a_{3n} \\ \vdots & & & & \vdots \\ a_{1n} & a_{2n} & a_{3n} & \cdots & a_{nn} \end{pmatrix}, \qquad \begin{aligned} \Delta_1 &= a_{11}, \\ \Delta_2 &= \det\begin{pmatrix} a_{11} & a_{12} \\ a_{12} & a_{22} \end{pmatrix}, \\ &\vdots \\ \Delta_n &= \det A. \end{aligned}$$

The general theorem can be formulated as follows:

(1.3.3) Theorem. *If A is an $n \times n$-symmetric matrix and if Δ_k is the* kth *principal minor of A for $1 \leq k \leq n$, then*:

(a) *A is positive definite if and only if $\Delta_k > 0$ for $k = 1, 2, \ldots, n$;*
(b) *A is negative definite if and only if $(-1)^k\Delta_k > 0$ for $k = 1, 2, \ldots, n$ (that is, the principal minors alternate in sign with $\Delta_1 < 0$).*

Mathematical induction can be used to establish this result. However, the formal inductive proof is somewhat complicated by the notation required for the step from $n = k$ to $n = k + 1$. It is quite illuminating to show how this inductive step works from $n = 2$ (Theorem (1.3.2)) to $n = 3$, because this step lays bare the essential features of the general inductive step. Consequently, we include the proof of this special case at this point.

PROOF FOR $n = 3$. Suppose that

$$A = \begin{pmatrix} a_{11} & a_{12} & a_{13} \\ a_{12} & a_{22} & a_{23} \\ a_{13} & a_{23} & a_{33} \end{pmatrix}$$

is a 3×3-symmetric matrix and that $\mathbf{x} = (x_1, x_2, x_3)$ is a nonzero vector in R^3. Then one of the following two cases must hold: Either $x_3 = 0$ or else $x_3 \neq 0$ and consequently $x_2 = tx_3$, $x_1 = sx_3$ for some real numbers s, t.

Case 1. If $x_3 = 0$, then a brief computation shows that

$$\mathbf{x} \cdot A\mathbf{x} = (x_1, x_2) \cdot \begin{pmatrix} a_{11} & a_{12} \\ a_{12} & a_{22} \end{pmatrix} \begin{pmatrix} x_1 \\ x_2 \end{pmatrix}$$

and $(x_1, x_2) \neq (0, 0)$, so (1.3.2) shows that:

(a) $\mathbf{x} \cdot A\mathbf{x} > 0$ for all $\mathbf{x} \neq \mathbf{0}$ such that $x_3 = 0$ if and only if $\Delta_1 > 0$, $\Delta_2 > 0$;
(b) $\mathbf{x} \cdot A\mathbf{x} < 0$ for all $\mathbf{x} \neq \mathbf{0}$ such that $x_3 = 0$ if and only if $\Delta_1 < 0$, $\Delta_2 > 0$;

Case 2. If $x_3 \neq 0$ and $x_2 = tx_3$, $x_1 = sx_3$ for real numbers s, t, then

$$\mathbf{x} \cdot A\mathbf{x} = x_3^2(a_{11}s^2 + a_{22}t^2 + a_{33} + 2a_{12}st + 2a_{13}s + 2a_{23}t).$$

Consequently, since $x_3 \neq 0$, it follows that $\mathbf{x} \cdot A\mathbf{x} > 0$ for all $\mathbf{x} \neq \mathbf{0}$ such that $x_3 \neq 0$ if and only if

$$\varphi(s, t) = a_{11}s^2 + a_{22}t^2 + a_{33} + 2a_{12}st + 2a_{13}s + 2a_{23}t > 0$$

for all real numbers s, t. In addition, $\mathbf{x} \cdot A\mathbf{x} < 0$ for all $\mathbf{x} \neq \mathbf{0}$ such that $x_3 \neq 0$ if and only if $\varphi(s, t) < 0$ for all real numbers s, t.

The critical points of $\varphi(s, t)$ are the solutions of the system

$$0 = \frac{\partial \varphi}{\partial s} = 2a_{11}s + 2a_{12}t + 2a_{13},$$

$$0 = \frac{\partial \varphi}{\partial t} = 2a_{12}s + 2a_{22}t + 2a_{23},$$

that is,

$$a_{11}s + a_{12}t = -a_{13},$$
$$a_{12}s + a_{22}t = -a_{23}.$$

This system has a unique solution (s^*, t^*) if and only if

$$\Delta_2 = \det\begin{pmatrix} a_{11} & a_{12} \\ a_{12} & a_{22} \end{pmatrix} \neq 0,$$

and this unique solution is given by Cramer's Rule as

$$s^* = \frac{1}{\Delta_2}\det\begin{pmatrix} -a_{13} & a_{12} \\ -a_{23} & a_{22} \end{pmatrix}, \qquad t^* = \frac{1}{\Delta_2}\det\begin{pmatrix} a_{11} & -a_{13} \\ a_{12} & -a_{23} \end{pmatrix}. \tag{1}$$

If we multiply the equation

$$a_{11}s^* + a_{12}t^* + a_{13} = 0$$

by s^*, and multiply the equation

$$a_{12}s^* + a_{22}t^* + a_{23} = 0$$

by t^* and add the results, we obtain

$$a_{11}(s^*)^2 + a_{22}(t^*)^2 + 2a_{12}s^*t^* + a_{13}s^* + a_{23}t^* = 0.$$

Consequently,

$$\varphi(s^*, t^*) = a_{13}s^* + a_{23}t^* + a_{33},$$

and so (1) implies that if $\Delta_2 \neq 0$, then

$$\varphi(s^*, t^*) = \frac{1}{\Delta_2}\det\begin{pmatrix} a_{11} & a_{12} & a_{13} \\ a_{12} & a_{22} & a_{23} \\ a_{13} & a_{23} & a_{33} \end{pmatrix} = \frac{\det A}{\Delta_2} = \frac{\Delta_3}{\Delta_2}. \tag{2}$$

(Just use the cofactor expansion of det A by the third column of A and basic properties of determinants.)

Since

$$H\varphi(s, t) = \det\begin{pmatrix} 2a_{11} & 2a_{12} \\ 2a_{12} & 2a_{22} \end{pmatrix} = 4\Delta_2,$$

it follows from (1.3.2) and (1.2.9) that (s^*, t^*) is a strict global minimizer for $\varphi(s, t)$ if and only if $\Delta_1 > 0, \Delta_2 > 0$. Similarly, (s^*, t^*) is a strict global maximizer for $\varphi(s, t)$ if and only if $\Delta_1 < 0, \Delta_2 > 0$.

If $\Delta_1 > 0, \Delta_2 > 0, \Delta_3 > 0$, then the conclusion (a) of Case 1 shows that if $\mathbf{x} \neq \mathbf{0}$ and $x_3 = 0$, then $\mathbf{x} \cdot A\mathbf{x} > 0$; on the other hand, the considerations in Case 2 show that if $\mathbf{x} \neq \mathbf{0}$, $x_3 \neq 0$, $x_2 = tx_3$, $x_1 = sx_3$, then

$$\mathbf{x} \cdot A\mathbf{x} = x_3^2\varphi(s, t) \geq x_3^2\varphi(s^*, t^*) = x_3^2\frac{\Delta_3}{\Delta_2} > 0.$$

Therefore $\mathbf{x} \cdot A\mathbf{x} > 0$ for all $\mathbf{x} \neq \mathbf{0}$ if $\Delta_1 > 0, \Delta_2 > 0, \Delta_3 > 0$.

On the other hand, if $\mathbf{x} \cdot A\mathbf{x} > 0$ for all $\mathbf{x} \neq \mathbf{0}$, then conclusion (a) of Case 1 shows that $\Delta_1 > 0$, $\Delta_2 > 0$. Also, if $\mathbf{x}^* = (s^*, t^*, 1)$, then (2) yields

$$\frac{\Delta_3}{\Delta_2} = \varphi(s^*, t^*) = \mathbf{x}^* \cdot A\mathbf{x}^* > 0,$$

so $\Delta_3 > 0$. This proves part (a) of (1.3.3) for $n = 3$. We can establish (b) by making obvious changes in this paragraph and the preceding one.

(1.3.4) Remarks

(a) If A is a 3×3-symmetric matrix such that $\Delta_1 > 0, \Delta_2 > 0, \Delta_3 = 0$, then a review of the proof of (1.3.3) for $n = 3$ shows that A is positive semidefinite. The proof of (1.3.3) for the general case shows that if $\Delta_1 > 0$, $\Delta_2 > 0, \ldots, \Delta_{n-1} > 0$, $\Delta_n = 0$, then A is positive semidefinite. Similarly, if $(-1)^k\Delta_k > 0$ for $k = 1, \ldots, n-1$ while $\Delta_n = 0$, then A is negative semidefinite.

(b) It is *not* true that if A is an $n \times n$-symmetric matrix, then A is positive semidefinite if and only if the principal minors $\Delta_1, \ldots, \Delta_n$ are all nonnegative. For example, if

$$A = \begin{pmatrix} 1 & 1 & 1 \\ 1 & 1 & 1 \\ 1 & 1 & \frac{1}{2} \end{pmatrix},$$

then all principal minors of A are nonnegative but A is *not* positive semidefinite since, for example, $\mathbf{x} \cdot A\mathbf{x} < 0$ for $\mathbf{x} = (1, 1, -2)$.

(c) Some principal minor criteria are available for indefinite matrices. For example, if A is a 2×2-symmetric matrix for which $\Delta_2 = \det A < 0$, then A is indefinite. In fact, if

$$Q_A(\mathbf{x}) = a_{11}x_1^2 + 2a_{12}x_1x_2 + a_{22}x_2^2,$$

and if $\Delta_2 = a_{11}a_{22} - a_{12}^2 < 0$, then either $a_{11} = a_{22} = 0$ or at least one of the numbers a_{11}, a_{22} is nonzero. In the former case, $Q_A(\mathbf{x}) = 2a_{12}x_1x_2$ assumes both positive and negative values. In the latter case, say $a_{11} \neq 0$, we can complete the square on x_1 to rewrite the quadratic form $Q_A(\mathbf{x})$ as follows:

$$\begin{aligned} Q_A(\mathbf{x}) &= a_{11}x_1^2 + 2a_{12}x_1x_2 + a_{22}x_2^2 \\ &= a_{11}\left(\left(x_1^2 + 2\frac{a_{12}}{a_{11}}x_1x_2 + \frac{a_{12}^2}{a_{11}^2}x_2^2\right) + \left(\frac{a_{22}}{a_{11}} - \frac{a_{12}^2}{a_{11}^2}\right)x_2^2\right) \\ &= \frac{1}{a_{11}}[(a_{11}x_1 + a_{12}x_2)^2 + \Delta_2 x_2^2]. \end{aligned}$$

Since $\Delta_2 < 0$, the final expression for $Q_A(\mathbf{x})$ makes it clear that Q_A has opposite signs at the points $(1, 0)$ and $(a_{12}, -a_{11})$, and therefore A is indefinite.

Now let us get back on track and apply what we have learned to the problems of global minimization. Here are four examples that summarize what we now know.

(1.3.5) Examples

(a) Minimize the function

$$f(x_1, x_2, x_3) = x_1^2 + x_2^2 + x_3^2 - x_1x_2 + x_2x_3 - x_1x_3.$$

The critical points of $f(x_1, x_2, x_3)$ are the solutions of the system

$$\begin{aligned} 2x_1 - x_2 - x_3 &= 0, \\ -x_1 + 2x_2 + x_3 &= 0, \\ -x_1 + x_2 + 2x_3 &= 0. \end{aligned}$$

This homogeneous system of linear equations has a coefficient matrix with a nonzero determinant, so $x_1 = 0, x_2 = 0, x_3 = 0$ is the one and only solution.

The Hessian of $f(x_1, x_2, x_3)$ is the constant matrix

$$Hf(x_1, x_2, x_3) = \begin{pmatrix} 2 & -1 & -1 \\ -1 & 2 & 1 \\ -1 & 1 & 2 \end{pmatrix}.$$

Note that $\Delta_1 = 2, \Delta_2 = 3, \Delta_3 = 4$, so $Hf(x_1, x_2, x_3)$ is positive definite everywhere on R^3. It follows for (1.2.9) that the critical point (0, 0, 0) is a strict global minimizer for $f(x_1, x_2, x_3)$.

Since $f(x_1, x_2, x_3)$ is defined and has continuous first partial derivatives everywhere on R^3 and since (0, 0, 0) is the only critical point of $f(x_1, x_2, x_3)$, it follows from (1.2.3) that $f(x_1, x_2, x_3)$ has no other minimizers or maximizers.

(b) Find the global minimizer of

$$f(x, y, z) = e^{x-y} + e^{y-x} + e^{x^2} + z^2.$$

To this end, compute

$$\nabla f(x, y, z) = \begin{pmatrix} e^{x-y} - e^{y-x} + 2xe^{x^2} \\ -e^{x-y} + e^{y-x} \\ 2z \end{pmatrix},$$

and

$$Hf(x, y, z) = \begin{pmatrix} e^{x-y} + e^{y-x} + 4x^2e^{x^2} + 2e^{x^2} & -e^{x-y} - e^{y-x} & 0 \\ -e^{x-y} - e^{y-x} & e^{x-y} + e^{y-x} & 0 \\ 0 & 0 & 2 \end{pmatrix}.$$

Clearly, $\Delta_1 > 0$ for all x, y, z because all the terms of it are positive. Also

$$\begin{aligned} \Delta_2 &= (e^{x-y} + e^{y-x})^2 + (e^{x-y} + e^{y-x})(4x^2e^{x^2} + 2e^{x^2}) - (e^{x-y} + e^{y-x})^2 \\ &= (e^{x-y} + e^{y-x})(4x^2e^{x^2} + 2e^{x^2}) > 0 \end{aligned}$$

because both factors are always positive. Finally, $\Delta_3 = 2\Delta_2 > 0$. Hence $Hf(x, y, z)$ is positive definite at all points. Therefore by Theorem (1.2.9), $f(x, y, z)$ is strictly globally minimized at any critical point (x^*, y^*, z^*). To find (x^*, y^*, z^*), solve

$$\mathbf{0} = \nabla f(x^*, y^*, z^*) = \begin{pmatrix} e^{x^*-y^*} - e^{y^*-x^*} + 2x^* e^{(x^*)^2} \\ -e^{x^*-y^*} + e^{y^*-x^*} \\ 2z^* \end{pmatrix}.$$

This leads to $z^* = 0$, $e^{x^*-y^*} = e^{y^*-x^*}$, hence $2x^* e^{(x^*)^2} = 0$. Accordingly, $x^* - y^* = y^* - x^*$; that is, $x^* = y^*$ and $x^* = 0$. Therefore $(x^*, y^*, z^*) = (0, 0, 0)$ is the strict global minimizer of $f(x, y, z)$.

(c) Find the global minimizers of

$$f(x, y) = e^{x-y} + e^{y-x}.$$

To this end, compute

$$\nabla f(x, y) = \begin{pmatrix} e^{x-y} - e^{y-x} \\ -e^{x-y} + e^{y-x} \end{pmatrix}$$

and

$$Hf(x, y) = \begin{pmatrix} e^{x-y} + e^{y-x} & -e^{x-y} - e^{y-x} \\ -e^{x-y} - e^{y-x} & e^{x-y} + e^{y-x} \end{pmatrix}.$$

Since $e^{x-y} + e^{y-x} > 0$ for all x, y and $\det Hf(x, y) = 0$, then, by Remark (1.3.4)(a), the Hessian $Hf(x, y)$ is positive semidefinite for all x, y. Therefore, by Theorem (1.2.9), $f(x, y)$ is minimized at any critical point (x^*, y^*) of $f(x, y)$. To find (x^*, y^*), solve

$$\mathbf{0} = \nabla f(x^*, y^*) = \begin{pmatrix} e^{x^*-y^*} - e^{y^*-x^*} \\ -e^{x^*-y^*} + e^{y^*-x^*} \end{pmatrix}.$$

This gives

$$e^{x^*-y^*} = e^{y^*-x^*}$$

or

$$x^* - y^* = y^* - x^*;$$

that is,

$$2x^* = 2y^*.$$

This shows that all points of the line $y = x$ are global minimizers of $f(x, y)$.

(d) Find the global minimizers of

$$f(x, y) = e^{x-y} + e^{x+y}.$$

In this case,

$$\nabla f(x, y) = \begin{pmatrix} e^{x-y} + e^{x+y} \\ -e^{x-y} + e^{x+y} \end{pmatrix}$$

$$Hf(x, y) = \begin{pmatrix} e^{x-y} + e^{x+y} & -e^{x-y} + e^{x+y} \\ -e^{x-y} + e^{x+y} & e^{x-y} + e^{x+y} \end{pmatrix}.$$

Since $e^{x-y} + e^{x+y} > 0$ for all x, y and det $Hf(x, y) > 0$, then by (1.3.2), $Hf(x, y)$ is positive definite for all x, y. Therefore, by Theorem (1.2.9), $f(x, y)$ is minimized at any critical point (x^*, y^*). To find (x^*, y^*), write

$$\mathbf{0} = \nabla f(x^*, y^*) = \begin{pmatrix} e^{x^*-y^*} + e^{x^*+y^*} \\ -e^{x^*-y^*} + e^{x^*+y^*} \end{pmatrix}.$$

Thus

$$e^{x^*-y^*} + e^{x^*+y^*} = 0$$

and

$$-e^{x^*-y^*} + e^{x^*+y^*} = 0.$$

But $e^{x^*-y^*} > 0$ and $e^{x^*+y^*} > 0$ for all x^*, y^*. Therefore the equality $e^{x^*-y^*} + e^{x^*+y^*} = 0$ is impossible. Thus $f(x, y)$ has no critical points and hence $f(x, y)$ has no global minimizers.

There is no disputing that global minimization is far more important than mere local minimization. Still there are certain situations in which scientists want knowledge of local minimizers of a function. Since we are in an excellent position to understand local minimization, let us get on with it. The basic fact to understand is the next theorem.

(1.3.6) Theorem. *Suppose that $f(\mathbf{x})$ is a function with continuous first and second partial derivatives on some set D in R^n. Suppose $\mathbf{x}^*$ is an interior point of D and that $\mathbf{x}^*$ is a critical point of $f(\mathbf{x})$. Then $\mathbf{x}^*$ is:*

(a) *a* strict local minimizer *of $f(\mathbf{x})$ if $Hf(\mathbf{x}^*)$ is positive definite*;
(b) *a* strict local maximizer *of $f(\mathbf{x})$ if $Hf(\mathbf{x}^*)$ is negative definite.*

PROOF. (a) Define $\Delta_k(\mathbf{x})$ to be the kth principal minor of $Hf(\mathbf{x})$. By hypothesis, we know $\Delta_k(\mathbf{x}^*) > 0$ for $k = 1, 2, \ldots, n$. Now because the second partials of $f(\mathbf{x})$ are continuous, each $\Delta_k(\mathbf{x})$ is a continuous function of $\mathbf{x}$. Since $\Delta_k(\mathbf{x}^*) > 0$, from continuity it follows that there exists for each k a number $r_k > 0$ such that $\Delta_k(\mathbf{x}) > 0$ if $\|\mathbf{x} - \mathbf{x}^*\| < r_k$. Set $r = \min\{r_1, \ldots, r_n\}$ and observe that for all $k = 1, \ldots, n$ we have $\Delta_k(\mathbf{x}) > 0$ if $\|\mathbf{x} - \mathbf{x}^*\| < r$. Therefore by Theorem (1.3.3), the matrix $Hf(\mathbf{x})$ is positive definite if $\|\mathbf{x} - \mathbf{x}^*\| < r$. Now if $0 < \|\mathbf{x} - \mathbf{x}^*\| < r$, then according to Theorem (1.2.4), we have

$$f(\mathbf{x}) = f(\mathbf{x}^*) + \nabla f(\mathbf{x}^*)\cdot(\mathbf{x} - \mathbf{x}^*) + \tfrac{1}{2}(\mathbf{x} - \mathbf{x}^*)\cdot Hf(\mathbf{z})(\mathbf{x} - \mathbf{x}^*),$$

where $\mathbf{z}$ is on the line segment from $\mathbf{x}$ to $\mathbf{x}^*$. Since $\|\mathbf{x} - \mathbf{x}^*\| < r$, it follows quickly that

$$\|\mathbf{z} - \mathbf{x}^*\| < r.$$

Hence $Hf(\mathbf{z})$ is positive definite. Consequently, if $0 < \|\mathbf{x} - \mathbf{x}^*\| < r$, then $f(\mathbf{x}) = f(\mathbf{x}^*) + 0 +$ a positive number. Thus $\|\mathbf{x} - \mathbf{x}^*\| < r$ and $\mathbf{x} \neq \mathbf{x}^*$ imply that

$$f(\mathbf{x}) > f(\mathbf{x}^*);$$

that is, $\mathbf{x}^*$ is a strict local minimizer of $f(\mathbf{x})$.

The proof of (b) is similar and will be omitted.

Let us briefly investigate the meaning of an indefinite Hessian at a critical point of a function before we go on. Suppose that $f(\mathbf{x})$ has continuous second partial derivatives on a set D in R^n, that $\mathbf{x}^*$ is an interior point of D which is a critical point of $f(\mathbf{x})$, and that $Hf(\mathbf{x}^*)$ is indefinite. This means that there are nonzero vectors $\mathbf{y}$, $\mathbf{w}$ in R^n such that

$$\mathbf{y} \cdot Hf(\mathbf{x}^*)\mathbf{y} > 0, \qquad \mathbf{w} \cdot Hf(\mathbf{x}^*)\mathbf{w} < 0.$$

Since $f(\mathbf{x})$ has continuous second partial derivatives on D, there is an $\varepsilon > 0$ such that

$$\mathbf{y} \cdot Hf(\mathbf{x}^* + t\mathbf{y})\mathbf{y} > 0, \qquad \mathbf{w} \cdot Hf(\mathbf{x}^* + t\mathbf{w})\mathbf{w} < 0$$

for all t with $|t| < \varepsilon$. But then if $Y(t)$, $W(t)$ are defined

$$Y(t) = f(\mathbf{x}^* + t\mathbf{y}), \qquad W(t) = f(\mathbf{x}^* + t\mathbf{w}),$$

then $Y'(0) = 0 = W'(0)$ and $Y''(0) = \mathbf{y} \cdot Hf(\mathbf{x}^*)\mathbf{y} > 0$, $W''(0) = \mathbf{w} \cdot Hf(\mathbf{x}^*)\mathbf{w} < 0$. Therefore, $t = 0$ is a strict local minimizer for $Y(t)$ and a strict local maximizer for $W(t)$.

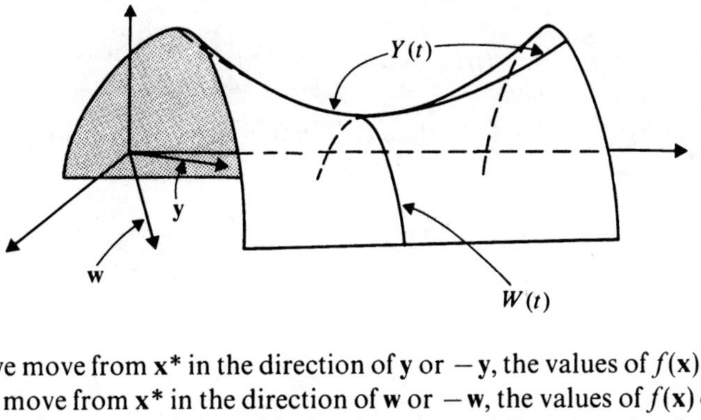

Thus if we move from $\mathbf{x}^*$ in the direction of $\mathbf{y}$ or $-\mathbf{y}$, the values of $f(\mathbf{x})$ increase, but if we move from $\mathbf{x}^*$ in the direction of $\mathbf{w}$ or $-\mathbf{w}$, the values of $f(\mathbf{x})$ decrease. For this reason, we call the critical point $\mathbf{x}^*$ a saddle point, that is, a *saddle point* for $f(\mathbf{x})$ is a critical point $\mathbf{x}^*$ for $f(\mathbf{x})$ such that there are vectors $\mathbf{y}$, $\mathbf{w}$ for which $t = 0$ is a strict local minimizer for $Y(t) = f(\mathbf{x}^* + t\mathbf{y})$ and a strict local maximizer for $W(t) = f(\mathbf{x}^* + t\mathbf{w})$.

The following result summarizes this little discussion:

(1.3.7) Theorem. *If $f(\mathbf{x})$ is a function with continuous second partial derivatives on a set D in R^n, if $\mathbf{x}^*$ is an interior point of D that is a critical point of $f(\mathbf{x})$, and if the Hessian $Hf(\mathbf{x}^*)$ is indefinite, then $\mathbf{x}^*$ is a saddle point for $f(\mathbf{x})$.*

(1.3.8) Example. Let us look for the global and local minimizers and maximizers (if any) of the function

$$f(x_1, x_2) = x_1^3 - 12x_1x_2 + 8x_2^3.$$

In this case, the critical points are the solutions of the system

$$0 = \frac{\partial f}{\partial x_1} = 3x_1^2 - 12x_2,$$

$$0 = \frac{\partial f}{\partial x_2} = -12x_1 + 24x_2^2.$$

This system can be readily solved to identify the critical points (2, 1) and (0, 0).

The Hessian of $f(x_1, x_2)$ is

$$Hf(x_1, x_2) = \begin{pmatrix} 6x_1 & -12 \\ -12 & 48x_2 \end{pmatrix}.$$

Since

$$Hf(2, 1) = \begin{pmatrix} 12 & -12 \\ -12 & 48 \end{pmatrix}$$

and since $\Delta_1 = 12$ and $\Delta_2 = 432$, it follows that the critical point (2, 1) is a strict local minimizer.

Now let us see whether (2, 1) is a global minimizer. Observe that $Hf(x_1, x_2)$ is not positive definite for all (x_1, x_2); for example,

$$Hf(0, 1) = \begin{pmatrix} 0 & -12 \\ -12 & 48 \end{pmatrix}$$

is indefinite by (1.3.4)(c). In view of (1.2.9), this leads us to suspect that (2, 1) may not be a global minimizer. The fact that

$$\lim_{x_1 \to -\infty} f(x_1, 0) = -\infty$$

shows conclusively that $f(x_1, x_2)$ has no global minimizer. Moreover, since

$$\lim_{x_1 \to +\infty} f(x_1, 0) = +\infty,$$

we see that there are no global maximizers or global minimizers.

How about the critical point (0, 0)? Well, since

$$Hf(0, 0) = \begin{pmatrix} 0 & -12 \\ -12 & 0 \end{pmatrix},$$

this matrix miserably fails the tests for positive definiteness and this leads us to expect trouble at (0, 0). Remark (1.3.4)(c) and Theorem (1.3.7) tell us that there is a saddle point at (0, 0). Alternatively, we can examine

$$f(x_1, x_2) = x_1^3 - 12x_1x_2 + 8x_2^3.$$

For example,

$$f(x_1, 0) = x_1^3,$$

which is positive for $x_1 > 0$, zero for $x_1 = 0$, and negative for $x_1 < 0$. Thus (0, 0) is not a local minimizer of $f(x_1, x_2)$.

One cautionary note is due here. If x^* is a critical point of $f(\mathbf{x})$ and $Hf(\mathbf{x}^*)$ is merely positive semidefinite, then nothing can be concluded in general. For instance, if $f(x, y) = x^4 - y^4$, then (0, 0) is the only critical point and $Hf(0, 0) = \begin{pmatrix} 0 & 0 \\ 0 & 0 \end{pmatrix}$, which is positive semidefinite. But plainly (0, 0) is neither a local maximizer nor a local minimizer of $f(x, y)$.

On the other hand, if $f(x, y) = x^4 + y^4$, the (0, 0) is the only critical point of $f(x, y)$ and $Hf(0, 0) = \begin{pmatrix} 0 & 0 \\ 0 & 0 \end{pmatrix}$, which is again positive semidefinite. This time it is clear that (0, 0) is the global minimizer of $f(x, y)$.

1.4. Coercive Functions and Global Minimizers

At this stage we can find global minimizers for $f(\mathbf{x})$ if $f(\mathbf{x})$ has a critical point and $Hf(\mathbf{x})$ is always positive definite. But what about global minimization for $f(\mathbf{x})$ in the case in which $Hf(\mathbf{x})$ is not known to be always positive definite? This short section is devoted to showing that this question sometimes has a very simple answer. The answer depends on the following theorem from calculus:

(1.4.1) Theorem. *Let D be a closed bounded subset of R^n. If $f(\mathbf{x})$ is a continuous function defined on D, then $f(\mathbf{x})$ has a global maximizer and a global minimizer on D.*[1]

Note that this theorem does not guarantee a global minimum on R^n, or on any other set that is either unbounded or not closed.

Now we isolate the type of function that is easily handled.

(1.4.2) Definition. A continuous function $f(\mathbf{x})$ that is defined on all of R^n is called *coercive* if

$$\lim_{\|\mathbf{x}\| \to \infty} f(\mathbf{x}) = +\infty.$$

This means that for any constant M there must be a positive number R_M such that $f(x) \geq M$ whenever $\|\mathbf{x}\| \geq R_M$. In particular, the values of $f(\mathbf{x})$ cannot remain bounded on a set A in R^n that is not bounded.

[1] For a proof see *Elements of Real Analysis* by R. G. Bartle.

(1.4.3) Examples

(a) Let $f(x, y) = x^2 + y^2 = \|\mathbf{x}\|^2$. Then

$$\lim_{\|\mathbf{x}\| \to \infty} f(\mathbf{x}) = \lim_{\|\mathbf{x}\| \to \infty} \|\mathbf{x}\|^2 = \infty.$$

Thus $f(x, y)$ is coercive.

(b) Let $f(x, y) = x^4 + y^4 - 3xy$. Note that

$$f(x, y) = (x^4 + y^4)\left(1 - \frac{3xy}{x^4 + y^4}\right).$$

If $\|\mathbf{x}\|$ is large, then $3xy/(x^4 + y^4)$ is very small. Hence

$$\lim_{\|(x,y)\| \to \infty} f(x, y) = \lim_{\|(x,y)\| \to \infty} (x^4 + y^4)\cdot(1 - 0) = +\infty.$$

Thus $f(x, y)$ is coercive.

(c) Let $f(x, y, z) =$

$$e^{x^2} + e^{y^2} + e^{z^2} - x^{100} - y^{100} - z^{100}.$$

Then because exponential growth is much faster than the growth of any polynomial, it follows that

$$\lim_{\|(x,y,z)\| \to \infty} f(x, y, z) = \infty.$$

Thus $f(x, y, z)$ is coercive.

(d) Linear functions on R^2 are never coercive. Such functions can be expressed as follows:

$$f(x, y) = ax + by + c,$$

where either $a \neq 0$ or $b \neq 0$. To see that $f(x, y)$ is not coercive, simply observe that $f(x, y)$ is constantly equal to c on the line

$$ax + by = 0.$$

Since this line is unbounded and $f(x, y)$ is not unbounded on this line, the function $f(x, y)$ is not coercive.

(e) If $f(x, y, z) = x^4 + y^4 + z^4 - 3xyz - x^2 - y^2 - z^2$, then as

$$\|(x, y, z)\| = \sqrt{x^2 + y^2 + z^2} \to \infty,$$

the higher degree terms dominate and force $\lim_{\|(x,y,z)\| \to \infty} f(x, y, z) = \infty$. Thus $f(x, y, z)$ is coercive. The following example helps us avoid some misunderstandings.

(f) Let $f(x, y) = x^2 - 2xy + y^2$. Then:

(i) for each fixed y_0, we have $\lim_{|x| \to \infty} f(x, y_0) = \infty$;
(ii) for each fixed x_0, we have $\lim_{|y| \to \infty} f(x_0, y) = \infty$;
(iii) but $f(x, y)$ is *not* coercive.

Properties (i) and (ii) above are more or less clear because in each case the quadratic term dominates. For example, in case (i), we have for a fixed y_0,

$$f(x, y_0) = x^2 - xy_0 + y_0^2.$$

This function of x is a parabola that opens upward. Therefore

$$\lim_{|x| \to \infty} f(x, y_0) = \infty.$$

To see that $f(x, y)$ is not coercive, factor to learn

$$f(x, y) = x^2 - 2xy + y^2 = (x - y)^2.$$

Therefore if $\|(x, y)\|$ goes to ∞ on the line $y = +x$, we see $f(x, y) = (x - x)^2 = 0$ and hence $f(x, y) = 0$ on the unbounded line $y = x$. Therefore, $\lim_{\|(x,y)\| \to \infty} f(x, y) \neq \infty$ so $f(x, y)$ is not coercive.

The point of this last example is very important. For $f(\mathbf{x})$ to be coercive, it is not sufficient that $f(\mathbf{x}) \to \infty$ as each coordinate tends to ∞. Rather $f(\mathbf{x})$ must become infinite along any path for which $\|\mathbf{x}\|$ becomes infinite. Exercise 31 contains a general result concerning functions of the sort discussed in Example (1.4.3)(a), (f) above.

The reason coercive functions are important is that they all have global minimizers.

(1.4.4) Theorem. *Let $f(\mathbf{x})$ be a continuous function defined on all R^n. If $f(\mathbf{x})$ is coercive, then $f(\mathbf{x})$ has at least one global minimizer.*

If, in addition, the first partial derivatives of $f(\mathbf{x})$ exist on all of R^n, then these global minimizers can be found among the critical points of $f(\mathbf{x})$.

PROOF. To prove the first statement, assume $\lim_{\|\mathbf{x}\| \to \infty} f(\mathbf{x}) = +\infty$. This means that if $\|\mathbf{x}\|$ is large, then so is $f(\mathbf{x})$. Accordingly, there is a number $r > 0$ such that if $\|\mathbf{x}\| > r$, then

$$f(\mathbf{x}) > f(\mathbf{0}).$$

Let $\bar{B}(\mathbf{0}, r)$ be the set $\{\mathbf{x}\colon \|\mathbf{x}\| \leq r\}$. The function $f(\mathbf{x})$ is continuous at each point of the set $\bar{B}(\mathbf{0}, r)$ and the set $\bar{B}(\mathbf{0}, r)$ is closed and bounded. From Theorem (1.4.1), it follows that $f(\mathbf{x})$ takes a minimum value on $\bar{B}(\mathbf{0}, r)$ at a point $\mathbf{x}^*$ in $\bar{B}(\mathbf{0}, r)$. In other words, $\mathbf{x} \in B(\mathbf{0}, r)$ implies $f(\mathbf{x}^*) \leq f(\mathbf{x})$. In particular, because $\mathbf{0} \in \bar{B}(\mathbf{0}, r)$, we see that

$$f(\mathbf{x}^*) \leq f(\mathbf{0}).$$

On the other hand, if $\mathbf{x} \notin \bar{B}(\mathbf{0}, r)$, then

$$f(\mathbf{x}) > f(\mathbf{0}) \geq f(\mathbf{x}^*).$$

Summarizing, we have seen that $\mathbf{x} \in \bar{B}(\mathbf{0}, r)$ implies $f(\mathbf{x}) \geq f(\mathbf{x}^*)$ and $\mathbf{x} \notin \bar{B}(\mathbf{0}, r)$ implies $f(\mathbf{x}) > f(\mathbf{x}^*)$. This shows $\mathbf{x}^*$ is a global minimizer of $f(\mathbf{x})$ and completes the proof of the first statement of the theorem.

The second statement holds because global minimizers on R^n are critical points by Theorem (1.2.3). This completes the proof.

This theorem sets up a method for trying to minimize coercive functions on R^n. If $f(\mathbf{x})$ is coercive and the first partial derivatives of $f(\mathbf{x})$ exist on R^n, then its minimizers are found among the critical points. Therefore to minimize $f(\mathbf{x})$ on R^n, merely list the critical points $\mathbf{x}^{(1)}, \ldots, \mathbf{x}^{(p)}$ of $f(\mathbf{x})$. Then choose the critical point $\mathbf{x}^{(i)}$ such that $f(\mathbf{x}^{(i)})$ is less than or equal to the other $f(\mathbf{x}^{(j)})$ for $j = 1, \ldots, p$. Theorem (1.4.4) guarantees that $f(\mathbf{x}^{(i)})$ is a global minimizer of $f(\mathbf{x})$ on R^n.

(1.4.5) Example. Minimize

$$f(x, y) = x^4 - 4xy + y^4$$

on R^2.

To this end, compute

$$\nabla f(x, y) = \begin{pmatrix} 4x^3 - 4y \\ -4x + 4y^3 \end{pmatrix}$$

and

$$Hf(x, y) = \begin{pmatrix} 12x^2 & -4 \\ -4 & 12y^2 \end{pmatrix}.$$

Note that

$$Hf(\tfrac{1}{2}, \tfrac{1}{2}) = \begin{pmatrix} 3 & -4 \\ -4 & 3 \end{pmatrix},$$

which is certainly not positive definite since $\det Hf(\frac{1}{2}, \frac{1}{2}) = 9 - 16 < 0$. Therefore the tests from the last section are not applicable. But all is not lost because $f(x, y)$ is coercive!

To see that $f(x, y)$ is coercive, note that

$$f(x, y) = x^4 + y^4\left(1 - \frac{4xy}{x^4 + y^4}\right).$$

As $\|(x, y)\| = \sqrt{x^2 + y^2} \to \infty$, the term $4xy/(x^4 + y^4) \to 0$. Hence

$$\lim_{\|(x,y)\| \to \infty} f(x, y) = \lim_{\|(x,y)\| \to \infty} (x^4 + y^4)(1 - 0) = +\infty.$$

Thus $f(x, y)$ is coercive. According to the last theorem $f(x, y)$ has a global minimizer at one of the critical points. Setting $\nabla f(x, y) = 0$, we get $y = x^3$, and $x = y^3$. Hence $x = x^9$ and $x(x^8 - 1) = 0$. This produces three critical points

$$(0, 0), (1, 1), (-1, -1).$$

Now

$$f(0, 0) = 0,$$

$$f(1, 1) = 1 - 4 + 1 = -2,$$

$$f(-1, -1) = 1 - 4 + 1 = -2.$$

Therefore $(-1, -1)$ and $(1, 1)$ are both global minimizers of $f(x, y)$.

1.5. Eigenvalues and Positive Definite Matrices

If the eigenvalues of a symmetric matrix are available, then they can easily be used to recognize definite, semidefinite, and indefinite matrices. The goal of this section is to see why this is so. Here is some background from linear algebra.

If A is an $n \times n$-matrix and if $\mathbf{x}$ is a nonzero vector in R^n such that $A\mathbf{x} = \lambda\mathbf{x}$ for some real or complex number λ, then λ is called an *eigenvalue* of A. If λ is an eigenvalue of A, then any nonzero vector $\mathbf{x}$ that satisfies the equation $A\mathbf{x} = \lambda\mathbf{x}$ is called an *eigenvector of A corresponding to* λ. Since λ is an eigenvalue of an $n \times n$-matrix A if and only if the homogeneous system $(A - \lambda I)\mathbf{x} = \mathbf{0}$ of n equations in n unknowns has a nonzero solution $\mathbf{x}$, it follows that the eigenvalues of A are just the roots of the *characteristic equation*

$$\det(A - \lambda I) = 0.$$

Since $\det(A - \lambda I)$ is a polynomial of degree n in λ, the characteristic equation has n real or complex roots if we count multiple roots according to their multiplicities, so an $n \times n$-matrix A has n real or complex eigenvalues counting multiplicities.

Symmetric matrices have the following special properties with respect to eigenvalues and eigenvectors:

(1) All of the eigenvalues of a symmetric matrix are real numbers.
(2) Eigenvectors corresponding to distinct eigenvalues of a symmetric matrix are orthogonal.
(3) If λ is an eigenvalue of multiplicity k for a symmetric matrix A (that is, λ is a root of characteristic equation $\det(A - \lambda I) = 0$, k times), there are k linearly independent eigenvectors corresponding to λ. By applying the Gram–Schmidt Orthogonalization Process, we can always replace these k linearly independent eigenvectors with a set of k mutually orthogonal eigenvectors of unit length. (For another view of this, see Example (7.2.4).)

By combining (2) and (3), we see that if A is an $n \times n$-symmetric matrix, then there are n mutually orthogonal unit eigenvectors $\mathbf{u}^{(1)}, \ldots, \mathbf{u}^{(n)}$ corresponding to the n eigenvalues $\lambda_1, \ldots, \lambda_n$ of A (with repeated eigenvalues listed according to their multiplicity). If P is the $n \times n$-matrix whose ith column is the unit eigenvector $\mathbf{u}^{(i)}$ corresponding to λ_i, and if D is the diagonal matrix with the eigenvalues $\lambda_1, \ldots, \lambda_n$ down the main diagonal, then the following matrix equation holds:

$$AP = PD,$$

because $A\mathbf{u}^{(i)} = \lambda_i \mathbf{u}^{(i)}$ for $i = 1, \ldots, n$. Since the matrix P is orthogonal (that is, its columns are mutually orthogonal unit vectors), P is invertible and the inverse P^{-1} of P is just the transpose P^{T} of P. It follows that

$$P^{\mathrm{T}}AP = D,$$

that is, the orthogonal matrix P *diagonalizes* A. If $Q_A(\mathbf{x}) = \mathbf{x} \cdot A\mathbf{x}$ is the quadratic form associated with the symmetric matrix A and if $\mathbf{x} = P\mathbf{y}$, then

$$\begin{aligned} Q_A(\mathbf{x}) &= \mathbf{x} \cdot A\mathbf{x} = (P\mathbf{y})^{\mathrm{T}}A(P\mathbf{y}) = \mathbf{y}^{\mathrm{T}}P^{\mathrm{T}}AP\mathbf{y} \\ &= \mathbf{y}^{\mathrm{T}}D\mathbf{y} = \lambda_1 y_1^2 + \lambda_2 y_2^2 + \cdots + \lambda_n y_n^2. \end{aligned}$$

Moreover, since P is invertible, $\mathbf{x} \neq \mathbf{0}$ if and only if $\mathbf{y} \neq \mathbf{0}$. Also, if $\mathbf{y}^{(i)}$ is the vector in R^n with the ith component equal to 1 and all other components equal to zero, and if $\mathbf{x}^{(i)} = P\mathbf{y}^{(i)}$, then

$$Q_A(\mathbf{x}^{(i)}) = \lambda_i$$

for $i = 1, 2, \ldots, n$. These considerations yield the following eigenvalue test for definite, semidefinite, and indefinite matrices.

(1.5.1) Theorem. *If A is a symmetric matrix, then*:

(a) *the matrix A is positive definite* (*resp. negative definite*) *if and only if all the eigenvalues of A are positive* (*resp. negative*);

(b) *the matrix A is positive semidefinite* (*resp. negative semidefinite*) *if and only if all of the eigenvalues of A are nonnegative* (*resp. nonpositive*);

(c) *the matrix A is indefinite if and only if A has at least one positive eigenvalue and at least one negative eigenvalue.*

The following example shows how the eigenvalue criteria in Theorem (1.5.1) can be applied to an optimization problem.

(1.5.2) Example. Let us locate all maximizers, minimizers, and saddle points of

$$f(x_1, x_2, x_3) = x_1^2 + x_2^2 + x_3^2 - 4x_1x_2.$$

The critical points of $f(x_1, x_2, x_3)$ are solutions of the system of equations

$$\begin{aligned} 0 &= \frac{\partial f}{\partial x_1} = 2x_1 - 4x_2, \\ 0 &= \frac{\partial f}{\partial x_2} = -4x_1 + 2x_2, \\ 0 &= \frac{\partial f}{\partial x_3} = 2x_3. \end{aligned}$$

It is easy to check that (0, 0, 0) is the one and only solution of this system. The Hessian of $f(x_1, x_2, x_3)$ is the constant matrix

$$Hf(x_1, x_2, x_3) = \begin{pmatrix} 2 & -4 & 0 \\ -4 & 2 & 0 \\ 0 & 0 & 2 \end{pmatrix}.$$

The eigenvalues of the Hessian are the solutions of the characteristic equation

$$0 = \det \begin{pmatrix} 2-\lambda & -4 & 0 \\ -4 & 2-\lambda & 0 \\ 0 & 0 & 2-\lambda \end{pmatrix} = (2-\lambda)[(2-\lambda)^2 - 16]$$

$$= (2-\lambda)[\lambda^2 - 4\lambda - 12].$$

Thus, the eigenvalues are $\lambda = 2, 6, -2$, so Theorem (1.3.7) shows that (0, 0, 0) is a saddle point. Since $f(x_1, x_2, x_3)$ has continuous first partial derivatives everywhere on R^3, it follows from (1.2.3) that $f(x_1, x_2, x_3)$ has no other minimizers, maximizers, or saddle points.

Exercises

1. Find the local and global minimizers and maximizers of the following functions:
   (a) $f(x) = x^2 + 2x$.
   (b) $f(x) = x^2 e^{-x^2}$.
   (c) $f(x) = x^4 + 4x^3 + 6x^2 + 4x$.
   (d) $f(x) = x + \sin x$.

2. Classify the following matrices according to whether they are positive or negative definite or semidefinite or indefinite:

   (a) $\begin{pmatrix} 1 & 0 & 0 \\ 0 & 3 & 0 \\ 0 & 0 & 5 \end{pmatrix}$. (b) $\begin{pmatrix} -1 & 0 & 0 \\ 0 & -3 & 0 \\ 0 & 0 & -2 \end{pmatrix}$.

   (c) $\begin{pmatrix} 7 & 0 & 0 \\ 0 & -8 & 0 \\ 0 & 0 & 5 \end{pmatrix}$. (d) $\begin{pmatrix} 3 & 1 & 2 \\ 1 & 5 & 3 \\ 2 & 3 & 7 \end{pmatrix}$.

   (e) $\begin{pmatrix} -4 & 0 & 1 \\ 0 & -3 & 2 \\ 1 & 2 & -5 \end{pmatrix}$. (f) $\begin{pmatrix} 2 & -4 & 0 \\ -4 & 8 & 0 \\ 0 & 0 & -3 \end{pmatrix}$.

3. Write the quadratic form $Q_A(\mathbf{x})$ associated with each of the following matrices A:

   (a) $A = \begin{pmatrix} -1 & 2 \\ 2 & 3 \end{pmatrix}$. (b) $A = \begin{pmatrix} 2 & -3 \\ -3 & 0 \end{pmatrix}$.

   (c) $A = \begin{pmatrix} 1 & -1 & 0 \\ -1 & -2 & 2 \\ 0 & 2 & 3 \end{pmatrix}$. (d) $A = \begin{pmatrix} -3 & 1 & 2 \\ 1 & 2 & -1 \\ 2 & -1 & 4 \end{pmatrix}$.

4. Write each of the quadratic forms in the form $\mathbf{x} \cdot A\mathbf{x}$ where A is an appropriate symmetric matrix:
   (a) $3x_1^2 - x_1x_2 + 2x_2^2$.
   (b) $x_1^2 + 2x_2^2 - 3x_3^2 + 2x_1x_2 - 4x_1x_3 + 6x_2x_3$.
   (c) $2x_1^2 - 4x_3^2 + x_1x_2 - x_2x_3$.

5. Suppose $f(\mathbf{x})$ is defined on R^3 by

$$f(\mathbf{x}) = c_1x_1^2 + c_2x_2^2 + c_3x_3^2 + c_4x_1x_2 + c_5x_1x_3 + c_6x_2x_3.$$

   Show that $f(\mathbf{x})$ is the quadratic form associated with $\frac{1}{2}Hf$. Discuss generalizations to higher dimensions.

6. Show that the principal minors of the matrix

$$A = \begin{pmatrix} 1 & -8 \\ 1 & 1 \end{pmatrix}$$

   are positive, but that there are $\mathbf{x} \neq \mathbf{0}$ in R^2 such that $\mathbf{x} \cdot A\mathbf{x} < 0$. Why does this not contradict Theorem (1.3.3)?

7. Use the principal minor criteria to determine (if possible) the nature of the critical points of the following functions:
   (a) $f(x_1, x_2) = x_1^3 + x_2^3 - 3x_1 - 12x_2 + 20$.
   (b) $f(x_1, x_2, x_3) = 3x_1^2 + 2x_2^2 + 2x_3^2 + 2x_1x_2 + 2x_2x_3 + 2x_1x_3$.
   (c) $f(x_1, x_2, x_3) = x_1^2 + x_2^2 + x_3^2 - 4x_1x_2$.
   (d) $f(x_1, x_2) = x_1^4 + x_2^4 - x_1^2 - x_2^2 + 1$.
   (e) $f(x_1, x_2) = 12x_1^3 - 36x_1x_2 - 2x_2^3 + 9x_2^2 - 72x_1 + 60x_2 + 5$.

8. Use the eigenvalue criteria on the Hessian matrix to determine the nature of the critical points for each of the functions in Exercise 7.

9. Show that the functions

$$f(x_1, x_2) = x_1^2 + x_2^3,$$

   and

$$g(x_1, x_2) = x_1^2 + x_2^4$$

   both have a critical point at (0, 0), both have positive semidefinite Hessians at (0, 0), but (0, 0) is a local minimizer for $g(x_1, x_2)$ but not for $f(x_1, x_2)$.

10. Find the global maximizers and minimizers, if they exist, for the following functions:
   (a) $f(x_1, x_2) = x_1^2 - 4x_1 + 2x_2^2 + 7$.
   (b) $f(x_1, x_2) = e^{-(x_1^2 + x_2^2)}$.
   (c) $f(x_1, x_2) = x_1^2 - 2x_1x_2 + \frac{1}{3}x_2^3 - 4x_2$.
   (d) $f(x_1, x_2, x_3) = (2x_1 - x_2)^2 + (x_2 - x_3)^2 + (x_3 - 1)^2$.
   (e) $f(x_1, x_2) = x_1^4 + 16x_1x_2 + x_2^8$.

11. Show that although (0, 0) is a critical point of $f(x_1, x_2) = x_1^5 - x_1x_2^6$, it is neither a local maximizer nor a local minimizer of $f(x_1, x_2)$.

12. Identify the coercive functions in the following list:
   (a) On R^3, let
$$f(x, y, z) = x^3 + y^3 + z^3 - xy.$$
   (b) On R^3, let
$$f(x, y, z) = x^4 + y^4 + z^2 - 3xy - z.$$
   (c) On R^3, let
$$f(x, y, z) = x^4 + y^4 + z^2 - 7xyz^2.$$
   (d) On R^3, let
$$f(x, y, z) = x^4 + y^4 - 2xy^2.$$
   (e) On R^3, let
$$f(x, y, z) = \ln(x^2y^2z^2) - x - y - z.$$
   (f) On R^3, let
$$f(x, y, z) = x^2 + y^2 + z^2 - \sin(xyz).$$

13. Define $f(x, y)$ on R^2 by
$$f(x, y) = x^4 + y^4 - 32y^2.$$
   (a) Find a point in R^2 at which Hf is indefinite.
   (b) Show $f(x, y)$ is coercive.
   (c) Minimize $f(x, y)$ on R^2.

14. Let $\mathbf{a} \in R^n$ be a fixed vector. Define $f(\mathbf{x})$ on R^n by
$$f(\mathbf{x}) = \mathbf{a} \cdot \mathbf{x}.$$
Show $f(\mathbf{x})$ is not coercive. Show that if ε is any positive number, then the function
$$g(\mathbf{x}) = \mathbf{a} \cdot \mathbf{x} + \varepsilon \|\mathbf{x}\|^2$$
is coercive.

15. Define $f(x, y, z)$ on R^3 by
$$f(x, y, z) = e^x + e^y + e^z + 2e^{-x-y-z}.$$
   (a) Show $Hf(x, y, z)$ is positive definite at all points of R^3.
   (b) Show (ln 2/4, ln 2/4, ln 2/4) is the strict global minimizer of $f(x, y, z)$ on R^3.

16. (a) Show that no matter what value of a is chosen, the function
$$f(x_1, x_2) = x_1^3 - 3ax_1x_2 + x_2^3$$
   has no global maximizers.
   (b) Determine the nature of the critical points of this function for all values of a.

17. If $\mathbf{x}^*$ is a critical point of a function $f(\mathbf{x})$, then $\mathbf{x}^*$ is a *weak saddle point* of $f(\mathbf{x})$ if there are points $\mathbf{x}$ arbitrarily close to $\mathbf{x}^*$ where $f(\mathbf{x}^*) > f(\mathbf{x})$ and other points $\mathbf{x}$ arbitrarily close to $\mathbf{x}^*$ where $f(\mathbf{x}) > f(\mathbf{x}^*)$.

(a) Show that the function

$$f(x_1, x_2) = (x_2 - x_1^2)(x_2 - 2x_1^2)$$

has a critical point at (0, 0) which is a weak saddle point. (Hint: Note that $f(x_1, x_2)$ is the product of two factors so that $f(x_1, x_2) > 0$ when both factors are positive or both factors are negative.)

(b) Show that $f(x_1, x_2)$ has a strict local minimizer along every line

$$\begin{cases} x_1 = at, \\ x_2 = bt, \end{cases}$$

through (0, 0) so that (0, 0) is not a saddle point.

18. (Linear Regression). Suppose that $(x_1, y_1), (x_2, y_2), \ldots, (x_n, y_n)$ are n points in the xy-plane and suppose that we want to "fit" a straight line $y = a + bx$ to these points in such a way that the sum of the squares of the vertical deviations of the given points from the line is as small as possible. In other words, we want to choose a, b so that

$$f(a, b) = \sum_{i=1}^{n} (a + bx_i - y_i)^2$$

is as small as possible. The resulting line is called the *linear regression line* for the points $(x_1, y_1), \ldots, (x_n, y_n)$.

(a) Show that the coefficients a and b of the linear regression line are given by

$$a = \bar{y} - b\bar{x}, \qquad b = \frac{n\bar{x}\bar{y} - \sum_{i=1}^{n} x_i y_i}{n(\bar{x})^2 - \sum_{i=1}^{n} x_i^2},$$

where

$$\bar{x} = \frac{1}{n}\sum_{i=1}^{n} x_i, \qquad \bar{y} = \frac{1}{n}\sum_{i=1}^{n} y_i$$

are the averages of the x- and y-coordinates of the n given points.

19. (a) Suppose that A is an arbitrary square matrix. Show that $\mathbf{x} \cdot A\mathbf{x} > 0$ for all $\mathbf{x} \neq \mathbf{0}$ if and only if the symmetric matrix $B = \frac{1}{2}(A + A^{\mathrm{T}})$ is positive definite. (Hint: Use the fact that

$$\mathbf{x} \cdot A\mathbf{x} = A^{\mathrm{T}}\mathbf{x} \cdot \mathbf{x} = \mathbf{x} \cdot A^{\mathrm{T}}\mathbf{x} \quad \text{for all } \mathbf{x}.)$$

(b) Show that if $A = \begin{pmatrix} 3 & 4 \\ 2 & 7 \end{pmatrix}$, then $\mathbf{x} \cdot A\mathbf{x} > 0$ for all $\mathbf{x} \neq \mathbf{0}$ in R^2.

20. Suppose that A is a square matrix and suppose that there is another matrix B such that $A = B^{\mathrm{T}}B$.

(a) Show that A is positive semidefinite.

(b) Show that if B has full column rank (that is, the rank of B is equal to the number of columns of B), then A is positive definite. (Hint: Recall that $\mathbf{y} \cdot B^{\mathrm{T}}\mathbf{x} = (B\mathbf{y}) \cdot \mathbf{x}$ for all $\mathbf{x}$, $\mathbf{y}$.)

21. Suppose that A is a positive definite matrix. Show that the diagonal elements a_{ii} of A are all positive. (Hint: Consider vectors with only one nonzero component.)

22. Suppose that $f(\mathbf{x})$ is a function with continuous second partial derivatives whose Hessian is positive definite at all points in R^n.
    (a) Show that $f(\mathbf{x})$ has at most one critical point in R^n.
    (b) Show, by example, that $f(\mathbf{x})$ may have no critical points.
    (c) Show that if $\mathbf{x}^*$ is a critical point of $f(\mathbf{x})$ then $\mathbf{x}^*$ is the unique strict global minimizer of $f(\mathbf{x})$.
    (d) Show that if $\nabla f(\mathbf{x}^{(1)}) = \nabla f(\mathbf{x}^{(2)})$, then $\mathbf{x}^{(1)} = \mathbf{x}^{(2)}$.

23. Suppose that $f(\mathbf{x})$ is a function with continuous second partial derivatives whose Hessian is positive semidefinite at all points of R^n. Prove that any critical point of $f(\mathbf{x})$ is a global minimizer of $f(\mathbf{x})$.

24. Suppose that A is a $n \times n$-symmetric matrix for which $a_{ii}a_{jj} - a_{ij}^2 < 0$ for some $i \neq j$. Show that A is indefinite. (See (1.3.4)(c).)

25. (a) Let A be a 3×3-symmetric matrix such that $\Delta_1 > 0$, $\Delta_2 > 0$, and $\Delta_3 = 0$. Prove A is positive semidefinite.
    (b) Give an example of a 3×3-symmetric matrix A such that $\Delta_1 > 0$, $\Delta_2 = 0$, and $\Delta_3 = 0$, but such that A is not positive semidefinite.
    (c) Show that if A is a 2×2-symmetric matrix with nonnegative entries such that $\Delta_1 \geq 0$ and $\Delta_2 \geq 0$, then A is positive semidefinite.

26. Show that the function

$$f(x, y, z) = e^{x^2+y^2+z^2} - x^4 - y^6 - z^6$$

has a global minimizer on R^3.

27. Let A be square $n \times n$-matrix. Show that $A + A^{\mathrm{T}}$ is symmetric. Show that

$$\mathbf{x} \cdot A\mathbf{x} = \mathbf{x} \cdot \left(\frac{A + A^{\mathrm{T}}}{2}\right)\mathbf{x}$$

for all $\mathbf{x}$ in R^n. Conclude that $\mathbf{x} \cdot A\mathbf{x} \geq 0$ for all $\mathbf{x}$ in R^n if and only if the symmetric matrix $A + A^{\mathrm{T}}$ is positive semidefinite.

28. Show that the Vandermonde matrix

$$\begin{pmatrix} x^4 & x^3 & x^2 \\ x^3 & x^2 & x \\ x^2 & x & 1 \end{pmatrix}$$

is positive semidefinite but is not positive definite.

29. Let $g(\mathbf{x})$ be a differentiable function on R^m with continuous first partial derivatives. Let A be a $m \times n$-matrix. Define f on R^n by

$$f(\mathbf{y}) = g(A\mathbf{y}).$$

Compute ∇f in terms of ∇g and A.

30. Let $\mathbf{a} = (a_i)$ and $\mathbf{b} = (b_j)$ be fixed vectors in R^n. Let $\mathbf{a} \otimes \mathbf{b}$ be the matrix whose entry on the ith row and jth column is $(a_i b_j)$.
    (a) Show $(\mathbf{a} \otimes \mathbf{b})\mathbf{x} = (\mathbf{b} \cdot \mathbf{x})\mathbf{a}$ for all $\mathbf{x}$ in R^n.
    (b) Show $\mathbf{a} \otimes \mathbf{a}$ is positive semidefinite but if $n \geq 2$, then $\mathbf{a} \otimes \mathbf{a}$ is not positive definite.

31. (a) Let A be an $n \times n$-symmetric matrix. Diagonalize A to show that

$$\frac{\mathbf{x} \cdot A\mathbf{x}}{\|\mathbf{x}\|^2}$$

is greater than or equal to the smallest eigenvalue of A for all $\mathbf{x} \neq \mathbf{0}$ in R^n.

(b) Show that the quadratic form $Q_A(\mathbf{x}) = \mathbf{x} \cdot A\mathbf{x}$ is coercive if and only if A is positive definite.

(c) Conclude from (b) that if

$$f(\mathbf{x}) = a + \mathbf{b} \cdot \mathbf{x} + \tfrac{1}{2}\mathbf{x} \cdot A\mathbf{x}$$

is any quadratic function where $a \in R$, $\mathbf{b} \in R^n$ and A is an $n \times n$-symmetric matrix, then $f(\mathbf{x})$ is coercive if and only if A is positive definite.

32. Find a function $f(x, y)$ on R^2 such that for each real number t, we have

$$\lim_{x \to \infty} f(x, tx) = \lim_{y \to \infty} f(ty, y) = \infty,$$

but such that $f(x, y)$ is not coercive.

33. Consider the function $f(\mathbf{x})$ defined on R^2 by

$$f(x, y) = x^3 + e^{3y} - 3xe^y.$$

Show that $f(\mathbf{x})$ has exactly one critical point and that this point is a local minimizer but not a global minimizer of $f(\mathbf{x})$.

CHAPTER 2

Convex Sets and Convex Functions

The study of convexity is a richly rewarding mathematical experience. Theorems dealing with convexity are invariably clean, easily understood statements such as: "Any critical point of a convex function is a global minimizer for that function." The proofs of convexity theorems are usually not difficult and are often suggested by the intuitive, geometric character of the concepts.

However, our interest in convexity here is not a result of its very appealing mathematical structure. Rather, it is driven by other important considerations. First, convex functions occur frequently and naturally in many optimization problems that arise in statistical, economic, or industrial applications. Second, convexity considerations often make it unnecessary to test the Hessians of functions for positive definiteness, a test which can be difficult in practice as we have seen in Chapter 1. Finally, convexity will be used to establish the entire mathematical basis for an important optimization procedure known as geometric programming, and will be central to our approach to nonlinear programming in general.

2.1. Convex Sets

Here is the basic definition.

(2.1.1) Definition. A set C in R^n is *convex* if for every $\mathbf{x}$ and $\mathbf{y}$ in C, the line segment joining $\mathbf{x}$ and $\mathbf{y}$ also lies in C.

The intuitive idea of a convex set is described in the following figures:

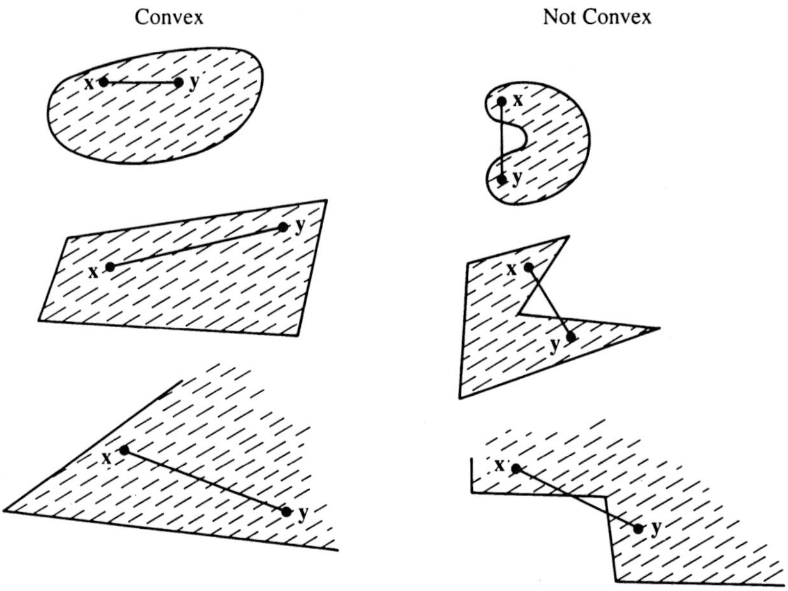

In Section 1.2 of Chapter 1, we defined the line segment $[\mathbf{x}, \mathbf{y}]$ joining $\mathbf{x}$ and $\mathbf{y}$ by

$$[\mathbf{x}, \mathbf{y}] = \{\mathbf{x} + \lambda(\mathbf{y} - \mathbf{x}): 0 \leq \lambda \leq 1\}.$$

Note that this set can also be described as follows:

$$[\mathbf{x}, \mathbf{y}] = \{\lambda\mathbf{y} + (1 - \lambda)\mathbf{x}: 0 \leq \lambda \leq 1\}.$$

Therefore, *a subset C of R^n is convex if and only if for every* $\mathbf{x}$ *and* $\mathbf{y}$ *in C and every λ with* $0 \leq \lambda \leq 1$, *the vector* $\lambda\mathbf{x} + (1 - \lambda)\mathbf{y}$ *is also in C.* (Since $\mathbf{x}$, $\mathbf{y}$ are arbitrary elements of C, they can be interchanged in the preceding statement.) It turns out that this latter description of a convex set is the easiest to use.

(2.1.2) Examples

(a) Lines in R^n can be described in a variety of ways. For example, if $\mathbf{x}$ and $\mathbf{v}$ are vectors in R^n, *the line L through* $\mathbf{x}$ *in the direction of* $\mathbf{v}$ is described below.

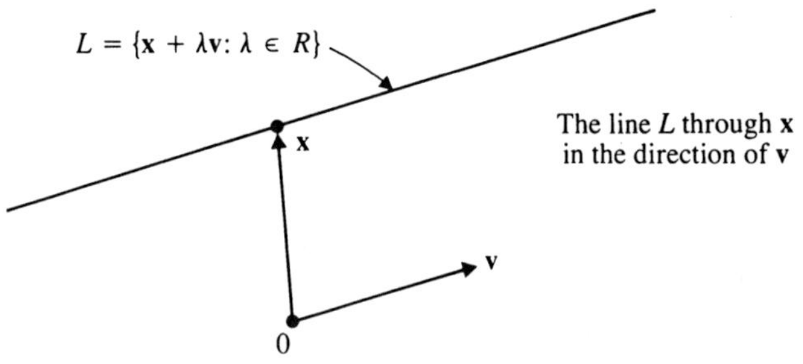

On the other hand, if $\mathbf{x}$ and $\mathbf{y}$ are vectors in R^n, the line L through $\mathbf{x}$ and $\mathbf{y}$ is the set described below.

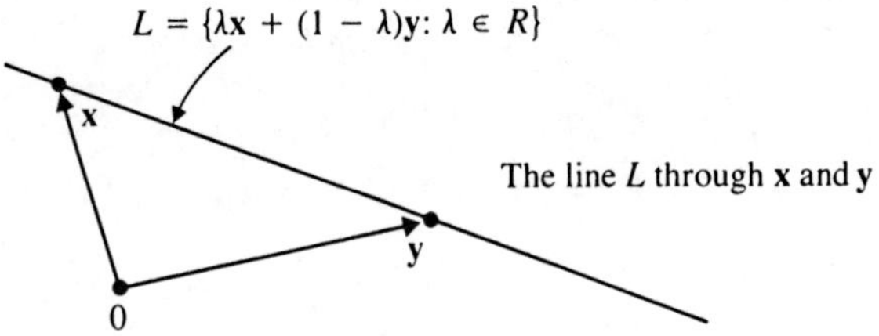

Lines in R^n can also be described as translates of one-dimensional linear subspaces. More precisely, if M is a one-dimensional linear subspace of R^n, if the vector **b** constitutes a basis for M, and if **x** is a given vector, then the *line obtained by translating the subspace M by the vector* **x** is

$$L = \{\mathbf{x} + \lambda \mathbf{b}: \lambda \in R\},$$

that is, L is just the line through **x** in the direction of the basis vector **b**.

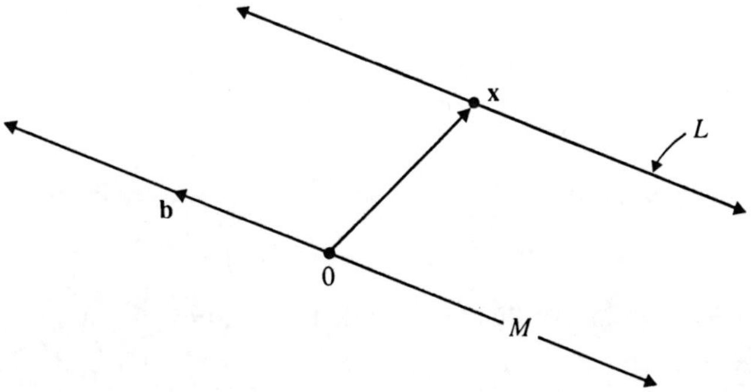

It is clear from any of the set descriptions given above that any line in R^n is a convex set.

(b) In our discussion of certain optimization methods in later chapters, we will often seek to minimize a function $f(\mathbf{x})$ defined on R^n along rays or half-lines in R^n. If **x** and **v** are given vectors in R^n, the *ray* (or *half-line*) *H from* **x** *in the direction of* **v** is described below.

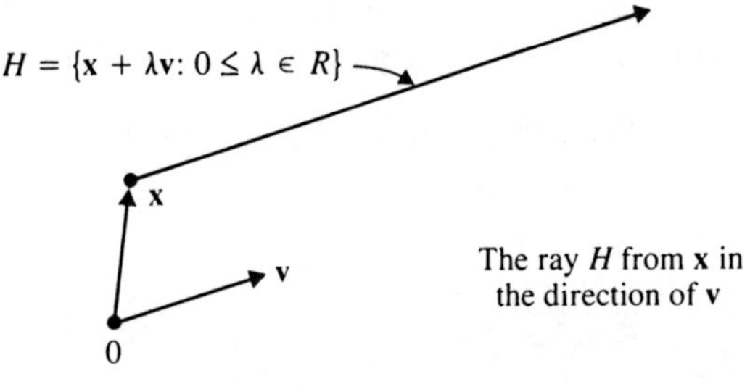

Clearly, any ray in R^n is a convex set.

(c) Any linear subspace M of R^n is a convex set since linear subspaces are closed under addition and scalar multiplication.

(d) If $\mathbf{x}^* \in R^n$ and if $\alpha \in R$, then the *closed half-spaces*

$$F^+ = \{\mathbf{y} \in R^n : \mathbf{x}^* \cdot \mathbf{y} \geq \alpha\},$$

$$F^- = \{\mathbf{y} \in R^n : \mathbf{x}^* \cdot \mathbf{y} \leq \alpha\},$$

and the *open half-spaces*

$$G^+ = \{\mathbf{y} \in R^n : \mathbf{x}^* \cdot \mathbf{y} > \alpha\},$$

$$G^- = \{\mathbf{y} \in R^n : \mathbf{x}^* \cdot \mathbf{y} < \alpha\},$$

determined by $\mathbf{x}^*$ and α are all convex sets. For example, if $\mathbf{y}$, $\mathbf{z}$ are in F^+ and if $0 \leq \lambda \leq 1$, then

$$\mathbf{x}^* \cdot [\lambda\mathbf{y} + (1 - \lambda)\mathbf{z}] = \lambda\mathbf{x}^* \cdot \mathbf{y} + (1 - \lambda)\mathbf{x}^* \cdot \mathbf{z} \geq \lambda\alpha + (1 - \lambda)\alpha = \alpha,$$

so

$$\lambda\mathbf{y} + (1 - \lambda)\mathbf{z} \in F^+.$$

(e) If $\mathbf{x}^* \in R^n$ and if $r > 0$, then the ball

$$B(\mathbf{x}^*, r) = \{\mathbf{y} \in R^n : \|\mathbf{y} - \mathbf{x}^*\| < r\}$$

centered at $\mathbf{x}^*$ of radius r is a convex set. In fact, if $\mathbf{y}$, $\mathbf{z}$ are in $B(\mathbf{x}, r)$, then

$$\|\mathbf{y} - \mathbf{x}^*\| < r, \qquad \|\mathbf{z} - \mathbf{x}^*\| < r.$$

For $0 \leq \lambda \leq 1$, we can apply the Triangle Inequality as follows:

$$\begin{aligned}\|\lambda\mathbf{y} + (1 - \lambda)\mathbf{z} - \mathbf{x}^*\| &= \|\lambda(\mathbf{y} - \mathbf{x}^*) + (1 - \lambda)(\mathbf{z} - \mathbf{x}^*)\| \\ &\leq \lambda\|\mathbf{y} - \mathbf{x}^*\| + (1 - \lambda)\|\mathbf{z} - \mathbf{x}^*\| < \lambda r + (1 - \lambda)r = r\end{aligned}$$

to conclude that $\lambda\mathbf{y} + (1 - \lambda)\mathbf{z} \in B(\mathbf{x}^*, r)$.

(f) If $C_1, C_2, C_3, \ldots, C_k, \ldots$ are convex sets in R^n, then the intersection $\bigcap C_i$ is also convex.[1] For if $\mathbf{y}$, $\mathbf{z}$ belong to this intersection and if $0 \leq \lambda \leq 1$, then $\mathbf{y}$, $\mathbf{z}$ belong to each C_i, so $\lambda\mathbf{y} + (1 - \lambda)\mathbf{z} \in C_i$ for each i since C_i is convex. But then $\lambda\mathbf{y} + (1 - \lambda)\mathbf{z} \in \bigcap C_i$ so that $\bigcap C_i$ is convex.

(g) If $A = (a_{ij})$ is an $m \times n$-matrix and if $\mathbf{b} \in R^m$, then the set S of all solutions $\mathbf{x} \in R^n$ of the system $A\mathbf{x} \leq \mathbf{b}$ of linear inequalities, that is,

$$\begin{aligned} a_{11}x_1 + a_{12}x_2 + \cdots + a_{1n}x_n &\leq b_1, \\ a_{21}x_1 + a_{22}x_2 + \cdots + a_{2n}x_n &\leq b_2, \\ \vdots \qquad\qquad & \quad \vdots \\ a_{m1}x_1 + a_{m2}x_2 + \cdots + a_{mn}x_n &\leq b_m, \end{aligned}$$

[1] Of course, the intersection may be empty but the empty set is convex by default!

is a convex set in R^n. (This fact is of fundamental importance in linear programming.) In fact, if $\mathbf{a}^{(i)} = (a_{i1}, a_{i2}, \ldots, a_{in}) \in R^n$ is the ith row of A and if F_i^- is the half-space

$$F_i^- = \{\mathbf{x} \in R^n: \mathbf{a}^{(i)} \cdot \mathbf{x} \leq b_i\}$$

for $i = 1, \ldots, m$, then the solution set S is just the intersection of the half-spaces $F_1^-, F_2^-, \ldots, F_m^-$, so the convexity of S follows from (d) and (f).

If $\mathbf{x}^{(1)}, \mathbf{x}^{(2)}, \ldots, \mathbf{x}^{(k)}$ are vectors in R^n, then a *weighted average* or *convex combination* of $\mathbf{x}^{(1)}, \ldots, \mathbf{x}^{(k)}$ is any vector

$$\lambda_1 \mathbf{x}^{(1)} + \lambda_2 \mathbf{x}^{(2)} + \cdots + \lambda_k \mathbf{x}^{(k)} = \sum_{i=1}^{k} \lambda_i \mathbf{x}^{(i)},$$

where $\lambda_1, \lambda_2, \ldots, \lambda_k$ are nonnegative numbers whose sum is 1. If C is a convex set in R^n, then, by definition of convexity, the set C contains any convex combination of *two* of its members. The following result shows that convex combinations of more than two vectors in a convex set also belong to the set.

(2.1.3) Theorem. *Let C be a convex set in R^n. Let $\mathbf{x}^{(1)}, \mathbf{x}^{(2)}, \ldots, \mathbf{x}^{(k)}$ be in C. If $\lambda_1, \lambda_2, \ldots, \lambda_k$ are nonnegative numbers whose sum is* 1, *then the convex combination $\sum_{i=1}^{k} \lambda_i \mathbf{x}^{(i)}$ is also in C.*

The essential idea needed for a formal proof of (2.1.3) by mathematical induction is not difficult. As we have already noted, the definition of convexity implies that any convex combination of $k = 2$ vectors in a convex set C also belongs to C. We will now show that this forces convex combinations of $k = 3$ vectors in C to belong to C as well.

Suppose that $\lambda_1, \lambda_2, \lambda_3$ are nonnegative numbers whose sum is 1 and that $\mathbf{x}^{(1)}, \mathbf{x}^{(2)}, \mathbf{x}^{(3)}$ belong to C. We want to show that the convex combination

$$\mathbf{x} = \lambda_1 \mathbf{x}^{(1)} + \lambda_2 \mathbf{x}^{(2)} + \lambda_3 \mathbf{x}^{(3)} = \sum_{i=1}^{3} \lambda_i \mathbf{x}^{(i)}$$

also belongs to C. This is clear if $\lambda_3 = 0$, since $\mathbf{x}$ is then a convex combination of *two* vectors $\mathbf{x}^{(1)}, \mathbf{x}^{(2)}$ in C. If $\lambda_3 \neq 0$, then

$$(*) \qquad \mathbf{x} = \lambda_1 \mathbf{x}^{(1)} + (\lambda_2 + \lambda_3)\left[\frac{\lambda_2}{\lambda_2 + \lambda_3}\mathbf{x}^{(2)} + \frac{\lambda_3}{\lambda_2 + \lambda_3}\mathbf{x}^{(3)}\right].$$

Since

$$\frac{\lambda_2}{\lambda_2 + \lambda_3} + \frac{\lambda_3}{\lambda_2 + \lambda_3} = 1,$$

the expression in square brackets in $(*)$ is a convex combination of *two* vectors in C and so it belongs to C as well. But then, since $\lambda_1 + (\lambda_2 + \lambda_3) = 1$, equation (*) shows that $\mathbf{x}$ is a convex combination of *two* vectors in C and so $\mathbf{x} \in C$.

The preceding argument demonstrates that if C contains any convex combination of two of its vectors, then it must also contain any convex combination of three of its vectors. It also suggests the correct procedure for the inductive step in a formal proof by mathematical induction. We will not supply the details of the formal proof here.

Given two vectors $\mathbf{x}^{(1)}, \mathbf{x}^{(2)}$ in R^n, the set of all convex combinations of $\mathbf{x}^{(1)}$, $\mathbf{x}^{(2)}$ is simply the line segment $[\mathbf{x}^{(1)}, \mathbf{x}^{(2)}]$.

For three vectors $\mathbf{x}^{(1)}, \mathbf{x}^{(2)}, \mathbf{x}^{(3)}$ in R^n, the set of all convex combinations of $\mathbf{x}^{(1)}$, $\mathbf{x}^{(2)}, \mathbf{x}^{(3)}$ is the triangular region determined by $\mathbf{x}^{(1)}, \mathbf{x}^{(2)}, \mathbf{x}^{(3)}$.

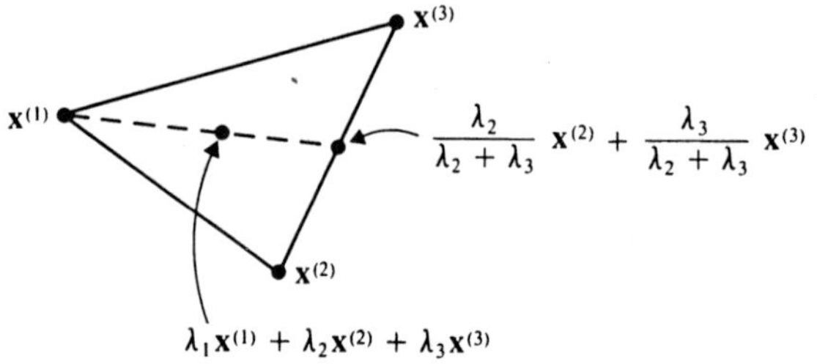

$$\lambda_1 + \lambda_2 + \lambda_3 = 1; \lambda_i > 0; i = 1, 2, 3.$$

More generally, the set of all convex combinations of k vectors $\mathbf{x}^{(1)}, \mathbf{x}^{(2)}, \ldots, \mathbf{x}^{(k)}$ in R^n is the convex polyhedral region determined by $\mathbf{x}^{(1)}, \ldots, \mathbf{x}^{(k)}$.

Given *any* subset D of R^n, there is a smallest convex set containing D, namely, the intersection of all convex sets in R^n that contain D. (R^n itself is a convex set containing D and (2.1.2)(f) shows that the intersection of convex sets is convex.) The smallest convex set containing D is called the *convex hull* of D and is denoted by co(D). The convex hull of two vectors $\mathbf{x}^{(1)}, \mathbf{x}^{(2)}$ in R^n is the line segment $[\mathbf{x}^{(1)}, \mathbf{x}^{(2)}]$ joining $\mathbf{x}^{(1)}$ and $\mathbf{x}^{(2)}$. The convex hull of k vectors $\mathbf{x}^{(1)}, \mathbf{x}^{(2)}, \ldots, \mathbf{x}^{(k)}$ in R^n is the convex polyhedron determined by $\mathbf{x}^{(1)}, \mathbf{x}^{(2)}, \ldots, \mathbf{x}^{(k)}$. These observations suggest the validity of the following result:

(2.1.4) Theorem. *If D is a subset of R^n, then the convex hull* co(D) *of D coincides with the set of all convex combinations of vectors in D.*

A procedure for establishing this result is outlined in Exercise 5.

*2.2. Some Illustrations of Convex Sets in Economics—Linear Production Models

We will now describe a simple and very basic economic model—the linear production model—from the point of view of the geometry of R^n. We will see that convexity plays a basic role in the geometric structure of this model. We will also see that natural economic terminology, assumptions, and conclusions related to the linear production model have equally natural mathematical counterparts. An understanding of the relationships between the economic and mathematical entities involved can help to shed light in both directions.

Let us begin our discussion with the description of the economic setting of the linear production model. Suppose that a certain commodity can be produced by n different production processes $P_1, P_2, \ldots, P_n$ that can operate simultaneously using some or all of m different inputs $I_1, I_2, \ldots, I_m$, each in limited supply. The inputs I_k might include labor, raw materials, energy, machine time, etc., while the production processes P_j can be thought of as "recipes" for combining the inputs to produce the commodity.

The basic economic assumption that underlies the linear production model is the so-called *Law of Constant Returns to Scale* which asserts that if the input levels required by any of the production processes P_j are increased or decreased by a certain multiple, the output level of the commodity is increased or decreased by the same multiple. As a rule, this assumption is not satisfied exactly by real production processes, but it is frequently a reasonable approximation to reality for limited ranges of the multipliers. As such, it provides a basis for a "first approximation" economic model of many "real world" production processes.

Next, we will formulate the mathematical model that corresponds to the economic description of the linear production model given above. Suppose that y_{ij} units of input I_i are rquired by the process P_j to produce x_j units of the output commodity. The corresponding *input coefficients* a_{ij} are defined by

$$y_{ij} = a_{ij}x_j; \qquad i = 1, \ldots, m; \quad j = 1, \ldots, n.$$

The *law of constant returns to scale* implies that the input coefficients a_{ij} depend only on the *type* of input I_i and the production process P_j and not on the output *level* x_j of the jth process. The *total output* of the produced commodity resulting from the simultaneous operation of the n production processes is

$$\text{(O)} \qquad x = x_1 + x_2 + \cdots + x_n = \sum_{j=1}^{n} x_j,$$

while the amount y_i of input I_i required for this production program is

$$y_i = y_{i1} + y_{i2} + \cdots + y_{in} = \sum_{j=1}^{n} y_{ij}.$$

Then y_i can be expressed in terms of the input coefficients as

(I) $$y_i = a_{i1}x_1 + \cdots + a_{in}x_n = \sum_{j=1}^{n} a_{ij}x_j; \qquad i = 1, \ldots, m,$$

so that the vector $\mathbf{y} = (y_1, y_2, \ldots, y_m)$ of input levels is related to the corresponding vector $\mathbf{x} = (x_1, x_2, \ldots, x_n)$ of output levels by the matrix equation $\mathbf{y} = A\mathbf{x}$ where $A = (a_{ij})$ is the matrix of input coefficients. Of course, the number x_j of output units produced by the jth process must be nonnegative, that is,

($\mathrm{P_O}$) $$x_j \geq 0; \qquad j = 1, \ldots, n,$$

and, similarly, the total number of units y_i of input I_i consumed by the production program must be nonnegative, that is,

($\mathrm{P_I}$) $$y_i \geq 0; \qquad i = 1, \ldots, m.$$

Finally, since the supply of the ith input I_i is limited, there are constants $b_1, \ldots, b_m$ such that

($\mathrm{B_I}$) $$y_i \leq b_i; \qquad i = 1, \ldots, m.$$

The conditions (O), (I), ($\mathrm{P_O}$), ($\mathrm{P_I}$), ($\mathrm{B_I}$) constitute the mathematical formulation of the *linear production model.* Economists use the term *technology set* to describe the set of all input levels $\mathbf{y} = (y_1, y_2, \ldots, y_m)$ corresponding to possible output levels $\mathbf{x} = (x_1, x_2, \ldots, x_n)$ satisfying conditions (I), ($\mathrm{P_O}$), ($\mathrm{P_I}$) of the model. (Notice that the total output (O) and the limitations on the supply of inputs ($\mathrm{B_I}$) do not enter into the description of the technology set of the model.)

What kind of set is the technology set T of the linear production model? Well, first, T is obviously a subset of R^m where m is the number of distinct production inputs. But T has a rather special geometric structure—it is a *cone* in R^m, that is, T is a convex set with the additional property that $\lambda\mathbf{y} \in T$ whenever $\mathbf{y} \in T$ and $\lambda \geq 0$. (See Exercise 6).

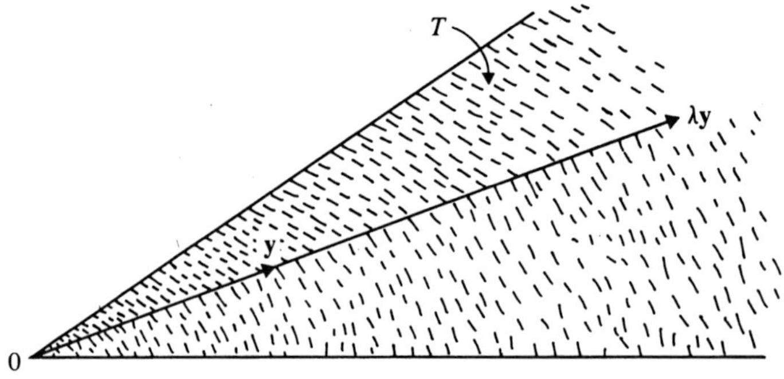

The geometric property that the technology set T is a cone corresponds to two basic economic properties of linear production models—the constant returns to scale discussed earlier and the absence of external discontinuities.

More precisely, the fact that $\lambda \mathbf{y} \in T$ whenever $\lambda \geq 0, \mathbf{y} \in T$ reflects the constant returns to scale, while the convexity of T implies that if a certain level of output is possible using two different sets of certain inputs, that same level of output can be maintained in infinitely many different ways by using different levels of the same inputs—no additional inputs are necessary. This latter economic feature—which economists refer to as the *absence of external discontinuities* —merely reflects that the technology set T has the property that $\lambda \mathbf{y}^{(1)} + (1 - \lambda)\mathbf{y}^{(2)} \in T$ whenever $\mathbf{y}^{(1)}, \mathbf{y}^{(2)} \in T$ and $0 \leq \lambda \leq 1$.

For a given output level $\bar{x}$ of production the *production isoquant* $Q_{\bar{x}}$ is the set of all input levels $\mathbf{y} = (y_1, y_2, \ldots, y_m)$ in the technology set T that result in the production level $\bar{x}$, that is,

$$Q_{\bar{x}} = \left\{ \mathbf{y} \in T : \mathbf{y} = A\mathbf{x} \text{ and } \bar{x} = \sum_{j=1}^{n} x_j \right\}.$$

It is straightforward to check that $Q_{\bar{x}}$ is a convex subset of R^m (Exercise 6). This geometric fact has its economic reflection in the absence of external discontinuities in the linear production model.

In our linear production model, there are n production processes $P_1, P_2, \ldots, P_n$ available to produce output. Given an output level $\bar{x}$ let $\mathbf{y}^{(k)}$ be the input levels required to produce the output level $\bar{x}$ with the *exclusive use of the* k*th production process* P_k, that is, $y_i^{(k)} = a_{ik}\bar{x}$ for $i = 1, \ldots, m$. The input vectors $\mathbf{y}^{(1)}, \ldots, \mathbf{y}^{(n)}$ are elements of the production isoquant $Q_{\bar{x}}$ that "generate" $Q_{\bar{x}}$ in the sense that $Q_{\bar{x}}$ is the convex hull of the set $\{\mathbf{y}^{(1)}, \mathbf{y}^{(2)}, \ldots, \mathbf{y}^{(n)}\}$ (Exercise 6).

The preceding discussion of linear production models illustrates two important points:

(1) Geometric conditions such as convexity can have economic interpretations and implications.
(2) Fancy terms from economics such as "constant returns to scale" and "absence of external discontinuities" often have simple and familiar mathemematical counterparts.

2.3. Convex Functions

Linear functions are very appealing because they are easy to manipulate and their graphs are especially simple (lines in the plane for one independent variable, planes in space for two independent variables, and so on). Anyone who has ever used linear programming knows that linear functions are important in applied mathematics.

In this section, we will begin study of a class of functions, called convex functions, which includes the class of linear functions but which has a much wider range of applications than the class of linear functions. As an added bonus, the mathematics of convex functions is quite beautiful and has a richly geometric and intuitive flavor. We will begin with an informal overview of convex functions of one variable.

A function $f(x)$ defined on an interval I of the real line is *convex* on I if the chord joining any pair of points $(x_1, f(x_1))$ and $(x_2, f(x_2))$ on the graph of $f(x)$ lies on or above the graph of $f(x)$.

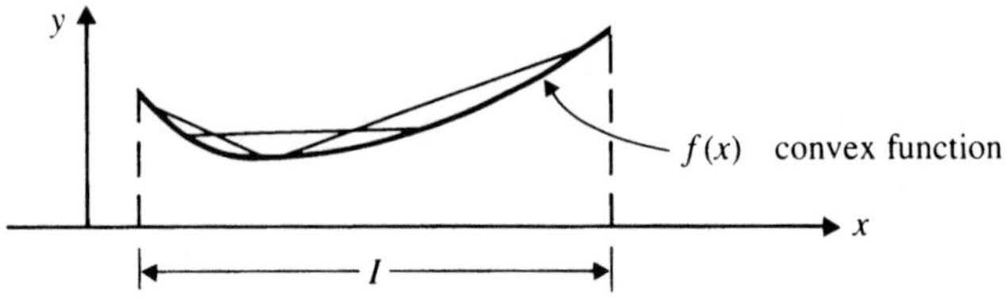

If all such chords lie on or below the graph of $f(x)$, then $f(x)$ is *concave* on I.

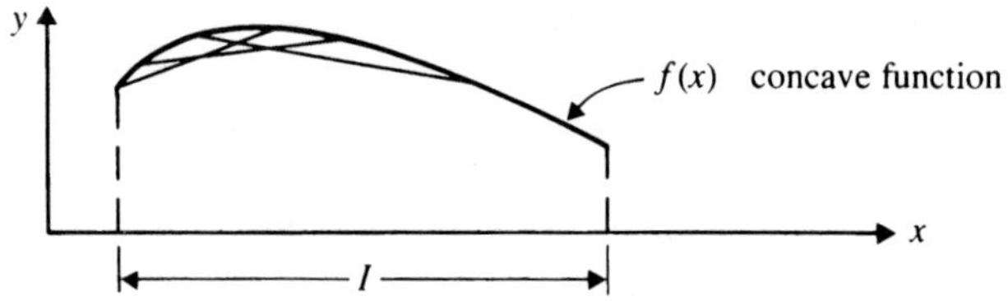

The graphs of convex or concave functions may have "flat spots" where chords joining pairs of points of the graph actually coincide with the corresponding segment of the graph. If no such flat spots occur, we say that $f(x)$ is *strictly convex* or *strictly concave*.

Let us find a more algebraic formulation of the preceding definitions. First, observe that if x_1, x_2 are real numbers, then a number u lies between x_1 and x_2 (inclusive) if and only if there is λ with $0 \leq \lambda \leq 1$ such that $u = \lambda x_1 + (1 - \lambda)x_2$.

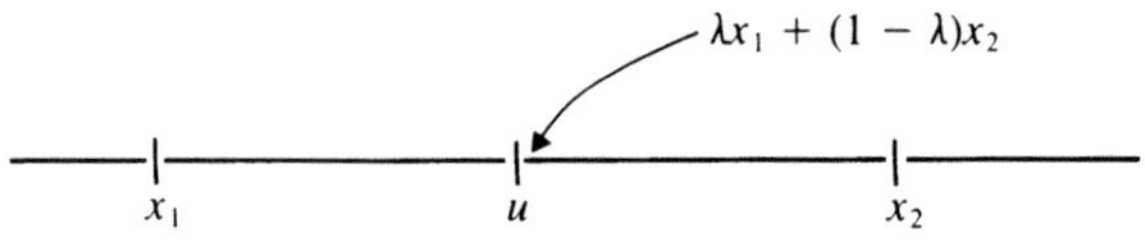

If we apply this observation to points on the x-axis and y-axis, then we see that a function $y = f(x)$ is convex on I if and only if

$$f(\lambda x_1 + [1 - \lambda]x_2) \leq \lambda f(x_1) + [1 - \lambda]f(x_2)$$

for all x_1, x_2 in I and all $0 \leq \lambda \leq 1$.

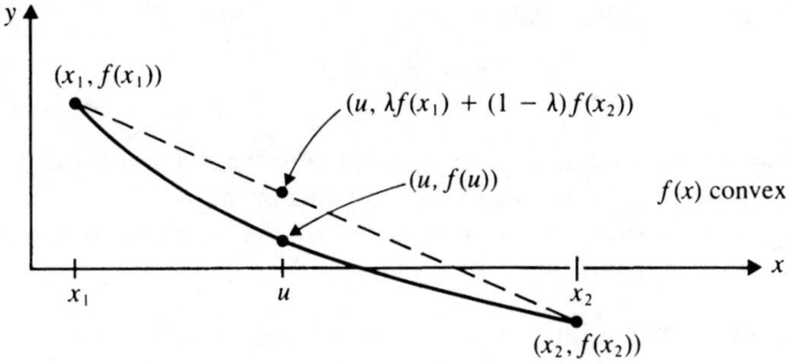

If strict inequality holds in the preceding inequality whenever x_1 and x_2 are distinct points of I and $0 < \lambda < 1$, then $f(x)$ is strictly convex. The corresponding descriptions of concave and strictly concave functions can be obtained by reversing inequalities.

The preceding description of convex function in terms of chords on their graphs can be used to establish the following remarkable fact:

(2.3.1) Theorem. *If $f(x)$ is a convex function defined on an open interval (a, b), then $f(x)$ is continuous on (a, b).*

In keeping with the informal nature of our discussion of convex functions of one variable, we shall present only an outline of the well-known and very elegant proof of (2.3.1). To prove that $f(x)$ is continuous from the right at a given point s in (a, b), choose two fixed points r and u in (a, b) such that $r < s < u$. Let t be a variable point in the interval (s, u) so that

$$r < s < t < u.$$

The convexity of $f(x)$ on (a, b) implies that the point $(t, f(t))$ must lie above or on the line through $(r, f(r))$ and $(s, f(s))$, but below or on the line through $(s, f(s))$ and $(u, f(u))$.

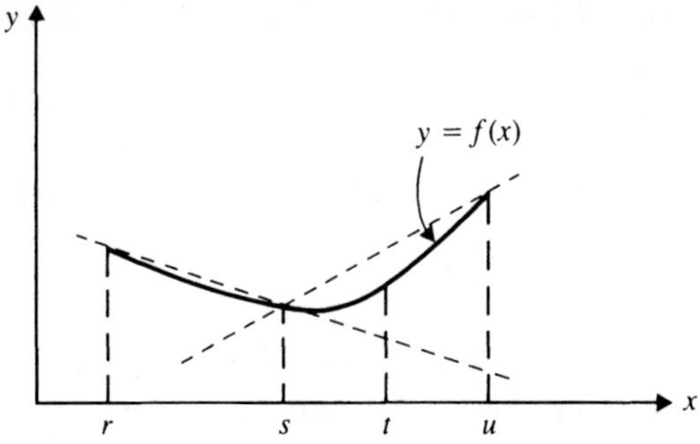

Therefore, by pinching, we see

$$\lim_{t \to s+} f(t) = f(s).$$

A similar argument shows that

$$\lim_{v \to s-} f(v) = f(s)$$

so $f(x)$ is continuous at s and hence on (a, b) since s is an arbitrary point of (a, b). (Exercise 4 deals with the details for this proof.)

There are two other useful descriptions of convex and concave functions that are differentiable or twice differentiable:

I. If $f(x)$ is differentiable on an interval I, then $f(x)$ is convex on I if and only if the tangent line to the graph of $f(x)$ always lies below or on the graph of $f(x)$, that is, for all x_1, x_2 in I

$$f(x_1) + f'(x_1)(x_2 - x_1) \leq f(x_2).$$

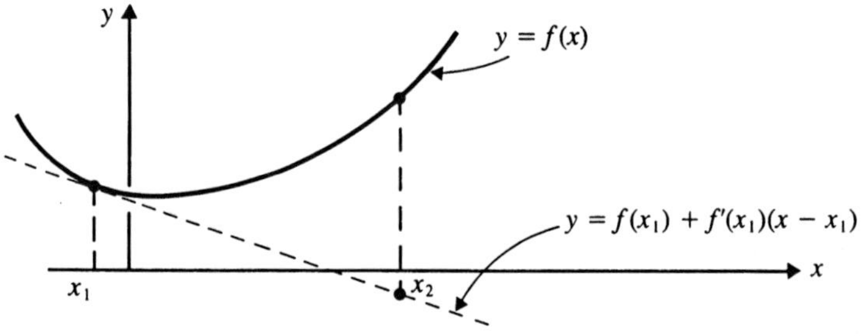

The function $f(x)$ is strictly convex on I if and only if all tangent lines to the graph of $f(x)$ lie below the graph of $f(x)$ and contact the graph only at the point of tangency, that is, for all x_1, x_2 in I with $x_1 \neq x_2$.

$$f(x_1) + f'(x_1)(x_2 - x_1) < f(x_2).$$

Similar descriptions of concave and strictly concave functions are obtained by making the obvious changes.

II. If $f(x)$ is twice differentiable on I, then $f(x)$ is convex on I if and only if $f''(x) \geq 0$ for all x in I. If $f''(x) > 0$ for all x in I, then $f(x)$ is strictly convex; however, $f(x)$ may be strictly convex on I and yet $f''(x)$ may be zero for some $x \in I$. (Think of an example to illustrate this possibility!) Of course, $f(x)$ is concave on I if and only if $f''(x) \leq 0$; moreover, if $f''(x) < 0$ for all x in I, then $f(x)$ is strictly concave on I.

Notice on the basis of the graphs that if $f(x)$ is a convex (resp. concave) function defined on an interval I then any local minimizer (resp. local maximizer) of $f(x)$ is a global minimizer (resp. global maximizer) of $f(x)$ on I. Also, if a strictly convex (resp. strictly concave) function has a local minimizer (resp. local maximizer) on I, then it has only one such minimizer (resp. maximizer) on I.

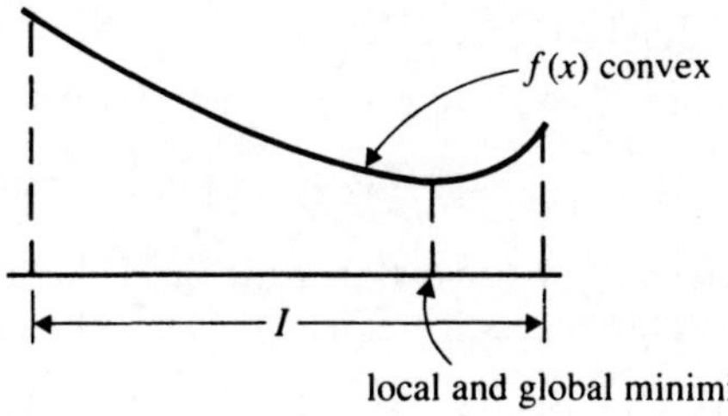

The preceding informal discussion of convex and concave functions can be generalized to functions of several variables. We will now proceed to this more general setting. Essentially all of the features of the one-dimensional case have appropriate counterparts in R^n.

(2.3.2) Definition. Suppose that $f(\mathbf{x})$ is a real-valued function defined on a convex set C in R^n. Then:

(a) the function $f(\mathbf{x})$ is *convex* on C if

$$f(\lambda\mathbf{x} + [1-\lambda]\mathbf{y}) \le \lambda f(\mathbf{x}) + [1-\lambda]f(\mathbf{y})$$

for all $\mathbf{x}$, $\mathbf{y}$ in C and all λ with $0 \le \lambda \le 1$;

(b) the function $f(\mathbf{x})$ is *strictly convex* on C if

$$f(\lambda\mathbf{x} + [1-\lambda]\mathbf{y}) < \lambda f(\mathbf{x}) + [1-\lambda]f(\mathbf{y})$$

for all $\mathbf{x}$, $\mathbf{y}$ in C with $\mathbf{x} \neq \mathbf{y}$ and all λ with $0 < \lambda < 1$. If the inequalities in the above definitions are reversed, we obtain the definitions of *concave* and *strictly concave* functions.

Note that $f(\mathbf{x})$ is convex (resp. strictly convex) on a convex set C if and only if $-f(\mathbf{x})$ is concave (resp. strictly concave) on C. Because of this close connection, we will formulate all results in terms of convex functions only. Corresponding results for concave functions should be clear.

Any linear function of n variables is both convex and concave on R^n. More precisely, if $\mathbf{a} \in R^n$ and $b \in R$, then the function $f(\mathbf{x})$ defined on R^n by

$$f(\mathbf{x}) = \mathbf{a}\cdot\mathbf{x} + b = a_1x_1 + a_2x_2 + \cdots + a_nx_n + b$$

satisfies

$$f(\lambda\mathbf{x} + [1-\lambda]\mathbf{y}) = \lambda f(\mathbf{x}) + [1-\lambda]f(\mathbf{y})$$

for all $\mathbf{x}$, $\mathbf{y}$ in R^n and all $\lambda \in R$, so $f(\mathbf{x})$ is convex and concave (but, of course, not strictly convex or strictly concave) on R^n.

Another example of a convex function on R^n is the function

$$f(\mathbf{x}) = (\mathbf{a}\cdot\mathbf{x})^2, \qquad \mathbf{x} \in R^n,$$

where $\mathbf{a}$ is a fixed vector in R^n. To verify that $f(\mathbf{x})$ is convex, let $\mathbf{x}$, $\mathbf{y}$ be vectors in R^n and let $0 < \lambda < 1$. Then

$$f(\lambda\mathbf{x} + [1-\lambda]\mathbf{y}) = [\mathbf{a}\cdot(\lambda\mathbf{x} + [1-\lambda]\mathbf{y})]^2$$
$$= [\lambda(\mathbf{a}\cdot\mathbf{x}) + [1-\lambda](\mathbf{a}\cdot\mathbf{y})]^2.$$

Since the function $\varphi(t) = t^2$ is convex on R (look at the graph or apply (II) above), the last term in the preceding equation is less than or equal to

$$\lambda(\mathbf{a}\cdot\mathbf{x})^2 + [1-\lambda](\mathbf{a}\cdot\mathbf{y})^2 = \lambda f(\mathbf{x}) + [1-\lambda]f(\mathbf{y}).$$

This shows that $f(\mathbf{x})$ is convex on R^n.

Other examples of convex functions on R^n are provided by those whose Hessians are always positive semidefinite. However, before we can verify this, we need to establish some important properties of convex functions.

(2.3.3) Theorem. *Suppose that $f(\mathbf{x})$ is a convex function defined on a convex subset C of R^n. If $\lambda_1, \lambda_2, \ldots, \lambda_k$ are nonnegative numbers with sum 1 and if $\mathbf{x}^{(1)}$, $\mathbf{x}^{(2)}, \ldots, \mathbf{x}^{(k)}$ are points of C, then*

$$(*) \qquad f\left(\sum_{i=1}^{k} \lambda_i \mathbf{x}^{(i)}\right) \leq \sum_{i=1}^{k} \lambda_i f(\mathbf{x}^{(i)}).$$

If $f(\mathbf{x})$ is strictly convex on C and if all of the λ_i's are positive, then equality holds in $(*)$ if and only if all of the $\mathbf{x}^{(i)}$'s are equal.

Notice that for $k = 2$, equation $(*)$ reduces to the definition of convex function and the second statement of (2.3.3) follows immediately from strict convexity. A formal proof of (2.3.3) can be obtained by mathematical induction on the number k of points of C. Rather than present such a proof here, we will demonstrate the essential part of the inductive step by working out the details for the case of $k = 3$ points of C.

To this end, let $\mathbf{x}^{(1)}, \mathbf{x}^{(2)}, \mathbf{x}^{(3)}$ be points of C and let $\lambda_1, \lambda_2, \lambda_3$ be nonnegative numbers whose sum is one. If $\lambda_3 = 0$, the conclusions of (2.3.3) follow immediately from the definition of convexity and strict convexity since $\lambda_1\mathbf{x}^{(1)} + \lambda_2\mathbf{x}^{(2)} + \lambda_3\mathbf{x}^{(3)}$ reduces to a convex combination of *two* points of C. Hence, we can assume that $\lambda_3 \neq 0$. In this case, observe that

$$\mathbf{z} = \frac{\lambda_2}{\lambda_2 + \lambda_3}\mathbf{x}^{(2)} + \frac{\lambda_3}{\lambda_2 + \lambda_3}\mathbf{x}^{(3)}$$

is a convex combination of *two* points in C and so

$$\lambda_1\mathbf{x}^{(1)} + (\lambda_2 + \lambda_3)\mathbf{z}$$

is also a convex combination of *two* points in C. Consequently, since we know that the conclusions of (2.3.3) hold for the case when $k = 2$, we can proceed as follows:

$$f(\lambda_1\mathbf{x}^{(1)} + \lambda_2\mathbf{x}^{(2)} + \lambda_3\mathbf{x}^{(3)})$$
$$= f(\lambda_1\mathbf{x}^{(1)} + (\lambda_2 + \lambda_3)\mathbf{z})$$

$$\leq \lambda_1 f(\mathbf{x}^{(1)}) + (\lambda_2 + \lambda_3) f\left(\frac{\lambda_2}{\lambda_2 + \lambda_3}\mathbf{x}^{(2)} + \frac{\lambda_3}{\lambda_2 + \lambda_3}\mathbf{x}^{(3)}\right)$$

$$\leq \lambda_1 f(\mathbf{x}^{(1)}) + (\lambda_2 + \lambda_3)\left(\frac{\lambda_2}{\lambda_2 + \lambda_3} f(\mathbf{x}^{(2)}) + \frac{\lambda_3}{\lambda_2 + \lambda_3} f(\mathbf{x}^{(3)})\right)$$

$$= \lambda_1 f(\mathbf{x}^{(1)}) + \lambda_2 f(\mathbf{x}^{(2)}) + \lambda_3 f(\mathbf{x}^{(3)}).$$

This establishes the first conclusion in (2.3.3) for the case when $k = 3$.

To establish the second statement in (2.3.3), observe that if $f(\mathbf{x})$ is strictly convex, then equality holds in the first of the preceding inequalities precisely when

$$\mathbf{x}^{(1)} = \mathbf{z} = \frac{\lambda_2}{\lambda_2 + \lambda_3}\mathbf{x}^{(2)} + \frac{\lambda_3}{\lambda_2 + \lambda_3}\mathbf{x}^{(3)},$$

and equality holds in the second inequality if and only if $\mathbf{x}^{(2)} = \mathbf{x}^{(3)}$. It now follows quickly that there is equality if and only if $\mathbf{x}^{(1)} = \mathbf{x}^{(2)} = \mathbf{x}^{(3)}$. A formal proof of (2.3.3) using mathematical induction follows very similar lines to the step from $k = 2$ to $k = 3$ points of C that we have just discussed.

The following result shows why convex functions are of interest in optimization problems.

(2.3.4) Theorem. *Any local minimizer of a convex function $f(\mathbf{x})$ defined on a convex subset C of R^n is also a global minimizer. Any local minimizer of a strictly convex function $f(\mathbf{x})$ defined on a convex set C in R^n is the unique strict global minimizer of $f(\mathbf{x})$ on C.*

PROOF. Suppose that $\mathbf{x}^*$ is a local minimizer for the convex function $f(\mathbf{x})$ on C. Then there is a positive number r such that $f(\mathbf{x}) \geq f(\mathbf{x}^*)$ whenever $\mathbf{x} \in C$ and $\|\mathbf{x} - \mathbf{x}^*\| < r$.

Given any $\mathbf{y} \in C$, we want to show that $f(\mathbf{y}) \geq f(\mathbf{x}^*)$. To this end, select λ with $0 < \lambda < 1$ and so small that $\mathbf{x}^* + \lambda(\mathbf{y} - \mathbf{x}^*) \in C$ and

$$\|\mathbf{x}^* + \lambda(\mathbf{y} - \mathbf{x}^*) - \mathbf{x}^*\| < r.$$

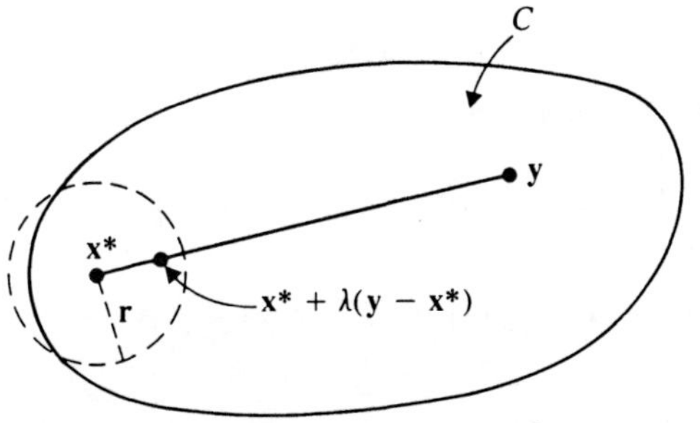

Accordingly,

$$f(\mathbf{x}^*) \leq f(\mathbf{x}^* + \lambda(\mathbf{y} - \mathbf{x}^*)) = f(\lambda\mathbf{y} + [1 - \lambda]\mathbf{x}^*) \leq \lambda f(\mathbf{y}) + [1 - \lambda] f(\mathbf{x}^*),$$

because $f(\mathbf{x})$ is convex on C. Notice that the last inequality is strict if $\mathbf{y} \neq \mathbf{x}^*$ and $f(\mathbf{x})$ is strictly convex. Now subtract $f(\mathbf{x}^*)$ from both sides of the preceding inequality and divide the result by λ to obtain $0 \leq f(\mathbf{y}) - f(\mathbf{x}^*)$ with strict inequality if $f(\mathbf{x})$ is strictly convex and $\mathbf{y} \neq \mathbf{x}^*$. This establishes the desired results.

One of our descriptions of convex differentiable functions of one variable stated that $y = f(x)$ is convex if and only if every tangent line to the graph of $f(x)$ lies on or below the graph of $f(x)$. For a function $f(\mathbf{x})$ with continuous first partial derivatives on some convex subset C of R^n, the object corresponding to a tangent line to the graph of a function of one variable is the *tangent hyperplane* $P_{\mathbf{x}}$ *to the graph of* $f(\mathbf{x})$ *at* $\mathbf{x}$ which is the subset of R^{n+1} defined by the equation

$$P_{\mathbf{x}} = \{(\mathbf{y}, y) \in R^{n+1}: y = f(\mathbf{x}) + \nabla f(\mathbf{x}) \cdot (\mathbf{y} - \mathbf{x})\}.$$

For example, if $f(\mathbf{x})$ is defined on R^2 by

$$f(\mathbf{x}) = f(x_1, x_2) = x_1^2 + 2x_2^2,$$

then the tangent hyperplane to the graph of $f(\mathbf{x})$ at the point $\mathbf{x} = (2, 1)$ is

$$P_{(2,1)} = \{\mathbf{y} \in R^3: y_3 = 6 + 4(y_1 - 2) + 4(y_2 - 1)\},$$

which is the usual tangent plane to the graph of this function at (2, 1) as computed in calculus.

The following result extends the tangent line characterization of the convex functions of one variable mentioned earlier. It says that convex functions are precisely those functions whose graphs are above their tangent hyperplanes.

(2.3.5) Theorem. *Suppose that* $f(\mathbf{x})$ *has continuous first partial derivatives on a convex set* D *in* R^n. *Then*:

(a) *the function* $f(\mathbf{x})$ *is convex if and only if*

$$f(\mathbf{x}) + \nabla f(\mathbf{x}) \cdot (\mathbf{y} - \mathbf{x}) \leq f(\mathbf{y})$$

for all $\mathbf{x}, \mathbf{y}$ *in* D;

(b) *the function* $f(\mathbf{x})$ *is strictly convex on* D *if and only if*

$$f(\mathbf{x}) + \nabla f(\mathbf{x}) \cdot (\mathbf{y} - \mathbf{x}) < f(\mathbf{y})$$

for all $\mathbf{x}, \mathbf{y}$ *in* D *with* $\mathbf{x} \neq \mathbf{y}$.

PROOF. Suppose that $f(\mathbf{x})$ is convex on D, that $\mathbf{x}, \mathbf{y}$ belong to D and that $0 < \lambda \leq 1$. Then

$$f(\mathbf{x} + \lambda(\mathbf{y} - \mathbf{x})) = f(\lambda\mathbf{y} + [1 - \lambda]\mathbf{x}) \leq \lambda f(\mathbf{y}) + [1 - \lambda]f(\mathbf{x}),$$

so that

$$\left(\frac{f(\mathbf{x} + \lambda(\mathbf{y} - \mathbf{x})) - f(\mathbf{x})}{\lambda}\right) \leq f(\mathbf{y}) - f(\mathbf{x}).$$

The expression on the left-hand side of the preceding inequality approaches $\|\mathbf{y} - \mathbf{x}\|$ times the directional derivative of $f(\mathbf{x})$ at $\mathbf{x}$ in the direction of $\mathbf{y} - \mathbf{x}$ as λ approaches 0. Letting $\lambda \to 0$, we obtain

$$\nabla f(\mathbf{x}) \cdot (\mathbf{y} - \mathbf{x}) \leq f(\mathbf{y}) - f(\mathbf{x}).$$

Therefore

$$f(\mathbf{x}) + \nabla f(\mathbf{x}) \cdot (\mathbf{y} - \mathbf{x}) \leq f(\mathbf{y})$$

for all $\mathbf{x}$, $\mathbf{y}$ in D.

Conversely, suppose that

$$f(\mathbf{x}) + \nabla f(\mathbf{x}) \cdot (\mathbf{y} - \mathbf{x}) \leq f(\mathbf{y}) \tag{1}$$

for all $\mathbf{x}$, $\mathbf{y}$ in D. Let $\mathbf{u}$, $\mathbf{v}$ belong to D and $0 < \lambda < 1$. If $\mathbf{w} = \lambda\mathbf{u} + [1 - \lambda]\mathbf{v}$, then $\mathbf{w}$ is in D and

$$\mathbf{v} = \frac{\mathbf{w} - \lambda\mathbf{u}}{1 - \lambda} = \mathbf{w} - \frac{\lambda}{1 - \lambda}(\mathbf{u} - \mathbf{w}),$$

so that

$$\mathbf{v} - \mathbf{w} = -\frac{\lambda}{1 - \lambda}(\mathbf{u} - \mathbf{w}).$$

Consequently, if we apply (1) to the pairs $\mathbf{w}$, $\mathbf{u}$ and $\mathbf{w}$, $\mathbf{v}$, we obtain

$$f(\mathbf{w}) + \nabla f(\mathbf{w}) \cdot (\mathbf{u} - \mathbf{w}) \leq f(\mathbf{u}),$$

$$f(\mathbf{w}) + \nabla f(\mathbf{w}) \cdot (\mathbf{u} - \mathbf{w}) \left(\frac{-\lambda}{1 - \lambda}\right) \leq f(\mathbf{v}).$$

Multiply the first of the preceding inequalities by λ, the second by $[1 - \lambda]$, and add the results to obtain

$$f(\lambda\mathbf{u} + [1 - \lambda]\mathbf{v}) = f(\mathbf{w}) \leq \lambda f(\mathbf{u}) + [1 - \lambda] f(\mathbf{v}).$$

It follows that $f(\mathbf{x})$ is convex on D. This proves (a).

To prove (b), suppose that $f(\mathbf{x})$ is strictly convex on D and that $\mathbf{x}$, $\mathbf{y}$ are distinct points of D. Glance at the first inequality in the proof of (a) and notice that under the assumption of strict convexity, this inequality is strict provided that $0 < \lambda < 1$, that is,

$$f(\mathbf{y}) - f(\mathbf{x}) > \frac{f(\mathbf{x} + \lambda(\mathbf{y} - \mathbf{x})) - f(\mathbf{x})}{\lambda}.$$

Since $f(\mathbf{x})$ is convex on D, part (a) implies that

$$\frac{f(\mathbf{x} + \lambda(\mathbf{y} - \mathbf{x})) - f(\mathbf{x})}{\lambda} \geq \frac{\nabla f(\mathbf{x}) \cdot [\lambda(\mathbf{y} - \mathbf{x})]}{\lambda} = \nabla f(\mathbf{x}) \cdot (\mathbf{y} - \mathbf{x}).$$

By combining the two preceding inequalities, we obtain

$$f(\mathbf{x}) + \nabla f(\mathbf{x})\cdot(\mathbf{y} - \mathbf{x}) < f(\mathbf{y})$$

for all $\mathbf{x}$, $\mathbf{y}$ in D with $\mathbf{x} \neq \mathbf{y}$.

Conversely, suppose this strict inequality holds for all $\mathbf{x}$, $\mathbf{y}$ in D with $\mathbf{x} \neq \mathbf{y}$. If $\mathbf{u}$, $\mathbf{v}$ are distinct points of D and if $0 < \lambda < 1$, then $\mathbf{w} = \lambda\mathbf{u} + [1 - \lambda]\mathbf{v}$ belongs to D and $\mathbf{w} \neq \mathbf{u}$, $\mathbf{w} \neq \mathbf{v}$. Consequently, as in the proof of (a),

$$f(\mathbf{w}) + \nabla f(\mathbf{w})\cdot(\mathbf{u} - \mathbf{w}) < f(\mathbf{u}),$$

$$f(\mathbf{w}) + \nabla f(\mathbf{w})\cdot(\mathbf{u} - \mathbf{w})\left(\frac{-\lambda}{1 - \lambda}\right) < f(\mathbf{v}),$$

which yields the strict inequality

$$f(\lambda\mathbf{u} + [1 - \lambda]\mathbf{v}) = f(\mathbf{w}) < \lambda f(\mathbf{u}) + [1 - \lambda]f(\mathbf{v}).$$

Thus, $f(\mathbf{x})$ is strictly convex on D.

The following striking result is an immediate consequence of (2.3.5). It is the most important and useful result in this chapter.

(2.3.6) Corollary. *If $f(\mathbf{x})$ is a convex function with continuous first partial derivatives on some convex set D, then any critical point of $f(\mathbf{x})$ in D is a global minimizer of $f(\mathbf{x})$.*

PROOF. Suppose that $\mathbf{x}$, $\mathbf{x}^*$ belong to D and that $\mathbf{x}^*$ is a critical point of D. Then $\nabla f(\mathbf{x}^*) = \mathbf{0}$ and so (2.3.5) implies that

$$f(\mathbf{x}^*) = f(\mathbf{x}^*) + \nabla f(\mathbf{x}^*)\cdot(\mathbf{x} - \mathbf{x}^*) \leq f(\mathbf{x}).$$

Consequently, $\mathbf{x}^*$ is a global minimizer of $f(\mathbf{x})$ on D.

Although the definitions of convex and strictly convex functions and their tangent hyperplane descriptions in (2.3.5) provide useful tools for deriving important information concerning their properties, they are not very useful for recognizing convex and strictly convex functions in concrete examples. For instance, the function $f(x) = x^2$ is certainly a convex (even strictly convex) function on R, yet it is cumbersome to verify this fact by using the definition or the tangent line description of convex function. The next two theorems will provide us with an effective means for recognizing convex functions in specific examples.

(2.3.7) Theorem. *Suppose that $f(\mathbf{x})$ has continuous second partial derivatives on some open convex set C in R^n. If the Hessian $Hf(\mathbf{x})$ of $f(\mathbf{x})$ is positive semidefinite (resp. positive definite) on C, then $f(\mathbf{x})$ is convex (resp. strictly convex) on C.*

PROOF. Suppose $\mathbf{x}$, $\mathbf{y}$ are arbitrary points of C. Since C is convex, the line segment $[\mathbf{x}, \mathbf{y}]$ joining $\mathbf{x}$ and $\mathbf{y}$ belongs to C and so (1.2.4) implies that there is a $\mathbf{z} \in [\mathbf{x}, \mathbf{y}]$ such that

$$f(\mathbf{y}) = f(\mathbf{x}) + \nabla f(\mathbf{x}) \cdot (\mathbf{y} - \mathbf{x}) + \tfrac{1}{2}(\mathbf{y} - \mathbf{x}) \cdot Hf(\mathbf{z})(\mathbf{y} - \mathbf{x}).$$

If $Hf(\mathbf{z})$ is positive semidefinite, this equation yields the inequality

$$f(\mathbf{y}) \geq f(\mathbf{x}) + \nabla f(\mathbf{x}) \cdot (\mathbf{y} - \mathbf{x}),$$

and this inequality is strict if $Hf(\mathbf{z})$ is positive definite and $\mathbf{y} \neq \mathbf{x}$. The desired conclusions now follow from (2.3.5).

The following example illustrates how (2.3.7) can be applied to test convexity.

(2.3.8) Example. Consider the function $f(\mathbf{x})$ defined on R^3 by

$$f(x_1, x_2, x_3) = 2x_1^2 + x_2^2 + x_3^2 + 2x_2x_3.$$

The Hessian of $f(\mathbf{x})$ is

$$Hf(\mathbf{x}) = \begin{pmatrix} 4 & 0 & 0 \\ 0 & 2 & 2 \\ 0 & 2 & 2 \end{pmatrix}.$$

The principal minors of $Hf(\mathbf{x})$ are $\Delta_1 = 4$, $\Delta_2 = 8$, $\Delta_3 = 0$, so (1.3.4)(a) shows that $Hf(\mathbf{x})$ is positive semidefinite, and so $f(\mathbf{x})$ is convex by (2.3.7). Since $Hf(\mathbf{x})$ is not positive definite (see (1.3.3)), it is not possible to conclude from (2.3.7) that $f(\mathbf{x})$ is strictly convex on R^3. As a matter of fact, since

$$f(x_1, x_2, x_3) = 2x_1^2 + (x_2 + x_3)^2,$$

we see that $f(\mathbf{x}) = 0$ for all $\mathbf{x}$ on the line where $x_1 = 0$ and $x_3 = -x_2$, so $f(\mathbf{x})$ is not strictly convex.

(2.3.9) Remark. If $f(\mathbf{x})$ is a convex function with continuous second partial derivatives on some convex set C in R^n, then the Hessian $Hf(\mathbf{x})$ can be shown to be positive semidefinite on C. (See Exercise 12.) However, the corresponding statement for strictly convex functions and positive definite Hessians is not valid even for functions of one variable. For example, $f(x) = x^4$ is strictly convex on R, yet its Hessian $12x^2$ is not positive definite at $x = 0$. Thus, the converse of (2.3.7) is valid for convex but not strictly convex functions.

The discussion above shows that many of the results of Chapter 1 are subsumed under the general heading of convex functions. But we must note that verifying that the Hessian is positive semidefinite is sometimes difficult. For instance, the function

$$f(x, y, z) = e^{x^2+y+z} - \ln(x + y) + 3^{z^2}$$

is convex on R^3 but its Hessian is a mess. Fortunately, there are ways other than checking the Hessian to show that a function is convex. The next group of results points in this direction. The following theorem shows that convex functions can be combined in a variety of ways to produce new convex functions.

(2.3.10) Theorem.

(a) *If $f_1(\mathbf{x}), \ldots, f_k(\mathbf{x})$ are convex functions on a convex set C in R^n, then*

$$f(\mathbf{x}) = f_1(\mathbf{x}) + f_2(\mathbf{x}) + \cdots + f_k(\mathbf{x})$$

is convex. Moreover, if at least one $f_i(\mathbf{x})$ is strictly convex on C, then the sum $f(\mathbf{x})$ is strictly convex.

(b) *If $f(\mathbf{x})$ is convex (resp. strictly convex) on a convex set C in R^n and if α is a positive number, then $\alpha f(x)$ is convex (resp. strictly convex) on C.*

(c) *If $f(\mathbf{x})$ is a convex (resp. strictly convex) function defined on a convex set C in R^n, and if $g(y)$ is an increasing (resp. strictly increasing) convex function defined on the range of $f(\mathbf{x})$ in R, then the composite function $g(f(\mathbf{x}))$ is convex (resp. strictly convex) on C.*

PROOF. (a) To show that any finite sum of convex functions on C is convex on C, it suffices to show that the sum $(f_1 + f_2)(\mathbf{x})$ of two convex functions $f_1(\mathbf{x})$ and $f_2(\mathbf{x})$ on C is again convex on C. If $\mathbf{y}, \mathbf{z}$ belong to C and $0 \leq \lambda \leq 1$, then

$$\begin{aligned}(f_1 + f_2)(\lambda\mathbf{y} + [1 - \lambda]\mathbf{z}) &= f_1(\lambda\mathbf{y} + [1 - \lambda]\mathbf{z}) + f_2(\lambda\mathbf{y} + [1 - \lambda]\mathbf{z}) \\ &\leq \lambda f_1(\mathbf{y}) + [1 - \lambda]f_1(\mathbf{z}) + \lambda f_2(\mathbf{y}) + [1 - \lambda]f_2(\mathbf{z}) \\ &= \lambda(f_1 + f_2)(\mathbf{y}) + [1 - \lambda](f_1 + f_2)(\mathbf{z}).\end{aligned}$$

Hence, $(f_1 + f_2)(\mathbf{x})$ is convex on C. Moreover, it is clear from this computation that if either $f_1(\mathbf{x})$ or $f_2(\mathbf{x})$ is strictly convex, then $(f_1 + f_2)(\mathbf{x})$ is strictly convex because strict convexity of either function introduces a strict inequality at the right place.

(b) This result follows by an argument similar to that used in (a).

(c) If $\mathbf{y}, \mathbf{z}$ belong to C and if $0 \leq \lambda \leq 1$, then

$$f(\lambda\mathbf{y} + [1 - \lambda]\mathbf{z}) \leq \lambda f(\mathbf{y}) + [1 - \lambda]f(\mathbf{z})$$

since $f(\mathbf{x})$ is convex on C. Consequently, since g is an increasing, convex function on the range of $f(\mathbf{x})$, it follows that

$$\begin{aligned}g(f(\lambda\mathbf{y} + [1 - \lambda]\mathbf{z})) &\leq g(\lambda f(\mathbf{y}) + [1 - \lambda]f(\mathbf{z})) \\ &\leq \lambda g(f(\mathbf{y})) + [1 - \lambda]g(f(\mathbf{z})).\end{aligned}$$

Thus, the composite function $g(f(\mathbf{x}))$ is convex on C. If $f(\mathbf{x})$ is strictly convex and g is strictly increasing, the first inequality in the preceding computation is strict for $\mathbf{y} \neq \mathbf{z}$ and $0 < \lambda < 1$, so $g(f(\mathbf{x}))$ is strictly convex on C.

(2.3.11) Examples

(a) The function $f(\mathbf{x})$ defined on R^3 by

$$f(x_1, x_2, x_3) = e^{x_1^2 + x_2^2 + x_3^2}$$

is strictly convex.

At first glance, it might seem that the most direct path to verify that $f(\mathbf{x})$ is strictly convex on R^3 would be to show that the Hessian $Hf(\mathbf{x})$ of $f(\mathbf{x})$ is positive definite on R^3. However, the Hessian turns out to be

$$Hf(x) = \begin{pmatrix} (2+4x_1^2)e^{x_1^2+x_2^2+x_3^2} & (4x_1x_2)e^{x_1^2+x_2^2+x_3^2} & (4x_1x_3)e^{x_1^2+x_2^2+x_3^2} \\ (4x_1x_2)e^{x_1^2+x_2^2+x_3^2} & (2+4x_2^2)e^{x_1^2+x_2^2+x_3^2} & (4x_2x_3)e^{x_1^2+x_2^2+x_3^2} \\ (4x_1x_3)e^{x_1^2+x_2^2+x_3^2} & (4x_2x_3)e^{x_1^2+x_2^2+x_3^2} & (2+4x_3^2)e^{x_1^2+x_2^2+x_3^2} \end{pmatrix}.$$

Obviously, proving that the Hessian is positive definite for all $x \in R^3$ will involve quite tedious algebra. No matter! There is a much simpler way to handle the problem.

First, note that

$$h(x_1, x_2, x_3) = x_1^2 + x_2^2 + x_3^2$$

is strictly convex since its Hessian

$$Hh(x_1, x_2, x_3) = \begin{pmatrix} 2 & 0 & 0 \\ 0 & 2 & 0 \\ 0 & 0 & 2 \end{pmatrix}$$

is obviously positive definite. Also, $g(t) = e^t$ is strictly increasing (since $g'(t) = e^t > 0$ for all $t \in R$) and (strictly) convex (since $g''(t) = e^t > 0$ for all $t \in R$). Therefore, by (2.3.10)(c), $f(\mathbf{x}) = g(h(\mathbf{x}))$ is strictly convex on R^3.

(b) Suppose that $\mathbf{a}^{(1)}, \mathbf{a}^{(2)}, \ldots, \mathbf{a}^{(k)}$ are fixed vectors in R^n and that $c_1, c_2, \ldots, c_k$ are positive real numbers. Then the function $f(\mathbf{x})$ defined on R^n by

$$f(\mathbf{x}) = \sum_{i=1}^{k} c_i e^{\mathbf{a}^{(i)} \cdot \mathbf{x}}$$

is convex.

To prove this statement, first observe that the functions $g_i(\mathbf{x})$ on R^n defined by

$$g_i(\mathbf{x}) = \mathbf{a}^{(i)} \cdot \mathbf{x}, \qquad i = 1, 2, \ldots, k$$

are linear and therefore convex on R^n. Since $h(t) = e^t$ is increasing and convex on R, it follows from (2.3.10)(c) that the functions

$$h(g_i(\mathbf{x})) = e^{\mathbf{a}^{(i)} \cdot \mathbf{x}}, \qquad i = 1, 2, \ldots, k$$

are all convex on R^n. Since $c_1, c_2, \ldots, c_k$ are positive real numbers, we can apply (2.3.10)(a), (b) to conclude that

$$f(\mathbf{x}) = \sum_{i=1}^{k} c_i e^{\mathbf{a}^{(i)} \cdot \mathbf{x}}$$

is convex on R^n.

(c) The function $f(\mathbf{x})$ defined on R^2 by

$$f(x_1, x_2) = x_1^2 - 4x_1x_2 + 5x_2^2 - \ln x_1x_2$$

is strictly convex on $C = \{\mathbf{x} \in R^2 : x_1 > 0, x_2 > 0\}$.

In fact, $f(\mathbf{x}) = g(\mathbf{x}) + h(\mathbf{x})$ where

$$g(x_1, x_2) = x_1^2 - 4x_1x_2 + 5x_2^2, \qquad h(x_1, x_2) = -\ln(x_1x_2),$$

so (2.3.10)(a) will imply that $f(\mathbf{x})$ is strictly convex once we show that $g(\mathbf{x})$ and $h(\mathbf{x})$ are convex and at least one of these functions is strictly convex on C. But the Hessian of $g(\mathbf{x})$ is

$$\begin{pmatrix} 2 & -4 \\ -4 & 10 \end{pmatrix}$$

and the principal minors of this matrix are $\Delta_1 = 2$, $\Delta_2 = 4$; hence, $g(\mathbf{x})$ is strictly convex on R^2. Consequently, all that we need to do now is to show that $h(\mathbf{x})$ is convex on C. But

$$h(x_1, x_2) = -\ln x_1 - \ln x_2$$

and the function $\varphi(t) = -\ln t$ $(t > 0)$ is (strictly) convex since $\varphi''(t) = 1/t^2$, so $h(\mathbf{x})$ is convex on C by (2.3.10)(c).

2.4. Convexity and the Arithmetic–Geometric Mean Inequality—An Introduction to Geometric Programming

Inequalities are very useful in almost every field of mathematics, and nonlinear programming is no exception. A number of important inequalities can be derived by applying the fact that suitably chosen functions are convex. One such inequality, the Arithmetic–Geometric Mean Inequality, is a very useful tool for the solution of certain practical optimization problems that are virtually intractable using methods based only on calculus. In fact, this inequality is the mathematical basis of an important formalized optimization procedure called geometric programming which we will introduce in this section and the next and study in greater detail in Chapter 5.

This section begins with the derivation of the Arithmetic–Geometric Mean Inequality. The most familiar special case of this inequality states that if x_1 and x_2 are positive numbers, then

$$\sqrt{x_1x_2} \le \tfrac{1}{2}x_1 + \tfrac{1}{2}x_2 \tag{1}$$

with equality in (1) if and only if $x_1 = x_2$. The right-hand side of (1) is the *arithmetic mean* and the left-hand side is called the *geometric mean* of the positive numbers x_1 and x_2. The inequality (1) is easy to verify by noting that this inequality is equivalent to

$$0 \le (\sqrt{x_1} - \sqrt{x_2})^2 = x_1 - 2\sqrt{x_1}\sqrt{x_2} + x_2,$$

which is obviously true. Note that there is equality in this inequality if and only if $x_1 = x_2$.

A more general version of the Arithmetic–Geometric Mean Inequality asserts that if $x_1, x_2, \ldots, x_n$ are positive real numbers, then

$$\sqrt[n]{x_1 x_2 \ldots x_n} \leq \frac{1}{n}x_1 + \frac{1}{n}x_2 + \cdots + \frac{1}{n}x_n, \tag{2}$$

with equality in (2) if and only if $x_1 = x_2 = \cdots = x_n$. A very simple proof of this version of the inequality can be based on the fact that the rectangle of maximum area with a given perimeter is a square. (See Exercise 20.)

Notice that the exponents of the variables on the left-hand sides of (1) and (2) are equal positive numbers whose sum is one (that is, there are two exponents equal to $\frac{1}{2}$ in (1) and n exponents equal to $1/n$ in (2)). These exponents, which also serve as multipliers of the variables on the right-hand sides of (1) and (2) are called *weights* of the associated variables. Although the weights of all the variables in (1) and (2) are equal, the general form of the Arithmetic–Geometric Mean Inequality allows these weights to vary from variable to variable so long as they are positive with sum equal to one. In the following statement of this inequality, we use the symbol $\prod$ to denote the *product* of the indexed terms that follow this symbol.

(2.4.1) Theorem (The Arithmetic–Geometric Mean Inequality (or (A–G) Inequality)). *If $x_1, x_2, \ldots, x_n$ are positive real numbers and if $\delta_1, \delta_2, \ldots, \delta_n$ are positive numbers whose sum is one, then*

$$\text{(A–G)} \qquad \prod_{i=1}^{n} (x_i)^{\delta_i} \leq \sum_{i=1}^{n} \delta_i x_i,$$

with equality in (A–G) *if and only if* $x_1 = x_2 = \cdots = x_n$.

The product of the left-hand side of (A–G) is called the *geometric mean* of $x_1, x_2, \ldots, x_n$ *with weights* $\delta_1, \delta_2, \ldots, \delta_n$ while the sum of the right of (A–G) is the *arithmetic mean* of $x_1, x_2, \ldots, x_n$ *with weights* $\delta_1, \delta_2, \ldots, \delta_n$. Notice that inequalities (1) and (2) are both special cases of (A–G) in which the weights are equal.

The (A–G) Inequality (2.4.1) is quite easy to establish by making use of convexity considerations in the following way. First, observe that the function $f(x)$ defined for $x > 0$ by

$$f(x) = -\ln x$$

is strictly convex since $f''(x) = 1/x^2 > 0$. Consequently, if $x_1, x_2, \ldots, x_n$ and $\delta_1, \delta_2, \ldots, \delta_n$ are positive numbers such that

$$\delta_1 + \delta_2 + \cdots + \delta_n = 1$$

then (2.3.3) implies that

$$-\ln\left(\sum_{i=1}^{n} \delta_i x_i\right) = f\left(\sum_{i=1}^{n} \delta_i x_i\right) \leq \sum_{i=1}^{n} \delta_i f(x_i) = -\sum_{i=1}^{n} \delta_i \ln x_i$$

with equality if and only if all of the x_i's are equal. The preceding inequality is equivalent to

$$\ln\left(\sum_{i=1}^{n} \delta_i x_i\right) \geq \sum_{i=1}^{n} \ln(x_i^{\delta_i}) = \ln\left(\prod_{i=1}^{n} x_i^{\delta_i}\right).$$

Consequently, since the logarithm function is strictly increasing, we obtain

$$\sum_{i=1}^{n} \delta_i x_i \geq \prod_{i=1}^{n} (x_i)^{\delta_i}$$

with equality in this inequality if and only if all of the x_i's are equal. This establishes the (A–G) Inequality (2.4.1).

As we shall see, the (A–G) Inequality is well suited for solving a fairly hefty class of nonlinear optimization problems. Before we attempt to identify this class of problems more precisely and formalize the optimization procedure, let us apply (A–G) to provide alternate solutions of some standard max–min problems from calculus.

(2.4.2) Examples

(a) Find the open rectangular box with a fixed surface area S_0 that has the largest volume.

SOLUTION. Refer to the figure to see that

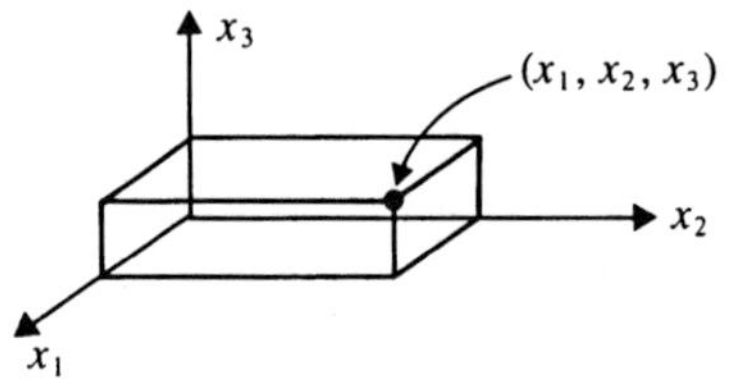

Volume $= V = x_1 x_2 x_3$.

Surface Area $= S_0 = x_1 x_2 + 2x_1 x_3 + 2x_2 x_3$.

Therefore, our job is to solve the following problem:

$$\begin{aligned} &\text{Maximize} \quad V(x_1, x_2, x_3) = x_1 x_2 x_3, \\ &\text{subject to} \quad x_1 x_2 + 2x_1 x_3 + 2x_2 x_3 = S_0. \end{aligned}$$

In this form, the problem is a natural for solution by the (A–G) Inequality. Let us see why. Note that

$$\begin{aligned} S_0 &= x_1 x_2 + 2x_1 x_3 + 2x_2 x_3 = 3\left(\frac{x_1 x_2 + 2x_1 x_3 + 2x_2 x_3}{3}\right) \\ &\overset{\text{(A-G)}}{\geq} 3((x_1 x_2)^{1/3}(2x_1 x_3)^{1/3}(2x_2 x_3)^{1/3}) \\ &= 3 \cdot 4^{1/3}(x_1^2 x_2^2 x_3^2)^{1/3} = 3 \cdot 4^{1/3} \cdot V^{2/3}. \end{aligned}$$

Therefore, V is largest when there is equality in this (A–G) Inequality, that is,

V is maximized when $x_1 = x_1^*$, $x_2 = x_2^*$, $x_3 = x_3^*$, where

$$x_1^* x_2^* = 2x_1^* x_3^* = 2x_2^* x_3^* = \frac{S_0}{3}.$$

A (very) little algebra shows that

$$x_1^* = x_2^* = \sqrt{\frac{S_0}{3}}, \qquad x_3^* = \frac{1}{2}\sqrt{\frac{S_0}{3}},$$

and therefore that the maximum volume of the box is

$$V_0 = x_1^* x_2^* x_3^* = \frac{S_0^{3/2}}{2 \cdot 3^{3/2}}.$$

(b) Find the open rectangular box with fixed volume V_0 that has the least surface area.

SOLUTION. Again refer to the figure in (a) to see that we want to solve the following problem:

$$\text{Minimize} \quad S(x_1, x_2, x_3) = x_1x_2 + 2x_1x_3 + 2x_2x_3$$
$$\text{subject to} \quad x_1x_2x_3 = V_0.$$

By proceeding exactly as in (a), we obtain

$$S = 3\left(\frac{x_1x_2 + 2x_1x_3 + 2x_2x_3}{3}\right) \overset{\text{(A-G)}}{\geq} 3 \cdot 4^{1/3} V_0^{2/3}.$$

Therefore, S is smallest when there is equality in the (A–G) Inequality, that is, S in minimized when $x_1 = x_1^*$, $x_2 = x_2^*$, $x_3 = x_3^*$ where

$$x_1^* x_2^* = 2x_1^* x_3^* = 2x_2^* x_3^* = 4^{1/3} V_0^{2/3}.$$

A short computation shows that

$$x_1^* = x_2^* = 4^{1/6} V_0^{1/3}, \qquad x_3^* = \frac{4^{1/6} V_0^{1/3}}{2}$$

are the dimensions of the box with the least surface area for a given volume V_0, and that the minimum surface area is

$$S_0 = x_1^* x_2^* + 2x_1^* x_3^* + 2x_2^* x_3^* = 3 \cdot 4^{1/3} V_0^{2/3}.$$

(2.4.3) Examples

(a) Maximize the volume of a cylindrical can of fixed cost c_0 cents if the cost of the top and bottom of the can is c_1 cents per square inch and the cost of the side of the can is c_2 cents per square inch.

SOLUTION. If r is the radius and h is the height of the can in inches, then the volume of the can is

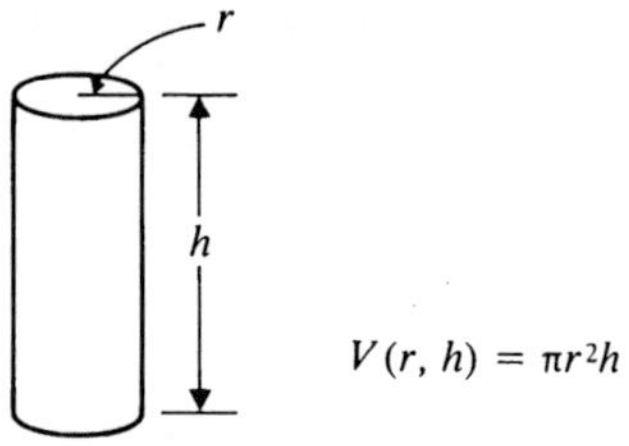

and the cost of the can is

$$c_0 = 2\pi r^2 c_1 + 2\pi r h c_2.$$

Now if we proceed as in Example (2.4.2), we obtain

$$c_0 = 2\pi r^2 c_1 + 2\pi r h c_2 = 4\left(\frac{\pi r^2 c_1}{2} + \frac{\pi r h c_2}{2}\right)$$
$$\overset{(A-G)}{\geq} 4(\pi r^2 c_1)^{1/2}(\pi r h c_2)^{1/2} = 4\pi r^{3/2} h^{1/2}(c_1 c_2)^{1/2}.$$

Unfortunately, unlike the situation in (2.4.2), the term on the right-hand side of the resulting inequality does not reduce to a constant multiple of a power of the volume V, so we cannot proceed as before. What we need to do is to "split" c_0 into a sum of terms in such a way that application of the (A–G) Inequality yields a constant multiple of a power of V on the right-hand side. A bit of experimentation will show that such a split can be accomplished as follows:

$$c_0 = 2\pi r^2 c_1 + \pi r h c_2 + \pi r h c_2 = 3\frac{2\pi r^2 c_1 + \pi r h c_2 + \pi r h c_2}{3}$$
$$\overset{(A-G)}{\geq} 3(2\pi r^2 c_1)^{1/3}(\pi r h c_2)^{1/3}(\pi r h c_2)^{1/3} = 3(2\pi)^{1/3}(c_1 c_2^2)^{1/3} V^{2/3}.$$

Now we can see that V is largest when there is equality in this (A–G) Inequality, that is, when

$$2\pi r^2 c_1 = \pi r h c_2 = \pi r h c_2 = \frac{c_0}{3}.$$

We will omit the easy calculation of the maximizing values of r and h. However, we want to point out an interesting cost analysis related to our solution of the problem: *Regardless of the values of c_1 and c_2, the optimal dimensions for the can will assign $\frac{1}{3}$ of the total cost to the top and bottom and $\frac{2}{3}$ of the total cost to the side.*

(b) Minimize the cost of a cylindrical can of fixed volume V_0 if the cost of the top and bottom of the can is c_1 cents per square foot and the cost of the side of the can is c_2 cents per square foot.

SOLUTION. If r and h denote the radius and height of the can, then our problem can be formulated as follows:

$$\text{Minimize} \quad c(r, h) = 2\pi r^2 c_1 + 2\pi r h c_2,$$
$$\text{subject to} \quad \pi r^2 h = V_0.$$

The same split of the cost function that we used in part (a), in conjunction with the (A–G) Inequality, yields

$$c(r, h) = 3\left(\frac{2\pi r^2 c_1 + \pi rhc_2 + \pi rhc_2}{3}\right)$$

$$\overset{(A-G)}{\geq} 3(2\pi)^{1/3}(c_1 c_2^2)^{1/3} V_0^{2/3}.$$

Therefore, the cost is smallest when there is equality in this (A–G) Inequality, that is, when

$$2\pi r^2 c_1 = \pi rhc_2 = \pi rhc_2 = (2\pi)^{1/3}(c_1 c_2^2)^{1/3} V_0^{2/3}.$$

For the minimizing values of r and h that solve these equations, we see again that *the optimal cost allocation is* $\frac{1}{3}$ *of the cost for the top and bottom and* $\frac{2}{3}$ *of the cost for the sides regardless of the values of* c_1 *and* c_2. We also see that the minimum cost is

$$3(2\pi)^{1/3}(c_1)^{1/3}(c_2)^{2/3} V_0^{2/3}.$$

Notice that if the price of the sides doubles, then the new minimum cost is

$$3(2\pi)^{1/3}(c_1)^{1/3}(2c_2)^{2/3} V_0^{2/3} = 2^{2/3} \cdot \text{old cost}.$$

On the other hand, if the cost of the top and bottom is doubled, the new cost is $2^{1/3}$ times the old cost, while if the volume V_0 is doubled, the optimal cost is increased by a factor of $2^{2/3}$.

The pairs of problems in (2.4.2) and (2.4.3) furnish our first examples of *dual problems.* In both examples, problems (a) and (b) deal with essentially the same functions except that the objective function in one problem is the constraint function in the other, and one problem is a minimization problem while the other is a maximization problem. The following example provides a further illustration of duality.

(2.4.4) Example. Consider the following problem:

(*P*) Maximize $f(x_1, x_2, x_3) = x_1 x_2^2 x_3$,

subject to $x_1 + x_2 + x_3^2 = k$,

where $k > 0$ is fixed and x_1, x_2, x_3 are positive real numbers.

In this case, the dual problem is

(*D*) Minimize $g(x_1, x_2, x_3) = x_1 + x_2 + x_3^2$,

subject to $x_1 x_2^2 x_3 = c$,

where $c > 0$ is fixed and x_1, x_2, x_3 are positive real numbers.

Of course, (*P*) is also the dual problem for (*D*).

To solve both problems, we need to split $x_1 + x_2 + x_3^2$ in such a way that application of the (A–G) Inequality will yield a constant multiple of a suitable power of $(x_1 x_2^2 x_3)$ at the lower end of the inequality. A bit of thought will show that this can be accomplished as follows:

$$x_1 + x_2 + x_3^2 = \frac{x_1}{2} + \frac{x_1}{2} + \frac{x_2}{4} + \frac{x_2}{4} + \frac{x_2}{4} + \frac{x_2}{4} + x_3^2$$

$$= 7\left(\frac{\frac{x_1}{2} + \frac{x_1}{2} + \frac{x_2}{4} + \frac{x_2}{4} + \frac{x_2}{4} + \frac{x_2}{4} + x_3^2}{7}\right)$$

$$\overset{\text{(A-G)}}{\geq} 7(\tfrac{1}{2})^{2/7}(\tfrac{1}{4})^{4/7}(x_1 x_2^2 x_3)^{2/7}.$$

Equality holds in this inequality precisely when

$$\frac{x_1}{2} = \frac{x_2}{4} = x_3^2 = \frac{k}{7}.$$

To maximize $x_1 x_2^2 x_3$ subject to the constraint $x_1 + x_2 + x_3^2 = k$, we choose the optimal values x_1^*, x_2^*, x_3^* that force equality in the (A–G) Inequality with the upper end of the inequality equal to k, that is,

$$x_1^* = \frac{2k}{7}, \qquad x_2^* = \frac{4k}{7}, \qquad x_3^* = \sqrt{\frac{k}{7}}.$$

To minimize $x_1 + x_2 + x_3^2$ subject to the constraint $x_1 x_2^2 x_3 = c$, choose the optimal values x_1^*, x_2^*, x_3^* that force equality in the (A–G) Inequality with the lower end of the inequality equal to $(1/2)^{2/7}(1/4)^{4/7}c^{2/7}$, that is,

$$x_1^* = 2(\tfrac{1}{2})^{2/7}(\tfrac{1}{4})^{4/7}c^{2/7}, \qquad x_2^* = 2x_1^*, \qquad x_3^* = \sqrt{\tfrac{1}{2}x_1^*}.$$

The (A–G) Inequality can also be used to solve some unconstrained minimization problems. The trick is to split the objective function in such a way that application of the (A–G) Inequality yields a constant at a lower end of the inequality. The following examples illustrate the procedure.

(2.4.5) Examples

(a) Find the values of $x > 0$ that minimize the function

$$f(x) = c_1 x^3 + \frac{c_2}{x},$$

where c_1, c_2 are positive constants.

To solve this problem with the (A–G) Inequality, we proceed as follows:

$$f(x) = c_1 x^3 + \frac{c_2}{x} = c_1 x^3 + \frac{1}{3}\frac{c_2}{x} + \frac{1}{3}\frac{c_2}{x} + \frac{1}{3}\frac{c_2}{x}$$

$$= 4\left(\frac{c_1 x^3 + \frac{1}{3}\frac{c_2}{x} + \frac{1}{3}\frac{c_2}{x} + \frac{1}{3}\frac{c_2}{x}}{4}\right)$$

$$\overset{\text{(A-G)}}{\geq} 4(\tfrac{1}{3})^{3/4}c_1^{1/4}c_2^{3/4}.$$

To minimize $f(x)$, force equality in (A–G). This amounts to choosing x^* so that

$$c_1(x^*)^3 = \frac{1}{3}\frac{c_2}{x^*} = (\tfrac{1}{3})^{3/4} c_1^{1/4} c_2^{3/4}.$$

This yields

$$x^* = (\tfrac{1}{3})^{1/4} c_1^{-1/4} c_2^{1/4}$$

as the minimizer.

Actually, the preceding example could have been solved easily as a calculus problem. The next example is not easy to do with calculus methods, but it is quite easy as an application of the (A–G) Inequality.

(b) Find the minimizers of

$$f(x_1, x_2) = 4x_1 + \frac{x_1}{x_2^2} + \frac{4x_2}{x_1}$$

for $x_1 > 0$, $x_2 > 0$.

This problem can be solved by the (A–G) Inequality in the following way:

$$f(x_1, x_2) = 4\left(\frac{4x_1 + \dfrac{x_1}{x_2^2} + \dfrac{2x_2}{x_1} + \dfrac{2x_2}{x_1}}{4}\right)$$

$$\overset{\text{(A-G)}}{\geq} 4(4^{1/4})(2^{2/4})\left[\frac{x_1^2 x_2^2}{x_2^2 x_1^2}\right]^{1/4} = 8.$$

If we force equality in the preceding inequality, we see that the values x_1^* and x_2^* that minimize $f(x_1, x_2)$ are given by

$$4x_1^* = \frac{x_1^*}{(x_2^*)^2} = \frac{2x_2^*}{x_1^*} = 2,$$

that is,

$$x_1^* = \tfrac{1}{2}, \qquad x_2^* = \tfrac{1}{2},$$

and that the minimum value is 8.

Note that this example is not easily attacked by the methods of Chapter 1. Indeed, the critical points are solutions of the system

$$\frac{\partial f}{\partial x_1} = 4 + \frac{1}{x_2^2} - \frac{4x_2}{x_1^2} = 0,$$

$$\frac{\partial f}{\partial x_2} = -\frac{2x_1}{x_2^3} + \frac{4}{x_1} = 0.$$

Solving this system is by no means easy and examination of the Hessian

$$Hf(x_1, x_2) = \begin{pmatrix} \dfrac{8x_2}{x_1^3} & -\dfrac{2}{x_2^3} - \dfrac{4}{x_1^2} \\ -\dfrac{2}{x_2^3} - \dfrac{4}{x_1^2} & \dfrac{6x_1}{x_2^4} \end{pmatrix}$$

is a frightening prospect!

The examples discussed in this section certainly indicate that the (A–G) Inequality can be an important tool for the solution of optimization problems. The only part of our approach that was not completely straightforward was "splitting" the function appropriately at the upper end of the (A–G) Inequality. We did this easily by inspection in the examples considered here because the functions under consideration involved three or fewer variables. However, it is apparent that the splitting process might be difficult to apply for complicated functions of a large number of variables. The other point in our approach that is a bit hazy at the moment is the scope of the method, that is, we have yet to identify precisely the types of optimization problems that are tractable via the (A–G) Inequality. We will address both of these issues in Section 2.5. In particular, we will develop a formal setting and a systematic procedure called *geometric programming* for applying the (A–G) Inequality to optimization problems. In geometric programming, the splitting process which was accomplished by inspection in the examples of this section is replaced by the solution of a certain system of linear equations determined by the problem at hand. This replacement results in a systematic, practical procedure that applies routinely to a wide class of problems. However, the essence of geometric programming is already contained in our informal solutions of the problems considered in this section—the rest is just a matter of making the procedure systematic and routine.

2.5. Unconstrained Geometric Programming

This section presents a systematic procedure for handling unconstrained geometric programming problems. As we demonstrated in the last section, it is often possible to solve these problems directly from the (A–G) Inequality and inspection. However, some problems are too complicated to be done in this way, and the procedure we are about to discuss then comes to the rescue.

We also have another objective in mind for this section. The alert student has already noted that we never proved that it was always possible to force equality in the (A–G) Inequality in the problems of the last section. Once we have established our systematic procedure for unconstrained geometric programming problems, we prove that it is always possible to force this equality.

(2.5.1) Definition. A function $g(\mathbf{t})$ defined for all $\mathbf{t} = (t_1, \ldots, t_m)$ in R^m with $t_j > 0$ for all $j = 1, \ldots, m$ is called a *posynomial* if $g(\mathbf{t})$ is of the form

$$g(\mathbf{t}) = \sum_{i=1}^{n} c_i \prod_{j=1}^{m} (t_j)^{\alpha_{ij}},$$

where the c_i's are *positive* constants and the α_{ij} are *arbitrary* real exponents.

Thus, a posynomial is a linear combination with nonnegative coefficients of terms that are products of real powers of the nonnegative variables $t_1, \ldots, t_m$. For example,

$$g(t_1, t_2) = 3t_1^{-1}t_2^{\sqrt{2}} + t_1^3 t_2^{1/2} + \sqrt{3}t_1^{-1/2}$$

is a posynomial defined on the interior of the first quadrant in R^2.

The goal of unconstrained geometric programming is to solve the following *primal geometric program*:

$$(GP) \quad \text{Minimize the posynomial} \quad g(\mathbf{t}) = \sum_{i=1}^{n} c_i \prod_{j=1}^{m} t_j^{\alpha_{ij}},$$

$$\text{where } t_1 > 0, \ldots, t_m > 0.$$

By a *solution* to (GP) we simply mean a global minimizer $\mathbf{t}^*$ for $g(\mathbf{t})$ on the set of vectors $\mathbf{t}$ in R^m with positive components.

We begin our attack on the program (GP) by observing that $g(\mathbf{t})$ can be rewritten as

$$g(\mathbf{t}) = \sum_{i=1}^{n} \delta_i \left(\frac{c_i \prod_{j=1}^{m} t_j^{\alpha_{ij}}}{\delta_i} \right),$$

where each δ_i is assumed to be a positive number (the Positivity Condition). If we add the restriction that

$$\sum_{i=1}^{n} \delta_i = 1 \qquad \text{(the Normality Condition)},$$

we can apply the (A–G) Inequality to this new expression for $g(\mathbf{t})$ to obtain

$$\begin{aligned} g(\mathbf{t}) &\overset{\text{(A-G)}}{\geq} \prod_{i=1}^{n} \left(\frac{c_i \prod_{j=1}^{m} t_j^{\alpha_{ij}}}{\delta_i} \right)^{\delta_i} \\ &= \prod_{i=1}^{n} \left(\frac{c_i}{\delta_i} \right)^{\delta_i} \left(\prod_{i=1}^{n} \prod_{j=1}^{m} t_j^{\alpha_{ij}\delta_i} \right) \\ &= \prod_{i=1}^{n} \left(\frac{c_i}{\delta_i} \right)^{\delta_i} \prod_{j=1}^{m} t_j^{\sum_i \alpha_{ij}\delta_i} \end{aligned}$$

Therefore, if we impose the additional rectriction that

$$\sum_{i=1}^{n} \alpha_{ij}\delta_i = 0; \qquad j = 1, \ldots, m \qquad \text{(the Orthogonality Condition)},$$

then the preceding inequality yields

$$g(\mathbf{t}) \geq \prod_{i=1}^{n} \left(\frac{c_i}{\delta_i}\right)^{\delta_i}.$$

Thus, if we define

$$v(\boldsymbol{\delta}) = \prod_{i=1}^{n} \left(\frac{c_i}{\delta_i}\right)^{\delta_i}$$

then the preceding computations show that

$$g(\mathbf{t}) \geq v(\boldsymbol{\delta}) \qquad \text{(the Primal–Dual Inequality)}$$

for any $\mathbf{t} \in R^m$ with positive components and any $\boldsymbol{\delta} \in R^n$ that satisfies the Positivity, Normality, and Orthogonality Conditions.

These considerations lead us to consider the following *dual geometric program*:

$$(DGP) \quad \text{Maximize} \quad v(\boldsymbol{\delta}) = \prod_{i=1}^{n} \left(\frac{c_i}{\delta_i}\right)^{\delta_i},$$

$$\text{subject to} \quad \delta_1 > 0, \ldots, \delta_n > 0 \qquad \text{(the Positivity Condition),}$$

$$\sum_{i=1}^{n} \delta_i = 1 \qquad \text{(the Normality Condition),}$$

$$\sum_{i=1}^{n} \alpha_{ij}\delta_i = 0; \quad \text{all } j \qquad \text{(the Orthogonality Condition).}$$

A vector $\boldsymbol{\delta} \in R^n$ that satisfies the Positivity, Normality, and Orthogonality Conditions is a *feasible* vector for (DGP). The dual program is *consistent* if the set of feasible vectors for (DGP) is nonempty. Finally, by a *solution* for the dual program (DGP) we mean a vector $\boldsymbol{\delta}^* \in R^n$ that is a global maximizer for $v(\boldsymbol{\delta})$ on the set of feasible vectors for (DGP).

Note that if $\mathbf{t}^*$ is a solution to the primal program (GP) and if $\boldsymbol{\delta}^*$ is a solution to the dual program (DGP), then

$$g(\mathbf{t}^*) \geq v(\boldsymbol{\delta}^*)$$

by the Primal–Dual Inequality. We will now show, among other things, that $g(\mathbf{t}^*)$ is actually equal to $v(\boldsymbol{\delta}^*)$ and that this equality yields a straightforward procedure for computing solutions of (GP) and (DGP) when these programs have solutions.

(2.5.2) Theorem. *If $\mathbf{t}^* = (t_1^*, \ldots, t_m^*)$ is a solution to the primal geometric program (GP), then the corresponding dual geometric program (DGP) is consistent. Moreover, the vector $\boldsymbol{\delta}^* = (\delta_1^*, \ldots, \delta_n^*)$ defined by*

$$\delta_i^* = \frac{u_i(t^*)}{g(t^*)}, \qquad i = 1, \ldots, n$$

(where $u_i(\mathbf{t}) = c_i t_1^{\alpha_{i1}} \ldots t_m^{\alpha_{im}}$ is the ith *term of $g(\mathbf{t})$) is a solution for (DGP) and equality holds in the Primal–Dual Inequality, that is,*

$$g(\mathbf{t}^*) = v(\boldsymbol{\delta}^*).$$

PROOF. A typical term of $g(\mathbf{t})$ is

$$u_i(\mathbf{t}) = c_i t_1^{\alpha_{i1}} \ldots t_m^{\alpha_{im}}.$$

Partial differentiation of $u_i(\mathbf{t})$ with respect to t_j has a very simple effect on $u_i(\mathbf{t})$—it simply multiplies $u_i(\mathbf{t})$ by α_{ij} and reduces the exponent of t_j by 1. This means that the following equation holds:

$$t_j \frac{\partial u_i}{\partial t_j} = \alpha_{ij} u_i.$$

Since $\mathbf{t}^* = (t_1^*, \ldots, t_m^*)$ is a minimizer for $g(\mathbf{t})$, it follows that

$$0 = \frac{\partial g}{\partial t_j}(\mathbf{t}^*) = \sum_{i=1}^{n} \frac{\partial u_i}{\partial t_j}(\mathbf{t}^*), \qquad j = 1, 2, \ldots, m.$$

But then the observation in the first paragraph in the proof shows that

$$0 = \sum_{i=1}^{n} \alpha_{ij} u_i(\mathbf{t}^*), \qquad j = 1, 2, \ldots, m.$$

Since $g(\mathbf{t}^*) > 0$ (Why?), we can divide both sides of the last equation by $g(\mathbf{t}^*)$ to obtain

$$0 = \sum_{i=1}^{n} \alpha_{ij} \left(\frac{u_i(\mathbf{t}^*)}{g(\mathbf{t}^*)} \right), \qquad j = 1, 2, \ldots, m.$$

Consequently, if we set

$$\delta_i^* = \frac{u_i(\mathbf{t}^*)}{g(\mathbf{t}^*)}, \qquad i = 1, \ldots, n,$$

then $\boldsymbol{\delta}^* = (\delta_1^*, \ldots, \delta_n^*)$ satisfies the Orthogonality Condition for the dual program (*DGP*). Also, $\delta_i^* > 0$ for $i = 1, \ldots, n$ so the Positivity Condition is satisfied. Finally,

$$\sum_{i=1}^{n} \delta_i^* = \sum_{i=1}^{n} \left(\frac{u_i(\mathbf{t}^*)}{g(\mathbf{t}^*)} \right) = \frac{g(\mathbf{t}^*)}{g(\mathbf{t}^*)} = 1$$

so the Normality Condition holds. We conclude that the vector $\boldsymbol{\delta}^*$ is feasible for the dual program, so the dual program (*DGP*) is consistent. Also

$$\begin{aligned} g(\mathbf{t}^*) &= g(\mathbf{t}^*)^{\delta_1^* + \cdots + \delta_n^*} = (g(\mathbf{t}^*))^{\delta_1^*} \cdots (g(\mathbf{t}^*))^{\delta_n^*} \\ &= \left(\frac{u_1(\mathbf{t}^*)}{\delta_1^*} \right)^{\delta_1^*} \cdots \left(\frac{u_n(\mathbf{t}^*)}{\delta_n^*} \right)^{\delta_n^*} \\ &= \left(\frac{c_1}{\delta_1^*} \right)^{\delta_1^*} \cdots \left(\frac{c_n}{\delta_n^*} \right)^{\delta_n^*} = v(\boldsymbol{\delta}^*), \end{aligned}$$

so equality holds in the Primal–Dual Inequality. This implies that $\boldsymbol{\delta}^*$ is a solution of the dual program (DGP). Since

$$\delta_i^* = \frac{u_i(\mathbf{t}^*)}{g(\mathbf{t}^*)}, \qquad i = 1, 2, \ldots, n,$$

the proof is complete.

With the preceding theorem in hand, we can formulate the following approach to geometric programming.

(2.5.3) The Geometric Programming Procedure. Given a primal geometric program

$$(DP) \quad \text{Minimize the posynomial} \quad g(\mathbf{t}) = \sum_{i=1}^{n} u_i(\mathbf{t}),$$

$$\text{where } u_i(\mathbf{t}) = c_i t_1^{\alpha_{i1}} \ldots t_m^{\alpha_{im}} \text{ and } t_i > 0, \ldots, t_m > 0;\ c_i > 0$$

we proceed as follows:

Step 1. Compute the set F of feasible vectors for the dual geometric program (DGP), that is, the set of all vectors $\boldsymbol{\delta}$ in R^n such that

$$\delta_1 > 0, \ldots, \delta_n > 0 \qquad \text{(the Positivity Condition)},$$

$$\sum_{i=1}^{n} \delta_i = 1 \qquad \text{(the Normality Condition)},$$

$$\sum_{i=1}^{n} \alpha_{ij}\delta_i = 0; \quad j = 1, \ldots, m \qquad \text{(the Orthogonality Condition)}.$$

Step 2. If the set F of feasible vectors for (DGP):

(a) is empty, then stop. The given program (GP) has no solution in this case;
(b) consists of a single vector $\boldsymbol{\delta}^*$, then $\boldsymbol{\delta}^*$ is a solution of (DGP). Proceed to Step 4;
(c) consists of more than one vector, then proceed to Step 3.

Step 3. Find a vector $\boldsymbol{\delta}^*$ that is a global maximizer for the dual function

$$v(\boldsymbol{\delta}) = \prod_{i=1}^{n} \left(\frac{c_i}{\delta_i}\right)^{\delta_i}$$

on the set F of feasible vectors for (DGP). Then $\boldsymbol{\delta}^*$ is a solution of (DGP). Proceed to Step 4.

Step 4. Given a solution $\boldsymbol{\delta}^*$ of (DGP), a solution $\mathbf{t}^*$ of the primal program is obtained by solving the equations

$$\delta_i^* = \frac{u_i(\mathbf{t}^*)}{v(\boldsymbol{\delta}^*)}, \qquad i = 1, \ldots, n,$$

for $t_1^*, \ldots, t_m^*$. The minimum value $g(\mathbf{t}^*)$ of $g(\mathbf{t})$ is equal to the maximum value $v(\delta^*)$ for the dual function $v(\boldsymbol{\delta})$.

(2.5.4) Remarks. Some comments are in order concerning the preceding systematic procedure for solving geometric programming problems.

(1) To find the set F of feasible vectors for the dual program, we first solve the system of *linear* equations consisting of the Normality and Orthogonality Conditions, and then impose the Positivity Condition on the resulting solutions.

(2) The statement in Step 2(a) that (GP) has no solution if the set F of feasible vectors for the dual program is empty follows from Theorem (2.5.2). For if (GP) has a solution $\mathbf{t}^*$, then (2.5.2) asserts that the dual program is consistent, that is, F is nonempty.

(3) The "difficult" alternative in Step 2 is (c), because we are required to find a maximizer $\boldsymbol{\delta}^*$ for $v(\boldsymbol{\delta})$ on the set F of feasible vectors for (DGP) by some means. As Example (2.5.5)(c) illustrates, the systematic procedure outlined in (2.5.3) can still be regarded as an effective method for solving (GP).

(4) The solution of the system of equations prescribed in Step 4 appears complicated because the equations are not linear in the variables $t_1^*, \ldots, t_m^*$. However, because

$$u_i(\mathbf{t}^*) = c_i(t_1^*)^{\alpha_{i1}} \ldots (t_m^*)^{\alpha_{im}},$$

we can obtain $t_1^*, \ldots, t_m^*$ by solving the system of *linear* equations

$$\alpha_{i1} \log t_1^* + \cdots + \alpha_{im} \log t_m^* = \log \delta_i^* - \log c_i + \log v(\delta^*), \qquad i = 1, \ldots, n.$$

Thus, the systematic procedure for geometric programming described in (2.5.3) is a highly *linear* process.

(5) The ith component δ_i^* of the solution $\boldsymbol{\delta}^*$ of the dual program (DGP) specifies the relative contribution of the ith term $u_i(t^*)$ to the minimum value $g(t^*)$ of (GP). (See Example (2.5.5)(a) for an illustration of this interpretation for δ_i^*.)

(2.5.5) Examples

(a) Let c_1, c_2, c_3, c_4 and V be positive constants. Consider the following geometric program:

$$\text{Minimize} \quad g(\mathbf{t}) = \frac{c_1 V}{t_1 t_2 t_3} + 2c_2 t_2 t_3 + 2c_3 t_1 t_3 + c_4 t_1 t_2,$$

$$\text{where} \qquad t_1 > 0, t_2 > 0, t_3 > 0.$$

Here is the dual problem

$$\text{Maximize} \quad V(\boldsymbol{\delta}) = \left(\frac{c_1 V}{\delta_1}\right)^{\delta_1} \left(\frac{2c_2}{\delta_2}\right)^{\delta_2} \left(\frac{2c_3}{\delta_3}\right)^{\delta_3} \left(\frac{c_4}{\delta_4}\right)^{\delta_4},$$

$$\text{subject to} \quad \delta_1 + \delta_2 + \delta_3 + \delta_4 = 1 \qquad \text{(the Normality Condition)},$$

$$\delta_i > 0, \qquad i = 1, 2, 3, 4 \qquad \text{(the Positivity Condition)},$$

$$\left.\begin{aligned} -\delta_1 + + \delta_3 + \delta_4 &= 0 \\ -\delta_1 + \delta_2 + \delta_4 &= 0 \\ -\delta_1 + \delta_2 + \delta_3 &= 0 \end{aligned}\right\} \qquad \text{(the Orthogonality Condition)}.$$

If we apply Step 1 of (2.5.3), we find that the vector

$$\boldsymbol{\delta}^* = (\tfrac{2}{5}, \tfrac{1}{5}, \tfrac{1}{5}, \tfrac{1}{5})$$

is the only feasible vector for the dual, so we follow alternative (b) in Step 2. Therefore, we can find the solution t_1^*, t_2^*, t_3^* by solving the system:

$$\frac{c_1 V}{t_1^* t_2^* t_3^*} = \delta_1^* v(\boldsymbol{\delta}^*) = \tfrac{2}{5} v(\boldsymbol{\delta}^*),$$

$$2c_2 t_2^* t_3^* = \delta_2^* v(\boldsymbol{\delta}^*) = \tfrac{1}{5} v(\boldsymbol{\delta}^*),$$

$$2c_3 t_1^* t_3^* = \delta_3^* v(\boldsymbol{\delta}^*) = \tfrac{1}{5} v(\boldsymbol{\delta}^*),$$

$$c_4 t_1^* t_2^* = \delta_4^* v(\boldsymbol{\delta}^*) = \tfrac{1}{5} v(\boldsymbol{\delta}^*).$$

We will omit the remaining computational details. Note that the relative contributions of the four terms in $g(t^*)$ are $\tfrac{2}{5}, \tfrac{1}{5}, \tfrac{1}{5}, \tfrac{1}{5}$ regardless of the values of c_1, c_2, c_3, c_4.

(b) Consider the geometric program

$$\text{Minimize} \quad g(t_1, t_2) = \frac{2}{t_1 t_2} + t_1 t_2 + t_1,$$

$$\text{where} \quad t_1 > 0, t_2 > 0.$$

The dual program is

$$\text{Maximize} \quad v(\boldsymbol{\delta}) = \left(\frac{2}{\delta_1}\right)^{\delta_1} \left(\frac{1}{\delta_2}\right)^{\delta_2} \left(\frac{1}{\delta_3}\right)^{\delta_3},$$

$$\begin{aligned} \text{subject to} \quad & \delta_1, \delta_2, \delta_3 > 0, \\ & \delta_1 + \delta_2 + \delta_3 = 1, \\ & -\delta_1 + \delta_2 + \delta_3 = 0, \\ & -\delta_1 + \delta_2 \qquad = 0. \end{aligned}$$

Solving these equations, we find that the only solution is $\delta_1 = \tfrac{1}{2}$, $\delta_2 = \tfrac{1}{2}$, and $\delta_3 = 0$. This vector is not feasible for the dual (because $\delta_3 = 0$) and hence there are no vectors feasible for the dual. Consequently, alternative(a) in Step 2 tells us the given program has no solution.

(c) Consider the geometric program

$$\text{Minimize} \quad g(t_1, t_2) = \frac{1}{t_1 t_2} + t_1 t_2 + t_1 + t_2,$$

$$\text{where} \quad t_1 > 0, t_2 > 0.$$

The dual program is

$$\text{Maximize} \quad v(\boldsymbol{\delta}) = \left(\frac{1}{\delta_1}\right)^{\delta_1}\left(\frac{1}{\delta_2}\right)^{\delta_2}\left(\frac{1}{\delta_3}\right)^{\delta_3}\left(\frac{1}{\delta_4}\right)^{\delta_4},$$

$$\begin{aligned} \text{subject to} \quad & \delta_1, \delta_2, \delta_3, \delta_4 > 0, \\ & \delta_1 + \delta_2 + \delta_3 + \delta_4 = 1, \\ & -\delta_1 + \delta_2 + \delta_3 \qquad = 0, \\ & -\delta_1 + \delta_2 + \qquad \delta_4 = 0. \end{aligned}$$

A little work shows that this system is equivalent to the following:

$$\begin{aligned} \delta_1 &= \delta_1 > 0, \\ \delta_2 &= 3\delta_1 - 1 > 0, \\ \delta_3 &= 1 - 2\delta_1 > 0, \\ \delta_4 &= 1 - 2\delta_1 > 0, \end{aligned}$$

and these inequalities restrict δ_1 to the range $\frac{1}{3} < \delta_1 < \frac{1}{2}$. Thus maximizing $v(\delta)$ amounts to maximizing

$$f(s) = \left(\frac{1}{s}\right)^{s}\left(\frac{1}{1-2s}\right)^{1-2s}\left(\frac{1}{1-2s}\right)^{1-2s}\left(\frac{1}{3s-1}\right)^{3s-1}$$

on the interval $\frac{1}{3} < s < \frac{1}{2}$. Taking logs, we maximize

$$\ln f(s) = -s\ln s - 2(1-2s)\ln(1-2s) - (3s-1)\ln(3s-1),$$

which is easily maximized with one of the numerical routines from Chapter 3.

*2.6. Convexity and Other Inequalities

In Section 2.4, the Arithmetic–Geometric Mean Inequality was derived on the basis of convexity considerations. Convexity can be exploited to prove a wide variety of other important inequalities that arise in many parts of pure and applied mathematics. Some of these inequalities including the Cauchy–Schwarz Inequality, Hölder's Inequality, and Minkowski's Inequality are derived in this section.

The following inequality, which follows from the Arithmetic–Geometric Mean Inequality, is fundamental to our derivations of the other inequalities mentioned above.

(2.6.1) Theorem (Young's Inequality). *Suppose that p and q are real numbers both greater than 1 such that $p^{-1} + q^{-1} = 1$. If x, y are positive real numbers,* *then*

$$xy \leq \frac{x^p}{p} + \frac{y^p}{q}.$$

Equality holds precisely when $x^p = y^q$.

Proof. If $s = x^p$ and $t = y^q$, then s, t are positive numbers and $p^{-1} + q^{-1} = 1$ so that the Arithmetic–Geometric Mean Inequality can be applied as follows:

$$xy = s^{1/p}t^{1/q} \overset{\text{(A-G)}}{\leq} \frac{s}{p} + \frac{t}{q} = \frac{x^p}{p} + \frac{y^q}{q}.$$

Moreover, (2.4.1) asserts that equality holds in the preceding inequality if and only if $s = t$, that is, if and only if $x^p = y^q$.

(2.6.2) Corollary (Hölder's Inequality). *Suppose that p and q are real numbers both greater than 1 such that $p^{-1} + q^{-1} = 1$. If $\mathbf{x} = (x_1, \ldots, x_n)$ and $\mathbf{y} = (y_1, \ldots, y_n)$ are vectors in R^n, then*

$$\sum_{i=1}^{n} |x_i y_i| \leq \left(\sum_{i=1}^{n} |x_i|^p\right)^{1/p} \left(\sum_{i=1}^{n} |y_i|^q\right)^{1/q}.$$

Proof. If $\mathbf{x} = \mathbf{0}$ or $\mathbf{y} = \mathbf{0}$, the inequality surely holds since both sides are equal to zero. If neither $\mathbf{x}$ nor $\mathbf{y}$ is $\mathbf{0}$, write

$$\|\mathbf{x}\|_p = \left(\sum_{i=1}^{n} |x_i|^p\right)^{1/p}, \qquad \|\mathbf{y}\|_q = \left(\sum_{i=1}^{n} |y_i|^q\right)^{1/q}.$$

From Young's Inequality, we have

$$\frac{|x_i y_i|}{\|\mathbf{x}\|_p \|\mathbf{y}\|_q} \leq \frac{1}{p}\frac{|x_i|^p}{\|\mathbf{x}\|_p^p} + \frac{1}{q}\frac{|y_i|^q}{\|\mathbf{y}\|_q^q}$$

for $i = 1, 2, \ldots, n$. Summing these inequalities over $i = 1, 2, \ldots, n$ yields

$$\frac{1}{\|\mathbf{x}\|_p^p \|\mathbf{y}\|_q^q} \sum_{i=1}^{n} |x_i y_i| \leq \frac{1}{p\|\mathbf{x}\|_p^p} \sum_{i=1}^{n} |x_i|^p + \frac{1}{q\|\mathbf{y}\|_q^q} \sum_{i=1}^{n} |y_i|^q = \frac{1}{p} + \frac{1}{q} = 1.$$

Now multiply the extremes of the preceding inequality by $\|\mathbf{x}\|_p \|\mathbf{y}\|_q$ to obtain

$$\sum_{i=1}^{n} |x_i y_i| \leq \|\mathbf{x}\|_p \|\mathbf{y}\|_q = \left(\sum_{i=1}^{n} |x_i|^p\right)^{1/p} \left(\sum_{i=1}^{n} |y_i|^q\right)^{1/q}$$

as promised.

Note that if we take $p = q = 2$ in Hölder's inequality, then $\|\mathbf{x}\|_p$ and $\|\mathbf{y}\|_q$ reduce the usual norms $\|\mathbf{x}\|$ and $\|\mathbf{y}\|$, and so we obtain the *Cauchy–Schwarz Inequality*

$$|\mathbf{x} \cdot \mathbf{y}| \leq \|\mathbf{x}\| \, \|\mathbf{y}\|, \qquad \mathbf{x} \in R^n, \quad \mathbf{y} \in R^n,$$

as a corollary.

In Section 1.2, we observed that the norm $\|\mathbf{x}\|$ on R^n has the following properties:

(1) $\|\mathbf{x}\| \geq 0$ for all vectors $\mathbf{x}$ in R^n.
(2) $\|\mathbf{x}\| = 0$ if and only if $\mathbf{x}$ is the zero vector $\mathbf{0}$.
(3) $\|\alpha\mathbf{x}\| = |\alpha|\,\|\mathbf{x}\|$ for all vectors $\mathbf{x}$ in R^n and all real numbers α.
(4) $\|\mathbf{x} + \mathbf{y}\| \leq \|\mathbf{x}\| + \|\mathbf{y}\|$ for all vectors $\mathbf{x}$, $\mathbf{y}$ in R^n (the Triangle Inequality).

Note that for any real number $p \geq 1$, the function defined on R^n by

$$\|\mathbf{x}\|_p = \left(\sum_{i=1}^{n} |x_i|^p\right)^{1/p}$$

obviously shares properties (1), (2), and (3) of the usual norm $\|\mathbf{x}\|$ on R^n. The following inequality asserts that $\|\mathbf{x}\|_p$ also satisfies the Triangle Inequality.

(2.6.3) Theorem (Minkowski's Inequality). *If* $\mathbf{x} = (x_1, x_2, \ldots, x_n)$ *and* $\mathbf{y} = (y_1, y_2, \ldots, y_n)$ *are vectors in* R^n *and if* $p \geq 1$, *then*

$$\left(\sum_{i=1}^{n} |x_i + y_i|^p\right)^{1/p} \leq \left(\sum_{i=1}^{n} |x_i|^p\right)^{1/p} + \left(\sum_{i=1}^{n} |y_i|^p\right)^{1/p},$$

that is,

$$\|\mathbf{x} + \mathbf{y}\|_p \leq \|\mathbf{x}\|_p + \|\mathbf{y}\|_p.$$

PROOF. If either $\mathbf{x}$ or $\mathbf{y}$ is the zero vector, then the two sides of the inequality are actually equal so the stated inequality surely holds. Hence, we can also suppose that $\mathbf{x} \neq \mathbf{0}$ and $\mathbf{y} \neq \mathbf{0}$.

If $p = 1$, then, since $|x_i + y_i| \leq |x_i| + |y_i|$ for each i, we can sum over i from 1 to n to obtain the desired inequality. Consequently, we can also suppose that $p > 1$.

Now consider the function φ defined for $t > 0$ by

$$\varphi(t) = t^p.$$

This function is (strictly) convex for $t > 0$ since

$$\varphi''(t) = p(p-1)t^{p-2} > 0$$

because $p > 1$. Consequently, since

$$\frac{\|\mathbf{x}\|_p}{\|\mathbf{x}\|_p + \|\mathbf{y}\|_p} + \frac{\|\mathbf{y}\|_p}{\|\mathbf{x}\|_p + \|\mathbf{y}\|_p} = 1,$$

it follows that

$$\left(\frac{\|\mathbf{x}\|_p}{\|\mathbf{x}\|_p + \|\mathbf{y}\|_p}\frac{|x_i|}{\|\mathbf{x}\|_p} + \frac{\|\mathbf{y}\|_p}{\|\mathbf{x}\|_p + \|\mathbf{y}\|_p}\frac{|y_i|}{\|\mathbf{x}\|_p}\right)^p$$

$$\leq \frac{\|\mathbf{x}\|_p}{\|\mathbf{x}\|_p + \|\mathbf{y}\|_p}\left(\frac{|x_i|}{\|\mathbf{x}\|_p}\right)^p + \frac{\|\mathbf{y}\|_p}{\|\mathbf{x}\|_p + \|\mathbf{y}\|_p}\left(\frac{|y_i|}{\|\mathbf{x}\|_p}\right)^p.$$

Therefore,

$$\begin{aligned}\sum_{i=1}^{n}\left(\frac{|x_i+y_i|}{\|\mathbf{x}\|_p+\|\mathbf{y}\|_p}\right)^p &\le \sum_{i=1}^{n}\left(\frac{|x_i|+|y_i|}{\|\mathbf{x}\|_p+\|\mathbf{y}\|_p}\right)^p\\ &= \sum_{i=1}^{n}\left(\frac{\|\mathbf{x}\|_p}{\|\mathbf{x}\|_p+\|\mathbf{y}\|_p}\left(\frac{|x_i|}{\|\mathbf{x}\|_p}\right)+\frac{\|\mathbf{y}\|_p}{\|\mathbf{x}\|_p+\|\mathbf{y}\|_p}\left(\frac{|y_i|}{\|\mathbf{y}\|_p}\right)^p\right)\\ &\le \frac{\|\mathbf{x}\|_p}{\|\mathbf{x}\|_p+\|\mathbf{y}\|_p}\sum_{i=1}^{n}\left(\frac{|x_i|}{\|\mathbf{x}\|_p}\right)^p+\frac{\|\mathbf{y}\|_p}{\|\mathbf{x}\|_p+\|\mathbf{y}\|_p}\sum_{i=1}^{n}\left(\frac{|y_i|}{\|\mathbf{y}\|_p}\right)^p\\ &= \frac{\|\mathbf{x}\|_p}{\|\mathbf{x}\|_p+\|\mathbf{y}\|_p}\cdot\frac{\|\mathbf{x}\|_p^p}{\|\mathbf{x}\|_p^p}+\frac{\|\mathbf{y}\|_p}{\|\mathbf{x}\|_p+\|\mathbf{y}\|_p}\cdot\frac{\|\mathbf{y}\|_p^p}{\|\mathbf{y}\|_p^p}=1,\end{aligned}$$

and so

$$\sum_{i=1}^{n}|x_i+y_i|^p \le (\|\mathbf{x}\|_p+\|\mathbf{y}\|_p)^p.$$

If we take the pth root of both sides of the preceding inequality, we obtain the desired result.

So far, we have made use of convexity to derive some important inequalities. Now we will turn the tables and show that these inequalities can help us to verify the convexity of certain functions.

(2.6.4) Example. If $\mathbf{a}_1, \ldots, \mathbf{a}_m$ are fixed vectors in R^n and if $c_1, \ldots, c_m$ are positive numbers, then the function

$$f(\mathbf{x}) = \ln\left(\sum_{i=1}^{m} c_i e^{\mathbf{a}_i\cdot\mathbf{x}}\right)$$

is convex on R^n.

The Hessian of $f(\mathbf{x})$ is complicated. Although it is possible to prove that it is positive semidefinite with a great deal of work and clever observation, it is much easier to prove that $f(\mathbf{x})$ is convex by making use of Hölder's Inequality.

We have to show that

$$f(\alpha\mathbf{x}+\beta\mathbf{y}) \le \alpha f(\mathbf{x}) + \beta f(\mathbf{y}),$$

or equivalently that

$$e^{f(\alpha\mathbf{x}+\beta\mathbf{y})} \le e^{\alpha f(\mathbf{x})+\beta f(\mathbf{y})} = (e^{f(\mathbf{x})})^\alpha (e^{f(\mathbf{y})})^\beta$$

for all $\mathbf{x}, \mathbf{y}$ in R^n and all $\alpha, \beta > 0$ with $\alpha + \beta = 1$. This amounts to showing that

$$\sum_{i=1}^{m} c_i e^{\mathbf{a}_i\cdot(\alpha\mathbf{x}+\beta\mathbf{y})} \le \left(\sum_{i=1}^{m} c_i e^{\mathbf{a}_i\cdot\mathbf{x}}\right)^\alpha \left(\sum_{i=1}^{m} c_i e^{\mathbf{a}_i\cdot\mathbf{y}}\right)^\beta,$$

which is the same as

$$\sum_{i=1}^{m}\left(c_i e^{\mathbf{a}_i\cdot\mathbf{x}}\right)^\alpha\left(c_i e^{\mathbf{a}_i\cdot\mathbf{y}}\right)^\beta \le \left(\sum_{i=1}^{m} c_i e^{\mathbf{a}_i\cdot\mathbf{x}}\right)^\alpha \left(\sum_{i=1}^{m} c_i e^{\mathbf{a}_i\cdot\mathbf{y}}\right)^\beta.$$

The form of the desired inequality suggests that this is a job for Hölder's Inequality! Just let

$$s_i = (c_i e^{\mathbf{a}_i \cdot \mathbf{x}})^{\alpha}, \qquad t_i = (c_i e^{\mathbf{a}_i \cdot \mathbf{y}})^{\beta}$$

for all $i = 1, \ldots, m$ and let $\alpha = 1/p$, $\beta = 1/q$. (This is a natural choice for p and q since we want $p^{-1} + q^{-1} = 1$.) With these choices, the inequality we want is just Hölder's Inequality

$$\sum_{i=1}^{m} s_i t_i \leq \left(\sum_{i=1}^{m} s_i^p\right)^{1/p} \left(\sum_{i=1}^{m} t_i^q\right)^{1/q}.$$

Exercises

1. Determine whether the given functions are convex, concave, strictly convex, or strictly concave on the specified intervals:
   (a) $f(x) = \ln x$ on $I = (0, +\infty)$.
   (b) $f(x) = e^{-x}$ on $I = (-\infty, +\infty)$.
   (c) $f(x) = |x|$ on $I = [-1, 1]$.
   (d) $f(x) = |x^3|$ on $I = (-\infty, +\infty)$.

2. Show that each of the following functions is convex or strictly convex on the specified convex set:
   (a) $f(x_1, x_2) = 5x_1^2 + 2x_1x_2 + x_2^2 - x_1 + 2x_2 + 3$ on $D = R^2$.
   (b) $f(x_1, x_2) = x_1^2/2 + 3x_2^2/2 + \sqrt{3}x_1x_2$ on $D = R^2$.
   (c) $f(x_1, x_2) = (x_1 + 2x_2 + 1)^8 - \ln(x_1x_2)^2$ on $D = \{(x_1, x_2) \in R^2 \colon x_1 > x_2 > 1\}$.
   (d) $f(x_1, x_2) = 4e^{3x_1 - x_2} + 5e^{x_1^2 + x_2^2}$ on $D = R^2$.
   (e) $f(x_1, x_2) = c_1x_1 + c_2/x_1 + c_3x_2 + c_4/x_2$ on $D = \{(x_1, x_2) \in R^2 \colon x_1 > 0, x_2 > 0\}$ where c_i is a positive number for $i = 1, 2, 3, 4$.

3. A *quadratic function in n variables* is any function defined on R^n which can be expressed in the form

$$f(\mathbf{x}) = a + \mathbf{b} \cdot \mathbf{x} + \mathbf{x} \cdot A\mathbf{x},$$

   where $a \in R$, $\mathbf{b} \in R^n$, and A is an $n \times n$-symmetric matrix.
   (a) Show that the function $f(\mathbf{x})$ defined on R^2 by

$$f(x_1, x_2) = (x_1 - x_2)^2 + (x_1 + 2x_2 + 1)^2 - 8x_1x_2$$

   is a quadratic function of two variables by finding the appropriate $a \in R$, $\mathbf{b} \in R^2$, and the 2×2-symmetric matrix A.
   (b) Compute the gradient $\nabla f(\mathbf{x})$ and the Hessian $Hf(\mathbf{x})$ of the quadratic function in (a) and express these quantities in terms of the $a \in R$, $\mathbf{b} \in R^2$, and the 2×2-symmetric matrix A computed in (a).
   (c) Show that a quadratic function $f(\mathbf{x})$ of n variables is convex if and only if the corresponding $n \times n$-symmetric matrix A is positive semidefinite, and is strictly convex if A is positive definite.
   (d) If $f(\mathbf{x})$ is a quadratic function of n variables such that the corresponding matrix A is positive definite, show that $\mathbf{0} = 2A\mathbf{x} + \mathbf{b}$ has a unique solution and that this solution is the strict global minimizer of $f(\mathbf{x})$.

4. Theorem (2.3.1) asserts that a convex function $f(x)$ defined on an open interval (a, b) is continuous. In the outline of a proof that follows that assertion, the following statements are used but are not verified:
   (a) If $r < s < t$, then the point $(t, f(t))$ must lie above or on the line through $(r, f(r))$ and $(s, f(s))$.
   (b) $\lim_{t \to s^+} f(t) = f(s)$.
   (c) $\lim_{v \to s^-} f(v) = f(s)$.
   Supply proofs for these statements and thereby complete the proof of (2.3.1).

5. Suppose that D is a subset of R^n. Show that the set C of all convex combinations of vectors in D coincides with the convex hull co(D) of D by the following procedure:
   (a) Show that C is a convex set containing D.
   (b) Show that if B is a convex set containing D then B contains C. (Hint: Use Theorem (2.1.3).)
   (c) Apply the conclusions of (a) and (b) to verify that $\text{co}(D) = C$.

6. The following exercises refer to the linear production model discussed in Section 2.2:
   (a) Show that the technology set of a linear production model is a cone.
   (b) Show that the production isoquant $Q_{\bar{x}}$ corresponding to a given output level $\bar{x}$ is a convex set.
   (c) If $\mathbf{y}^{(k)}$ is a vector of input levels required to produce the output level $\bar{x}$ with the exclusive use of the kth producing process P_k for $k = 1, 2, \ldots, n$, show that the production isoquant $Q_{\bar{x}}$ is the convex hull of $\{\mathbf{y}^{(1)}, \mathbf{y}^{(2)}, \ldots, \mathbf{y}^{(n)}\}$.

7. Consider the function $f(\mathbf{x})$ defined on the set

$$D = \{\mathbf{x} \in R^3 : x_1 > 0, x_2 > 0, x_3 > 0\}$$

   by

$$f(\mathbf{x}) = (x_1)^{r_1} + (x_2)^{r_2} + (x_3)^{r_3},$$

   where $r_i > 0$ for $i = 1, 2, 3$.
   (a) Show that $f(\mathbf{x})$ is strictly convex on D if $r_i > 1$ for $i = 1, 2, 3$.
   (b) Show that $f(\mathbf{x})$ is strictly concave on D if $0 < r_i < 1$ for $i = 1, 2, 3$.

8. (a) Show that

$$\frac{x}{4} + \frac{3y}{4} \leq \sqrt{\ln\left(\frac{e^{x^2}}{4} + \tfrac{3}{4}e^{y^2}\right)}$$

   for all positive numbers x and y. (Hint: The desired inequality follows from the convexity of an appropriate function.)
   (b) Show that

$$\left(\frac{x}{2} + \frac{y}{3} + \frac{z}{12} + \frac{w}{12}\right)^4 \leq \tfrac{1}{2}x^4 + \tfrac{1}{3}y^4 + \tfrac{1}{12}z^4 + \tfrac{1}{12}w^4$$

   with equality if and only if $x = y = z = w$.

9. Suppose that $f(\mathbf{x})$ and $g(\mathbf{x})$ are convex functions defined on a convex set C in R^n and that

$$h(\mathbf{x}) = \max\{f(\mathbf{x}), g(\mathbf{x})\}, \qquad \mathbf{x} \in C.$$

   Show that $h(\mathbf{x})$ is convex on C.

10. Suppose that $f(\mathbf{x})$ is a convex function with continuous first partials defined on a convex set C in R^n. Prove that a point $\mathbf{x}^*$ in C is a global minimizer of $f(\mathbf{x})$ on C if and only if $\nabla f(\mathbf{x}^*)\cdot(\mathbf{x}-\mathbf{x}^*)\geq 0$ for all $\mathbf{x}$ in C.

11. Suppose that $f(\mathbf{x})$ is a function defined on a convex subset D of R^n. The *epigraph* of $f(\mathbf{x})$ is the set in R^{n+1} defined by

$$\text{epi}(f)=\{(\mathbf{x},\alpha)\colon \mathbf{x}\in D, \alpha\in R, f(\mathbf{x})\leq\alpha\}.$$

(a) Sketch the epigraph of the function

$$f(\mathbf{x})=e^x \quad \text{for } x\in R.$$

(b) Sketch the epigraph of the function

$$f(x_1,x_2)=x_1^2+x_2^2 \quad \text{for } \mathbf{x}=(x_1,x_2)\in R^2.$$

(c) Show that $f(\mathbf{x})$ is convex if and only if the epigraph of $f(\mathbf{x})$ is convex.
(d) Prove Theorem (2.3.3) by using part (c) and Theorem (2.1.3).
(e) Show that the epigraph of a continuous convex function on R^n is a closed set. (Hint: Show that if $(\mathbf{x}_n,\alpha_n)\in\text{epi}(f)$ and $\mathbf{x}_n\to\mathbf{x}^*$, $\alpha_n\to\alpha$, then $(\mathbf{x}^*,\alpha)\in\text{epi}(f)$.
(f) Give an alternative derivation of the result in Exercise 9 by showing that

$$\text{epi}(\max\{f(x),g(x)\})=\text{epi}(f(x))\cap\text{epi}(g(x)).$$

12. Suppose that $f(\mathbf{x})$ is a function with continuous second partial derivatives on R^n. Show that if there is a $\mathbf{z}$ such that the Hessian $Hf(\mathbf{z})$ of $f(\mathbf{x})$ at $\mathbf{z}$ is not positive semidefinite, then $f(\mathbf{x})$ is not convex.

13. Suppose that $f(\mathbf{x})$ is a function with continuous second partial derivatives on R^n. Denote by $\nabla f(\mathbf{x})\otimes\nabla f(\mathbf{x})$ the $n\times n$-matrix whose (i,j)th entry is

$$\frac{\partial f}{\partial x_i}(\mathbf{x})\cdot\frac{\partial f}{\partial x_j}(\mathbf{x}),$$

that is, $\nabla f(\mathbf{x})\otimes\nabla f(\mathbf{x})$ is just the matrix product

$$(\text{column vector } \nabla f(\mathbf{x})\cdot(\text{row vector } \nabla f(\mathbf{x})).$$

(a) Show that $\nabla f(\mathbf{x})\otimes\nabla f(\mathbf{x})$ is always positive semidefinite.
(b) Show that if α is a positive number the function $e^{\alpha f(\mathbf{x})}$ is convex on R^n if

$$Hf(\mathbf{x})+\alpha(\nabla f(\mathbf{x})\otimes\nabla f(\mathbf{x}))$$

is positive semidefinite for all $\mathbf{x}$ in R^n.

14. Show that the matrix

$$A(x)=\begin{pmatrix} x^4 & x^3 & x^2\\ x^3 & x^2 & x\\ x^2 & x & 1\end{pmatrix}$$

is positive semidefinite for all $x\in R$. (Hint: Show that there is a column vector $\mathbf{u}(x)\in R^3$ such that $\mathbf{u}(x)\mathbf{u}(x)^{\mathrm{T}}=A(x)$, that is, $\mathbf{u}(x)\otimes\mathbf{u}(x)=A(x)$.) (See Exercise 13.)

15. Use the Arithmetic–Geometric Mean Inequality to solve the following optimization problem:
(a) Minimize x^2+y+z subject to $xyz=1$ and $x,y,z>0$.
(b) Maximize xyz subject to $3x+4y+12z=1$ and $x,y,z>0$.

(c) Minimize $3x + 4y + 12z$ subject to $xyz = 1$ and $x, y, z > 0$.
(d) Maximize xy^2z^3 subject to $x^3 + y^2 + z = 39$ and $x, y, z > 0$.
(e) Minimize $x^3 + y^2 + z$ subject to $xy^2z^3 = 39$.

16. Suppose that c_1, c_2, c_3 are positive constants and that

$$f(x, y) = c_1 x + c_2 x^{-2} y^{-3} + c_3 y^4.$$

(a) Minimize $f(x, y)$ over all $x, y > 0$.
(b) Find the relative contributions of each term to the minimum.

17. Solve the following "classical" calculus problems by making use of the Arithmetic–Geometric Mean Inequality.
(a) Find the largest circular cylinder that can be inscribed in a sphere of a given radius.
(b) Find the smallest radius r such that a circular cylinder of volume 8 cubic units can be inscribed in the sphere of radius r.

18. (a) Show that if c_1, c_2, c_3, c_4 are positive numbers, then

$$g(x_1, x_2, x_3, x_4) = c_1 x_3^{-1} + c_2 x_2^{-6} x_4^2 + c_3 x_1^3 x_4^2 + c_4 x_1^{-1} x_2^4 x_3^2 x_4^2$$

has no minimum on the set

$$D = \{(x_1, x_2, x_3, x_4): x_i > 0 \text{ for } i = 1, 2, 3, 4\}.$$

(b) Set up the log-linear equations whose solution produces the minimizer of

$$g(t_1, t_2, t_3, t_4) = c_1 t_3^{-1} t_4^{-2} + c_2 t_2^{-6} t_4^2 + c_3 t_1^3 t_4^2 + c_4 t_1^{-1} t_2^4 t_3^2 t_4^2$$

over all $t_1, t_2, t_3, t_4 > 0$. Give the relative contributions of each term to the minimum.

19. Find necessary and sufficient conditions for equality in:
(a) Young's Inequality;
(b) Hölder's Inequality;
(c) the Cauchy–Schwarz Inequality;
(d) Minkowski's Inequality.

20. Prove the special case (2) in Section 2.4 of the Arithmetic–Geometric Mean Inequality. (Hint: Show that for a fixed p the solution of the maximization problem.)

Maximize $\prod_{k=1}^{n} v_k$ subject to $v_k > 0$, $\sum_{k=1}^{n} v_k = p$ is $v_k = p/n$ for $k = 1, \ldots, n$. (Hint: Suppose that $v_1, v_2, \ldots, v_n$ are positive and that $\sum_{k=1}^{n} v_k = p$. Show that if $v_1 \neq v_2$, then setting $\bar{v}_1 = \bar{v}_2 = (v_1 + v_2)/2$ makes $\bar{v}_1 + \bar{v}_2 + v_3 + \cdots + v_n = p$ and $v_1 v_2 < \bar{v}_1 \bar{v}_2$.)

21. Let $f(x, y, z) = x^2 + y - 3z$. Show $f(x, y, z)$ is convex on R^3. Is it strictly convex?

22. (a) Let $g(\mathbf{x})$ be a convex function on R^n and suppose $g(\mathbf{x}) \geq 3$ for all $\mathbf{x}$ in R^n. Show

$$f(\mathbf{x}) = (g(\mathbf{x}) - 3)^2$$

is convex on R^n.
(b) Give an example of a convex function $g(x)$ on R^1 such that $(g(\mathbf{x}) - 3)^2$ is not convex.

23. Let $f(\mathbf{x})$ be a continuously differentiable function on R^n. Show that if $f(\mathbf{x})$ is strictly convex, then $\nabla f(\mathbf{x}) = \nabla f(\mathbf{y})$ if and only if $\mathbf{x} = \mathbf{y}$.

24. Let $f(x)$ be a convex function on R^1. Let $\mathbf{a}$ be a fixed vector on R^n and let α be a fixed real number. Define $g(\mathbf{x})$ on R^n by

$$g(\mathbf{x}) = f(\mathbf{a} \cdot \mathbf{x} + \alpha).$$

Show $g(\mathbf{x})$ is convex on R^n but that if $n \geq 2$, then $g(\mathbf{x})$ is *not* strictly convex. Deduce that

$$g(x, y, z) = (4x + 5y - 8z + 17)^8$$

is convex but not strictly convex on R^3. State why this does not follow from Theorem (2.3.10)(c).

25. Let $f(\mathbf{x})$ be a strictly convex function on R^n. Suppose $\mathbf{x}$ and $\mathbf{y}$ are distinct points in R^n such that $f(\mathbf{x}) = f(\mathbf{y}) = 0$. Show that there is a $\mathbf{z}$ in R^n such that $f(\mathbf{z}) < 0$.

26. (Calculator Exercise). Use geometric programming to solve the following program:

$$\text{Minimize} \quad g(x, y) = \frac{1000}{xy} + 2x + 2y + xy$$

over all $x, y > 0$. (*Note*: The set of feasible vectors for the dual contains more than one vector.)

27. Consider the function defined for $x > 0$ by

$$f(x) = x + \frac{1}{x}.$$

(a) Show that $x^* = 1$ is a strict global minimizer of $f(x)$ for $x > 0$.
(b) Use the result of (a) to minimize the function g defined on R^2 by

$$g(x_1, x_2) = 2x_1^2 + x_2^2 + \frac{1}{2x_1^2 + x_2^2}.$$

(c) Use the result of (a) to minimize the function h defined on R^3 by

$$h(x_1, x_2, x_3) = e^{x_1 - x_2 + x_3} + e^{-x_1 + x_2 - x_3}.$$

28. Let $f(\mathbf{x})$ be a convex function defined on R^n. Show that if $f(\mathbf{x})$ has continuous first partial derivatives and if $\varepsilon > 0$, then

$$g(\mathbf{x}) = f(\mathbf{x}) + \varepsilon \|\mathbf{x}\|^2$$

is both strictly convex and coercive.

CHAPTER 3

Iterative Methods for Unconstrained Optimization

Chapters 1 and 2 developed the theoretical basis for the solution of optimization problems. In this chapter, we confront the problem of finding numerical methods to solve optimization problems that arise in practice. Briefly put, we want iterative methods suitable for computer implementation that will permit us to compute minimizers for a given function $f(\mathbf{x})$ on R^n. By an iterative method, we mean a computational routine that produces a sequence $\{\mathbf{x}^{(k)}\}$ in R^n that we can expect to converge to a minimizer of $f(\mathbf{x})$ under reasonably broad conditions on $f(\mathbf{x})$. To be of practical interest, the routine should produce an approximating sequence $\{\mathbf{x}^{(k)}\}$ that is numerically reliable and that is not too costly to compute.

The need for iterative methods derives from the fact that it is usually not practical and often impossible to locate the critical points of $f(\mathbf{x})$ by attempting to find the exact solutions of the system

$$\nabla f(\mathbf{x}) = \mathbf{0}.$$

It is even more difficult in practice to numerically analyze the Hessian $Hf(\mathbf{x})$ to test the nature of the critical points. Instead, iterative methods are used to search out the minimizer $\mathbf{x}^*$ by means of an approximating sequence $\{\mathbf{x}^{(k)}\}$ whose points are generated in some computationally acceptable way from $f(\mathbf{x})$. The value of such a method is measured in terms of the computational cost of obtaining the successive terms of the corresponding sequence $\{\mathbf{x}^{(k)}\}$, the convergence rate, and the convergence guarantees of this sequence.

The obvious practical importance of solving unconstrained optimization problems has spawned the development of an extensive variety of iterative methods. Research and practice continue to refine and improve these methods. In this presentation, we will begin by investigating two methods that are historically important and convey general principles from which new methods can be derived. We will then develop newer techniques based on these general principles that are now "methods of choice" in practical applications.

We will not discuss the mechanics of the computer implementation of the methods we describe. However, we will be vitally interested in the mathematics that underlies these computer implementations. Those readers who are interested in the details of computer implementation of iterative minimization methods are advised to master the ideas in this chapter, and then proceed to a more advanced text such as *Numerical Methods for Unconstrained Optimization and Nonlinear Equations* by J. E. Dennis and R. B. Schnabel (Prentice-Hall, Englewood Cliffs, NJ, 1983).

Virtually every successful method for unconstrained minimization has its origins in one or both of the classical methods known as Newton's Method and the Method of Steepest Descent. Both of these methods have features that are desirable enough to retain, and both have features that are to be avoided in more practical and modern methods. For this reason, the study of Newton's Method and the Method of Steepest Descent is an ideal way to set the stage for our discussion of the more modern methods in this chapter.

3.1. Newton's Method

Given a differentiable function $f(\mathbf{x})$ on R^n, we know from (1.2.3) that any minimizers of $f(\mathbf{x})$ will be found among the solutions of the system of equations

$$\nabla f(\mathbf{x}) = \mathbf{0}. \tag{1}$$

Since this system of equations is in general nonlinear, it is usually not easy to find exact solutions even when the number of variables is small. Instead, it is usually more effective and practical to use some iterative method to find "zeros" of (1), that is, solutions of $\nabla f(\mathbf{x}) = \mathbf{0}$, to any desired degree of accuracy.

Newton's method is such a "zero-finder." It actually applies more generally to any system of the form

$$\mathbf{g}(\mathbf{x}) = \mathbf{0},$$

where $\mathbf{g}(\mathbf{x})$ is a differentiable function on R^n with values in R^n. It seeks to find solutions of $\mathbf{g}(\mathbf{x}) = \mathbf{0}$. When applied to $\mathbf{g}(\mathbf{x}) = \nabla f(\mathbf{x})$, it therefore seeks out zeros of $\nabla f(\mathbf{x}) = \mathbf{0}$. Newton's Method for the single variable case is often discussed in introductory calculus. We will begin our presentation of the method at that point.

For a given real-valued differentiable function $g(x)$ defined on R, Newton's Method seeks solutions of the equation

$$g(x) = 0$$

by beginning with an initial guess $x^{(0)}$ and generating successive terms of a sequence $\{x^{(k)}\}$ according to the formula

$$x^{(k+1)} = x^{(k)} - \frac{g(x^{(k)})}{g'(x^{(k)})}, \qquad k \geq 0. \tag{2}$$

The geometric basis for the iteration formula (2) is easy to describe: The point $x^{(k+1)}$ is merely the x-intercept of the tangent line to $y = g(x)$ at the point $(x^{(k)}, g(x^{(k)}))$.

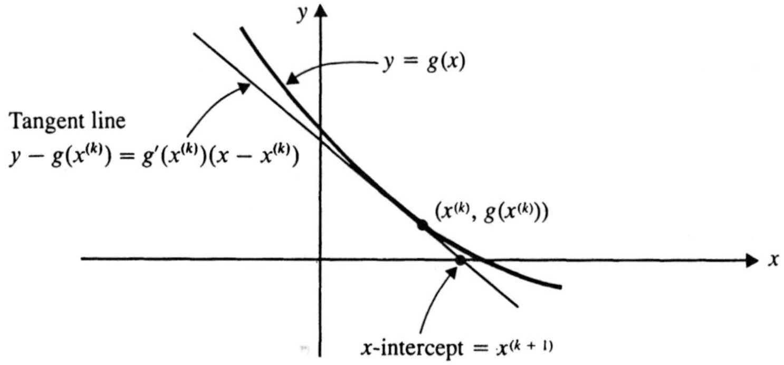

The equation of this tangent line is

$$y - g(x^{(k)}) = g'(x^{(k)})(x - x^{(k)}).$$

If we find the x-intercept of this line by setting $y = 0$ and solving for x, the solution $x^{(k+1)}$ is given by the iterative formula (2).

Most texts in numerical analysis show that the sequence $\{x^{(k)}\}$ produced by Newton's Method can be trusted to converge to a solution x^* of $g(x)$ provided that:

(1) the initial guess $x^{(0)}$ is not too far from x^*;
(2) the graph of $y = g(x)$ is not too wobbly.

(A precise statement of one sufficient condition for convergence of $\{x^{(k)}\}$ is given in Exercise 1.) It is not hard to come up with a function that will fool Newton's Method (see Exercise 1) but for most well-behaved functions Newton's Method works very well.

The idea underlying Newton's Method for solving the equation

$$\mathbf{g}(\mathbf{x}) = \mathbf{0}$$

for a differentiable function $\mathbf{g}(\mathbf{x})$ of n variables with values in R^n is basically the same as the single variable case. The tangent line at $(x^{(k)}, g(x^{(k)}))$ of the single variable case is replaced by the tangent hyperplanes at $\mathbf{x}^{(k)}$ to the graphs of the n component functions $g_1(\mathbf{x}), \ldots, g_n(\mathbf{x})$ of $\mathbf{g}(\mathbf{x})$. (See the discussion preceding (2.3.5).) More precisely, if $\mathbf{x}^{(k)}$ is the current point, the next term $\mathbf{x}^{(k+1)}$ of the sequence produced by Newton's Method is the solution of the system

$$g_1(\mathbf{x}^{(k)}) + \nabla g_1(\mathbf{x}^{(k)})\cdot(\mathbf{x} - \mathbf{x}^{(k)}) = 0,$$
$$\vdots$$
$$g_n(\mathbf{x}^{(k)}) + \nabla g_n(\mathbf{x}^{(k)})\cdot(\mathbf{x} - \mathbf{x}^{(k)}) = 0,$$

where $g_i(\mathbf{x})$ is the ith component function for $\mathbf{g}(\mathbf{x})$. This system can be expressed in matrix notation as

$$\mathbf{g}(\mathbf{x}^{(k)}) + \nabla \mathbf{g}(\mathbf{x}^{(k)})(\mathbf{x} - \mathbf{x}^{(k)}) = \mathbf{0},$$

where $\nabla \mathbf{g}(\mathbf{x})$ is the *Jacobian matrix* for $\mathbf{g}(\mathbf{x})$ at $\mathbf{x}$, that is, $\nabla \mathbf{g}(\mathbf{x})$ is the $n \times n$-matrix whose (i, j) entry is

$$\frac{\partial g_i}{\partial x_j}(\mathbf{x})$$

for $i, j = 1, 2, \ldots, n$. This yields the following iteration formulas for Newton's Method in the n-variable case.

(3.1.1) Newton's Method for Solving Equations. Suppose that $\mathbf{g}$ is a differentiable function of n variables with values in R^n and that $\mathbf{x}^{(0)} \in R^n$. Then the *Newton's Method sequence* $\{\mathbf{x}^{(k)}\}$ (*with initial point* $\mathbf{x}^{(0)}$ *for solving* $\mathbf{g}(\mathbf{x}) = \mathbf{0}$) is defined by the following recurrence formula:

(A) $$[\nabla \mathbf{g}(\mathbf{x}^{(k)})](\mathbf{x}^{(k+1)} - \mathbf{x}^{(k)}) = -\mathbf{g}(\mathbf{x}^{(k)}) \quad \text{for } k \geq 0,$$

or alternatively by

(B) $$\mathbf{x}^{(k+1)} = \mathbf{x}^{(k)} - [\nabla \mathbf{g}(\mathbf{x}^{(k)})]^{-1}\mathbf{g}(\mathbf{x}^{(k)}) \quad \text{for } k \geq 0.$$

A few words of caution and explanation are in order in conjunction with the definition (3.1.1):

(a) The recurrence formula (B) reduces to the classical formula (2) when g is a real-valued differentiable function of a single variable.
(b) In general, the system of linear equations in (A) may not have a unique solution for some value of k. In that case, the Newton's Method sequence is undefined.
(c) Even if the Newton's Method sequence is defined, it may not converge to a solution of $\mathbf{g}(\mathbf{x}) = \mathbf{0}$.
(d) Recurrence formula (B) provides a clear, explicit description of $\mathbf{x}^{(k+1)}$ in terms of $\mathbf{x}^{(k)}$ which is quite useful for theoretical purposes. However, formula (A) is better for computation than (B) since it does not require the explicit calculation of the inverse of the Jacobian matrix. In practice, we always solve the linear system (A).

Let us have a look at the iteration process prescribed by Newton's Method for the solution of a nonlinear system of three equations in three unknowns.

(3.1.2) Example. Consider the nonlinear system of equations

$$\begin{aligned} x^2 + y^2 + z^2 &= 3, \\ x^2 + y^2 - z &= 1, \\ x + y + z &= 3. \end{aligned}$$

Since solutions of this system are precisely the points of intersection of the sphere $x^2 + y^2 + z^2 = 3$, the paraboloid $x^2 + y^2 - z = 1$, and the plane $x + y + z = 3$, it is easy to see that the system has one and only one solution: $x = 1, y = 1, z = 1$.

This system can be expressed in the form

$$\mathbf{g}(\mathbf{x}) = \mathbf{0},$$

provided that $\mathbf{g}$ is taken to be the function with domain and range in R^3 defined by

$$\mathbf{g}(x, y, z) = (x^2 + y^2 + z^2 - 3, x^2 + y^2 - z - 1, x + y + z - 3).$$

Let us compute the first three terms of the sequence produced by Newton's Method for the initial point $x^{(0)} = (1, 0, 1)$. The Jacobian matrix of the system is

$$\nabla g(\mathbf{x}) = \begin{pmatrix} 2x & 2y & 2z \\ 2x & 2y & -1 \\ 1 & 1 & 1 \end{pmatrix}.$$

Therefore, given that $x^{(0)} = (1, 0, 1)$, we can find $\mathbf{x}^{(1)}$ by solving the system (A), that is,

$$\begin{aligned} 2(x - 1) \quad\quad + 2(z - 1) &= 1, \\ 2(x - 1) \quad\quad - \ (z - 1) &= 1, \\ (x - 1) + y + \ (z - 1) &= 1. \end{aligned}$$

It is easy to check that this system has a unique solution $x = \frac{3}{2}, y = \frac{1}{2}, z = 1$, so $\mathbf{x}^{(1)} = (\frac{3}{2}, \frac{1}{2}, 1)$.

We compute $\mathbf{x}^{(2)}$ from $\mathbf{x}^{(1)}$ using (A) again, that is, $\mathbf{x}^{(2)}$ is the solution of the system

$$\begin{aligned} 3(x - \tfrac{3}{2}) + (y - \tfrac{1}{2}) + 2(z - 1) &= -\tfrac{1}{2}, \\ 3(x - \tfrac{3}{2}) + (y - \tfrac{1}{2}) - \ (z - 1) &= -\tfrac{1}{2}, \\ (x - \tfrac{3}{2}) + (y - \tfrac{1}{2}) + \ (z - 1) &= 0. \end{aligned}$$

This system has the unique solution $x = \frac{5}{4}, y = \frac{3}{4}, z = 1$ so $\mathbf{x}^{(2)} = (\frac{5}{4}, \frac{3}{4}, 1)$. A similar computation shows that $\mathbf{x}^{(3)} = (\frac{9}{8}, \frac{7}{8}, 1)$.

Suppose that we had started with the initial point $\mathbf{x} = (0, 0, 0)$. Then the system (A) would be:

$$\begin{aligned} 0x + 0y + 0z &= 3, \\ 0x + 0y - z &= 1, \\ x + y + z &= 3, \end{aligned}$$

which has no solution. Thus, the Newton's Method sequence is undefined for that choice of initial point.

A general technique for proving convergence for Newton's Method was devised by the Soviet mathematician L. V. Kantorovitch.[1] He first observed that the Newton's Method sequence $\{t^{(k)}\}$ with *any* initial point $t^{(0)}$ for the quadratic real-valued function of a single variable

$$\bar{g}(t) = \frac{a}{2}t^2 - \frac{1}{b}t + \frac{c}{b}$$

always converges to the unique solution

$$t^* = \frac{1}{ab}(1 - \sqrt{1 - 2abc})$$

of $\bar{g}(t) = 0$ provided that the constants a, b, c satisfy $abc \leq \frac{1}{2}$. He then used this rather special one-dimensional result to obtain a convergence theorem for Newton's Method in the n-dimensional case. More precisely, he showed that if $\mathbf{g}(\mathbf{x})$ is a differentiable function of n variables, and if the initial point $\mathbf{x}^{(0)}$ is chosen so that constants a, b, c exist for which $abc \leq \frac{1}{2}$, and if:

(1) $\|[\nabla \mathbf{g}(\mathbf{x}^{(0)})]^{-1}\| \leq b$;
(2) $\|[\nabla \mathbf{g}(\mathbf{x}^{(0)})]^{-1}\mathbf{g}(\mathbf{x}^{(0)})\| \leq c$;
(3) there is a $\delta > 0$ so that

$$\|\nabla \mathbf{g}(\mathbf{x}) - \nabla \mathbf{g}(\mathbf{y})\| \leq a\|\mathbf{x} - \mathbf{y}\|,$$

whenever $\|\mathbf{x} - \mathbf{x}^{(0)}\| < \delta$, $\|\mathbf{y} - \mathbf{x}^{(0)}\| < \delta$;

then the Newton's Method sequence $\{\mathbf{x}^{(k)}\}$ with initial point $\mathbf{x}^{(0)}$ converges to a solution $\mathbf{x}^*$ of $\mathbf{g}(\mathbf{x})$ at a rate governed by the inequality

$$\|\mathbf{x}^{(k)} - \mathbf{x}\| \leq |t^{(k)} - t^*|,$$

and $t^{(k)}$, t^* correspond to the quadratic function $\bar{g}$ of one variable as indicated earlier. Thus, when the initial point $\mathbf{x}^{(0)}$ can be chosen so that (1), (2), and (3) are satisfied for constants a, b, c with $abc \leq \frac{1}{2}$, then the convergence of Newton's Method for the quadratic function $\bar{g}(t)$ *majorizes* the convergence of Newton's Method for the function $\mathbf{g}(\mathbf{x})$ of n variables.

[1] See Section 5.3 of *Numerical Methods for Unconstrained Optimization and Nonlinear Equations* by J. E. Dennis and R. B. Schnabel (Prentice-Hall, Englewood Cliffs, NJ, 1983.)

The preceding convergence theorem of Kantorovitch and Example (3.1.2) show that the choice of a "good" initial point $\mathbf{x}^{(0)}$ is crucial to the success of Newton's Method. If $\mathbf{x}^{(0)}$ is too far away from a solution $\mathbf{x}^*$ of $\mathbf{g}(\mathbf{x}) = \mathbf{0}$, the Newton's Method sequence $\{\mathbf{x}^{(k)}\}$ with initial point $\mathbf{x}^{(0)}$ may not be defined or it may not converge to $\mathbf{x}^*$.

It is usually not practical to attempt to choose $\mathbf{x}^{(0)}$ so that the hypotheses of a convergence theorem such as the Kantorovitch Theorem are satisfied. Instead, a choice of $\mathbf{x}^{(0)}$ is made on the basis of educated guesswork and calculations begin. If these calculations with Newton's Method do not produce a suitably accurate approximation to a solution of $\mathbf{g}(\mathbf{x}) = \mathbf{0}$, we try to make a "better" choice of the initial point $\mathbf{x}^{(0)}$ and try again. The role of convergence theorems in this process is that they assure us that the successive refinement of the choice of $\mathbf{x}^{(0)}$ will eventually lead to a Newton's Method solution of $\mathbf{g}(\mathbf{x}) = \mathbf{0}$ if such a solution exists and $\mathbf{g}(\mathbf{x})$ is a "reasonable" function.

We were led to Newton's Method for solving the equation $\mathbf{g}(\mathbf{x}) = \mathbf{0}$ because of our interest in finding a minimizer $\mathbf{x}^*$ for a real-valued, differentiable function $f(\mathbf{x})$ defined on R^n. The connection between these two problems is that if $\mathbf{x}^*$ is a minimizer of $f(\mathbf{x})$ then $\mathbf{x}^*$ is necessarily a solution of the equation $\mathbf{g}(\mathbf{x}) = \mathbf{0}$ where $\mathbf{g}(\mathbf{x}) = \nabla f(\mathbf{x})$ for $\mathbf{x} \in R^n$. Thus, when Newton's Method is applied to function minimization, the function whose zeros we seek is actually the gradient of the function to be minimized. We will now investigate how the special features of a gradient function are reflected in Newton's Method when it is applied to function minimization.

If $f(\mathbf{x})$ is the function on R^n to be minimized and $\mathbf{g}(\mathbf{x}) = \nabla f(\mathbf{x})$, then the Jacobian matrix $\nabla \mathbf{g}(\mathbf{x}^{(k)})$ which appears in the defining equations for Newton's Method (see (3.1.1)) is the matrix whose ith row is the gradient of the ith component $\partial f/\partial x_i$ of $\nabla f(\mathbf{x})$. If we think about that for a moment, we see that this Jacobian matrix is just the Hessian matrix of $f(\mathbf{x})$, that is,

$$\nabla \mathbf{g}(\mathbf{x}^{(k)}) = Hf(\mathbf{x}^{(k)}).$$

This observation leads to the following statement of Newton's Method for function minimization.

(3.1.3) Newton's Method for Function Minimization. Suppose that $f(\mathbf{x})$ is a twice continuously differentiable, real-valued function of n variables and suppose that $\mathbf{x}^{(0)} \in R^n$. Then the *Newton's Method sequence* $\{\mathbf{x}^{(k)}\}$ *with initial point* $\mathbf{x}^{(0)}$ *for minimizing* $f(\mathbf{x})$ is defined by the following recurrence formula:

$$\text{(A)}' \qquad [Hf(\mathbf{x}^{(k)})](\mathbf{x}^{(k+1)} - \mathbf{x}^{(k)}) = -\nabla f(\mathbf{x}^{(k)}) \quad \text{for } k \geq 0,$$

or alternatively by

$$\text{(B)}' \qquad \mathbf{x}^{(k+1)} = \mathbf{x}^{(k)} - [Hf(\mathbf{x}^{(k)})]^{-1}\nabla f(\mathbf{x}^{(k)}) \quad \text{for } k \geq 0.$$

Note that, unlike the situation in (1.3.1), the matrices that appear in the defining equations (A)′ and (B)′ are necessarily symmetric since the Hessian matrix for a twice continuously differentiable function on R^n is always symmetric.

There is another approach to the definition (3.1.3) of Newton's Method for function minimization that is quite illuminating. According to Taylor's Theorem (1.2.4), the function

$$f_k(\mathbf{x}) = f(\mathbf{x}^{(k)}) + \nabla f(\mathbf{x}^{(k)})\cdot(\mathbf{x} - \mathbf{x}^{(k)}) + \tfrac{1}{2}(\mathbf{x} - \mathbf{x}^{(k)})\cdot Hf(\mathbf{x}^{(k)})(\mathbf{x} - \mathbf{x}^{(k)})$$

is the quadratic function that "best fits" the function $f(\mathbf{x})$ at the current point $\mathbf{x}^{(k)}$ in the sense that $f_k(\mathbf{x})$ and $f(\mathbf{x})$, as well as their first and second derivatives, agree at $\mathbf{x}^{(k)}$.

Given that $\mathbf{x}^{(k)}$ is the kth term of a sequence that we hope will converge to a minimizer $\mathbf{x}^*$ of the given function $f(\mathbf{x})$, it would be a reasonable strategy to take as the next term of this sequence the point $\mathbf{x}^{(k+1)}$ that represents the minimizer of the quadratic function $f_k(\mathbf{x})$ provided that $f_k(\mathbf{x})$ has such a minimizer. Let us see how this strategy works out.

If the quadratic function $f_k(\mathbf{x})$ has a minimizer $\mathbf{y}^*$, then $\mathbf{y}^*$ must be a critical point of $f_k(\mathbf{x})$, that is,

$$0 = \nabla f_k(\mathbf{y}^*) = \nabla f(\mathbf{x}^{(k)}) + Hf(\mathbf{x}^{(k)})\cdot(\mathbf{y}^* - \mathbf{x}^{(k)}).$$

Moreover, if the Hessian $Hf(\mathbf{x})$ of the function $f(\mathbf{x})$ to be minimized is positive definite at the current point $\mathbf{x}^{(k)}$, then the function $f_k(\mathbf{x})$ is strictly convex and does have a strict global minimizer at the unique solution $\mathbf{y}^*$ of the equation $\nabla f_k(\mathbf{y}^*) = \mathbf{0}$. This latter equation can be written as

$$[Hf(\mathbf{x}^{(k)})](\mathbf{y}^* - \mathbf{x}^{(k)}) = -\nabla f(\mathbf{x}^{(k)}),$$

which is exactly the recurrence formula (A)′ if we replace y^* by $\mathbf{x}^{(k+1)}$. Thus, the *Newton's Method sequence* $\{\mathbf{x}^{(k)}\}$ *for minimizing* $f(\mathbf{x})$ *can be looked at in the following way: Given the current point* $\mathbf{x}^{(k)}$ *of this sequence the next point* $\mathbf{x}^{(k+1)}$ *is just the critical point of the quadratic function* $f_k(\mathbf{x})$ *that best fits* $f(\mathbf{x})$ *at* $\mathbf{x}^{(k)}$.

The preceding observation yields the following interesting result concerning the application of Newton's Method to minimize quadratic functions.

(3.1.4) Theorem. *Suppose that* A *is a positive definite* $n \times n$*-matrix, that* $\mathbf{b} \in R^n$, $\mathbf{x}^{(0)} \in R^n$, *and that* $a \in R$. *Then the quadratic function* $f(\mathbf{x})$ *of* n *variables defined by*

$$f(\mathbf{x}) = a + \mathbf{b}\cdot\mathbf{x} + \tfrac{1}{2}\mathbf{x}\cdot A\mathbf{x}$$

is strictly convex and has a unique global minimizer at the point $\mathbf{x}^*$ *that is the unique solution of the system*

$$A\mathbf{x} = -\mathbf{b}.$$

Moreover, the Newton's Method sequence $\{\mathbf{x}^{(k)}\}$ *with initial point* $\mathbf{x}^{(0)}$ *for minimizing* $f(\mathbf{x})$ *reaches* $\mathbf{x}^*$ *in one step, that is,*

$$\mathbf{x}^{(1)} = \mathbf{x}^*.$$

PROOF. Note that

$$\nabla f(\mathbf{x}) = \mathbf{b} + A\mathbf{x}, \qquad Hf(\mathbf{x}) = A$$

for any $\mathbf{x} \in R^n$. Hence, since A is positive definite, (2.3.4) and (2.3.7) imply that $f(\mathbf{x})$ is strictly convex and has a unique global minimizer at the unique solution $\mathbf{x}^*$ of the system

$$A\mathbf{x} = -\mathbf{b}.$$

To prove that Newton's Method reaches $\mathbf{x}^*$ in just one step from any initial point $\mathbf{x}^{(0)}$, we simply note that

$$\begin{aligned}\mathbf{x}^{(1)} &= \mathbf{x}^{(0)} - [Hf(\mathbf{x}^{(0)})]^{-1}\nabla f(\mathbf{x}^{(0)}) \\ &= \mathbf{x}^{(0)} - A^{-1}[\mathbf{b} + A\mathbf{x}^{(0)}] \\ &= \mathbf{x}^{(0)} - A^{-1}[-A\mathbf{x}^* + A\mathbf{x}^{(0)}] \\ &= \mathbf{x}^{(0)} + \mathbf{x}^* - \mathbf{x}^{(0)} = \mathbf{x}^*.\end{aligned}$$

Thus, $\mathbf{x}^{(1)} = \mathbf{x}^*$, so the proof is complete.

For nonquadratic functions $f(\mathbf{x})$ of n variables, the Newton's Method sequence $\{\mathbf{x}^{(k)}\}$ for $f(\mathbf{x})$ does not usually converge in a single step or even a finite number of steps. In fact, it may not converge at all even when $f(\mathbf{x})$ has a unique global minimizer $\mathbf{x}^*$, and the initial point $\mathbf{x}^{(0)}$ is arbitrarily close but not coincident with $\mathbf{x}^*$. (See Exercise 2 for a one-dimensional example of this sort.)

The hypotheses of (3.1.4) require that the Hessian of the quadratic function $f(\mathbf{x})$ is positive definite, and this requirement is crucial to the conclusion that $f(\mathbf{x})$ has a unique global minimizer $\mathbf{x}^*$ and that Newton's Method reaches $\mathbf{x}^*$ in a single step starting from any initial point $\mathbf{x}^{(0)}$. For instance, if A is negative definite, then Newton's Method finds the unique global *maximizer* $\mathbf{x}^*$ of the quadratic function $f(\mathbf{x})$. The method simply does not discriminate among maximizers, minimizers, or mere saddle points of a function $f(\mathbf{x})$ whether it is quadratic or not.

For an arbitrary (that is, not necessarily quadratic) function $f(\mathbf{x})$, the assumption that $Hf(\mathbf{x})$ is positive definite does yield the conclusion that Newton's Method at least heads in the direction of decreasing function values.

(3.1.5) Theorem. *Suppose that $\{\mathbf{x}^{(k)}\}$ is the Newton's Method sequence for minimizing a function $f(\mathbf{x})$. If the Hessian $Hf(\mathbf{x}^{(k)})$ of $f(\mathbf{x})$ at $\mathbf{x}^{(k)}$ is positive definite and if $\nabla f(\mathbf{x}^{(k)}) \neq \mathbf{0}$, then the direction*

$$\mathbf{p}^{(k)} = -[Hf(\mathbf{x}^{(k)})]^{-1}\nabla f(\mathbf{x}^{(k)})$$

from $\mathbf{x}^{(k)}$ to $\mathbf{x}^{(k+1)}$ is a descent direction for $f(\mathbf{x})$ in the sense that there is an $\varepsilon > 0$ such that

$$f(\mathbf{x}^{(k)} + t\mathbf{p}^{(k)}) < f(\mathbf{x}^{(k)})$$

for all t such that $0 < t < \varepsilon$.

PROOF. Define $\varphi(t)$ on R by

$$\varphi(t) = f(\mathbf{x}^{(k)} + t\mathbf{p}^{(k)}).$$

Note that $\varphi(t)$ is the restriction of $f(\mathbf{x})$ to the line through $\mathbf{x}^{(k)}$ in the direction of $\mathbf{p}^{(k)}$. Then

$$\varphi'(t) = \nabla f(\mathbf{x}^{(k)} + t\mathbf{p}^{(k)})\cdot\mathbf{p}^{(k)},$$

and so

$$\begin{aligned}\varphi'(0) &= \nabla f(\mathbf{x}^{(k)})\cdot\mathbf{p}^{(k)}\\ &= -\nabla f(\mathbf{x}^{(k)})[Hf(\mathbf{x}^{(k)})]^{-1}\nabla f(\mathbf{x}^{(k)}) < 0,\end{aligned}$$

since $[Hf(\mathbf{x}^{(k)})]^{-1}$ is positive definite and $\nabla f(\mathbf{x}^{(k)}) \neq \mathbf{0}$. Therefore, there is an $\varepsilon > 0$ such that $\varphi(t) < \varphi(0)$ for all t for which $0 < t < \varepsilon$, that is,

$$f(\mathbf{x}^{(k)} + t\mathbf{p}^{(k)}) < f(\mathbf{x}^{(k)})$$

for all t such that $0 < t < \varepsilon$. This completes the proof.

The following example illustrates the Newton's Method iteration procedure for a simple nonquadratic function.

(3.1.6) Example. Let us construct a Newton's Method sequence $\{\mathbf{x}^{(k)}\}$ for minimizing the function

$$f(x_1, x_2) = x_1^4 + 2x_1^2x_2^2 + x_2^4.$$

We will take advantage of the symmetry of $f(x_1, x_2)$ with respect to the variables x_1, x_2 by computing the next iterate for a current point of the form (a, a). Note that

$$\nabla f(x_1, x_2) = (4x_1^3 + 4x_1x_2^2, 4x_1^2x_2 + 4x_2^3),$$

$$Hf(x_1, x_2) = \begin{pmatrix} 12x_1^2 + 4x_2^2 & 8x_1x_2 \\ 8x_1x_2 & 4x_1^2 + 12x_2^2 \end{pmatrix},$$

so that

$$\nabla f(a, a) = (8a^3, 8a^3), \qquad Hf(a, a) = \begin{pmatrix} 16a^2 & 8a^2 \\ 8a^2 & 16a^2 \end{pmatrix}.$$

Therefore, equation (A)′ takes the following form at the current point (a, a):

$$\begin{aligned} 16a^2(x_1 - a) + 8a^2(x_2 - a) &= -8a^3,\\ 8a^2(x_1 - a) + 16a^2(x_2 - a) &= -8a^3.\end{aligned}$$

This system reduces to

$$\begin{aligned} 2x_1 + 1x_2 &= 2a,\\ 1x_1 + 2x_2 &= 2a,\end{aligned}$$

which has the solution $(x_1, x_2) = (\frac{2}{3}a, \frac{2}{3}a)$. Thus, for example, the Newton's Method sequence with initial point $\mathbf{x}^{(0)} = (1, 1)$ for minimizing $f(x_1, x_2)$ is given by

$$\mathbf{x}^{(k)} = ((\tfrac{2}{3})^k, (\tfrac{2}{3})^k).$$

Evidently, this sequence converges to $\mathbf{x}^* = (0, 0)$ which is the global minimizer of this function since

$$f(x_1, x_2) = (x_1^2 + x_2^2)^2.$$

The symmetry and simplicity of the objective function in the preceding example allowed us to compute the general term of the Newton's Method sequence for minimizing the function. It is usually neither possible nor necessary to compute this general term in practice. Rather, we proceed from one iterate to the next by computing the solution or approximate solution of the system (A)′ of linear equations. We terminate this iteration process when the current iterate satisfies the equation $\nabla f(\mathbf{x}) = \mathbf{0}$ to within some acceptable error bound. In this connection, we point out that the alternate formula (B)′ is never used for computing iterates in Newton's Method because it is computationally too expensive to compute the inverse of the Hessian matrix as (B)′ requires.

Several problems can occur in the implementation of Newton's Method. One of the most obvious is that the sequence of iterates may converge to a point that is not a global minimizer for the function. This can occur if the objective function is "wobbly" as the one-dimensional example pictured below illustrates.

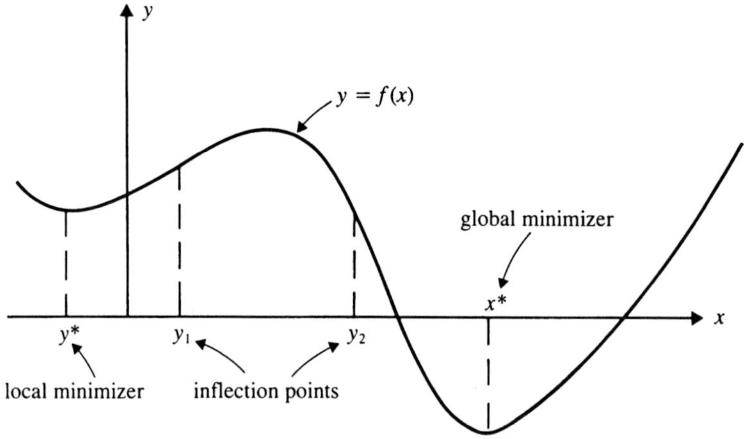

It seems rather clear (and it can be proved) that if we choose an initial point $x^{(0)} < y_1$, then the Newton's Method sequence with initial point $x^{(0)}$ will converge to the local minimizer y^* of $f(x)$. On the other hand, if we choose an initial point $x^{(0)} > y_2$, then the corresponding Newton's Method sequence will find the global minimizer x^* for $f(x)$. If the stopping condition for Newton's Method simply measures the accuracy with which the current iterate satisfies the critical point condition $f'(x) = 0$, y^* and x^* appear to be equally good limits for a Newton's Method sequence even though we really want to find x^* and not y^*. Of course, the way to avoid y^* is to pick $x^{(0)}$ as close to x^* as we can. Thus, if we find that a given initial point leads to a local rather than global minimizer, the remedy is to pick a new initial point and try again.

Newton's Method can also produce a sequence of iterates that diverges. For example, if we apply Newton's Method to minimize the function $f(x) = \frac{2}{3}|x|^{3/2}$ starting with the initial point $x^{(0)} = 1$,

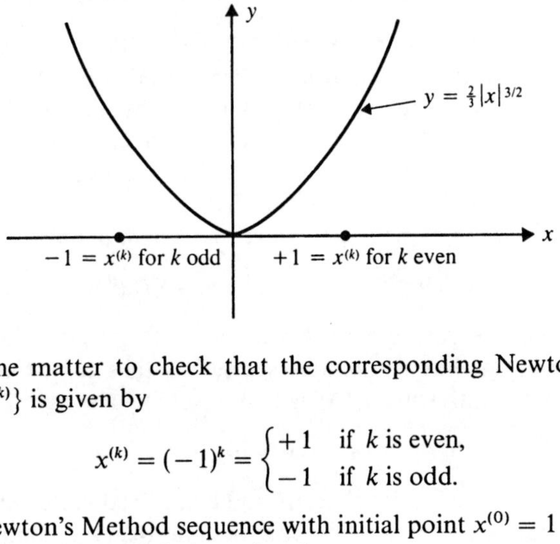

it is a routine matter to check that the corresponding Newton's Method sequence $\{\mathbf{x}^{(k)}\}$ is given by

$$x^{(k)} = (-1)^k = \begin{cases} +1 & \text{if } k \text{ is even,} \\ -1 & \text{if } k \text{ is odd.} \end{cases}$$

Thus, the Newton's Method sequence with initial point $x^{(0)} = 1$ never comes close to the true global minimizer $x^* = 0$.

More trouble can occur with Newton's Method when the Hessian $Hf(\mathbf{x})$ fails to be positive definite or even invertible because the Newton's Method sequence may fail to be defined (cf. (3.1.2)) or it may wander away from the global minimizer. The following example illustrates this latter phenomenon.

(3.1.7) Example. The function of a single variable

$$f(x) = x^4 - 32x^2$$

can readily be shown to have two global minimizers at $x = \pm 4$, no global maximizers, and a local maximizer at $x = 0$.

If we begin with the initial point $x^{(0)} = 1$ and construct the first few terms of the Newton's Method sequence for minimizing $f(x)$, we find that $x^{(1)} = -0.153846$, $x^{(2)} = 0.000629$, and that $x^{(3)}$ is zero to six decimal places. Thus, instead of approaching the nearest global minimizer at $x = +4$, the minimizing sequence $\{x^{(k)}\}$ appears to (and in fact does) converge to the local maximizer at $x = 0$. The problem here is that the Hessian

$$Hf(x) = 12x^2 - 64$$

is negative on the interval $[-4/\sqrt{3}, 4/\sqrt{3}]$ which includes the initial point $x^{(0)}$.

The apparent rapidity with which the Newton's Method sequence $\{x^{(k)}\}$ converges to the critical point $x^* = 0$ of $f(x)$ in the preceding example is not an accidental feature of the example. One can show that if $\mathbf{x}^*$ is a critical point of a function on R^n and if $\{\mathbf{x}^{(k)}\}$ is the Newton's Method sequence for minimizing $f(\mathbf{x})$ starting from the initial point $\mathbf{x}^{(0)}$, then there is a constant $C > 0$ such that

$$(*) \qquad \|\mathbf{x}^{(k+1)} - \mathbf{x}^*\| \leq C\|\mathbf{x}^{(k)} - \mathbf{x}^*\|^2$$

under suitable conditions on $f(\mathbf{x})$ and $\{\mathbf{x}^{(k)}\}$. (See, for example, page 156 of *Introduction to Linear and Non-Linear Programming* by D. G. Luenberger. (Addison-Wesley, Reading, MA., 1973).) The inequality $(*)$ shows that if a given $\mathbf{x}^{(k)}$ is sufficiently close to $\mathbf{x}^*$, then $\mathbf{x}^{(k+1)}$ is much closer since the square of a small positive number is much smaller than the given number.

Now a final word about computation. To compute the successive terms of the Newton's Method sequence, we solve the system of linear equations

$$(A)' \qquad Hf(\mathbf{x}^{(k)})(\mathbf{x}^{(k+1)} - \mathbf{x}^{(k)}) = -\nabla f(\mathbf{x}^{(k)})$$

for $\mathbf{x}^{(k+1)}$ given the current $\mathbf{x}^{(k)}$. Now, the matrix $Hf(\mathbf{x}^{(k)})$ of this system is symmetric if $f(\mathbf{x})$ has continuous second partials so the computational problem involved with solving (A)′ is that of solving a system

$$(*) \qquad A\mathbf{y} = \mathbf{b},$$

where A is an $n \times n$-symmetric matrix and $\mathbf{b}$ is a fixed vector in R^n.

Of course, Gaussian elimination provides one means to compute solutions to a system $(*)$. However, there is a more efficient procedure that can be used. Suppose that the matrix A has a factorization

$$A = LU,$$

where L is a lower triangular matrix (that is, $L = (l_{ij})$ where $l_{ij} = 0$ for $j > i$) and U is an upper triangular matrix and the diagonal elements of both L and U are nonzero. Then if $\mathbf{z} \in R^n$ is a solution of the system

$$L\mathbf{z} = \mathbf{b},$$

and $\mathbf{y}$ is a solution of

$$U\mathbf{y} = \mathbf{z},$$

it follows that $\mathbf{y}$ is a solution of $(*)$ since

$$A\mathbf{y} = LU\mathbf{y} = L\mathbf{z} = \mathbf{b}.$$

Because the matrices L and U are triangular and have nonzero diagonal elements, the solution procedure for the systems

$$L\mathbf{z} = \mathbf{b}, \qquad U\mathbf{y} = \mathbf{z}$$

is especially simple. In fact, if we write $L\mathbf{z} = \mathbf{b}$ as a system of linear equations

$$\begin{aligned} l_{11}z_1 \qquad\qquad\qquad\qquad &= b_1, \\ l_{21}z_1 + l_{22}z_2 \qquad\qquad\quad &= b_2, \\ \vdots \qquad\qquad\qquad\qquad\quad &\;\;\vdots \\ l_{n1}z_1 + l_{n2}z_2 + \cdots + l_{nn}z_n &= b_n, \end{aligned}$$

then obviously $z_1 = b_1/l_{11}$, we can solve for z_2 by substituting z_1 into the second equation, then we substitute z_1, z_2 into the third equation to find z_3, and so on. This solution procedure is referred to as *forward-substitution.* Once a solution $\mathbf{z}$ to $L\mathbf{z} = \mathbf{b}$ has been computed, we proceed to solve the upper triangular system $U\mathbf{y} = \mathbf{z}$, that is,

$$\begin{aligned} u_{11}y_1 + u_{12}y_2 + \cdots + u_{1n}y_n &= z_1, \\ u_{22}y_2 + \cdots + u_{2n}y_n &= z_2, \\ &\;\;\vdots \\ u_{nn}y_n &= z_n. \end{aligned}$$

This system is solved by *back-substitution*, that is, we use the last equation to find y_n, then substitute into the next to last to find y_{n-1}, and so on, until we find y_1 from the first equation. The fact that the diagonal elements of L and U are nonzero assures us that the forward- and back-substitution procedures can be carried to completion (that is, without encountering divisors that are zero.)

In the above discussion, we have assumed that the given symmetric matrix A has a factorization

$$A = LU, \tag{3}$$

where L is a lower triangular matrix and U is an upper triangular matrix and the diagonal elements of L and U are nonzero. We will now discuss some conditions that assure the existence of such a factorization as well as a procedure for computing L and U.

If A is a positive definite matrix (a case of special interest for Newton's Method in view of (3.1.5)), it can be shown that A has a unique factorization of the form

$$A = \hat{L}D\hat{L}^{\mathrm{T}}, \tag{4}$$

where $\hat{L}$ is a lower triangular matrix with diagonal entries equal to 1 and D is a diagonal matrix with positive diagonal entries. If $D^{1/2}$ is the *square root* of the matrix D, that is, $D^{1/2}$ is the diagonal matrix whose diagonal elements are just the square roots of the corresponding elements of D, then clearly $D^{1/2} \cdot D^{1/2} = D$ so

$$A = \hat{L}D^{1/2}D^{1/2}\hat{L}^{\mathrm{T}} = (\hat{L}D^{1/2})(\hat{L}D^{1/2})^{\mathrm{T}}.$$

Therefore, if $L = \hat{L}D^{1/2}$, then L is lower triangular with positive diagonal entries and

$$A = LL^{\mathrm{T}}.$$

This factorization of a positive definite matrix A is called the *Cholesky factorization* of A. It clearly provides a factorization of the sort we need to solve the system $A\mathbf{x} = \mathbf{b}$ by forward- and back-substitution since $U = L^{\mathrm{T}}$ is upper triangular with positive diagonal elements.

The matrix L in the Cholesky factorization of a positive definite matrix A can be computed directly by equating the corresponding elements on the two sides of the matrix equation $A = LL^{\mathrm{T}}$

$$\begin{pmatrix} a_{11} & a_{21} & \cdots & a_{n1} \\ a_{21} & a_{22} & \cdots & a_{n2} \\ \vdots & & & \\ a_{n1} & a_{n2} & \cdots & a_{nn} \end{pmatrix} = \begin{pmatrix} l_{11} & 0 & \cdots & 0 \\ l_{21} & l_{22} & \cdots & 0 \\ \vdots & & & \\ l_{n1} & l_{n2} & \cdots & l_{nn} \end{pmatrix} \begin{pmatrix} l_{11} & l_{21} & \cdots & l_{n1} \\ 0 & l_{22} & \cdots & l_{n2} \\ \vdots & & & \\ 0 & 0 & \cdots & l_{nn} \end{pmatrix}.$$

For the elements of the first columns, this yields

$$a_{11} = l_{11}^2, \qquad a_{21} = l_{11}l_{21}, \ldots, a_{n1} = l_{11}l_{n1},$$

which allows us to compute the elements of the first column of L since we know that $l_{11} \neq 0$. Next, we see that

$$a_{22} = l_{21}^2 + l_{22}^2,$$

so we can compute l_{22} since l_{21} is now known. If we equate the remaining elements of the second column, we obtain

$$a_{23} = l_{31}l_{21} + l_{32}l_{22}, \ldots, a_{n2} = l_{n1}l_{21} + l_{n2}l_{22},$$

so the elements of the second column of L can now be computed. Evidently, we can continue in this way until all elements of L are determined.

There are numerical advantages associated with solving the system

$$A\mathbf{x} = \mathbf{b}$$

by first computing the Cholesky factorization

$$A = LL^{\mathrm{T}},$$

and then solving the systems

$$L\mathbf{z} = \mathbf{b}, \qquad L^{\mathrm{T}}\mathbf{y} = \mathbf{z},$$

by forward- and back-substitution rather than solving the given system by Gaussian elimination. One such advantage is computational accuracy. It turns out that the Cholesky factorization procedure, coupled with forward- and back-substitution, is less influenced by round-off error than Gaussian elimination.

For more information concerning triangular factorizations of matrices and the solution of systems of equations by triangular factorization, see, for example, *Introduction to Matrix Computations* by G. W. Stewart (Academic Press, New York, 1973).

3.2. The Method of Steepest Descent

This classical iteration method for minimizing functions on R^n is just one of many contributions of the French mathematician Augustin-Louis Cauchy. After its discovery in 1847, it became the minimization method most frequently used by mathematicians and scientists because it is relatively easy to implement by hand for small-scale problems.

With the advent of the electronic computer, the Method of Steepest Descent and Newton's Method were eclipsed by iterative methods that were algorithmically more complicated but computationally more efficient than these classical methods. This was due to the fact that high-speed computers made it possible to attack large-scale applications that were previously considered intractable. This, in turn, stimulated the development of a new generation of computer-based algorithms that were more effective than these classical algorithms. New algorithms often evolved from efforts to remedy undesirable characteristics of these classical methods.

The Method of Steepest Descent rests on the following important property of the gradient of a differentiable function on R^n:

At a given point $\mathbf{x}^{(0)}$, the vector $\mathbf{v} = -\nabla f(\mathbf{x}^{(0)})$ points in the direction of most rapid decrease for $f(\mathbf{x})$ and the rate of decrease of $f(\mathbf{x})$ at $\mathbf{x}^{(0)}$ in this direction is $-\|\nabla f(\mathbf{x}^{(0)})\|$.

This result is a simple consequence of the Chain Rule. For if $\mathbf{u}$ is a given vector of unit length and if

$$\varphi_{\mathbf{u}}(t) = f(\mathbf{x}^{(0)} + t\mathbf{u}), \qquad t \geq 0,$$

then $\varphi_{\mathbf{u}}(t)$ is the restriction of $f(\mathbf{x})$ to the ray from $\mathbf{x}^{(0)}$ in the direction of $\mathbf{u}$ (see (2.1.2)(b)). By the Chain Rule, we obtain

$$\varphi'_{\mathbf{u}}(t) = \nabla f(\mathbf{x}^{(0)} + t\mathbf{u}) \cdot \mathbf{u},$$

so that

$$\varphi'_{\mathbf{u}}(0) = \nabla f(\mathbf{x}^{(0)}) \cdot \mathbf{u}$$

measures the rate of change of $f(\mathbf{x})$ at $\mathbf{x}^{(0)}$ in the direction of $\mathbf{u}$. For this reason, $\varphi'_{\mathbf{u}}(0)$ is usually called the *directional derivative of* $f(\mathbf{x})$ *at* $\mathbf{x}^{(0)}$ *in the direction* $\mathbf{u}$ and is denoted by

$$\nabla_{\mathbf{u}} f(\mathbf{x}^{(0)}).$$

The question we want to answer is: Which direction u makes $\nabla_{\mathbf{u}} f(\mathbf{x}^{(0)})$ as small as possible? The direction $\mathbf{u}$ that answers this question is the direction of greatest instantaneous decrease of $f(\mathbf{x})$ at $\mathbf{x}^{(0)}$; in plain language, the direction of *steepest descent* of $f(\mathbf{x})$ at $\mathbf{x}^{(0)}$.

Computing the direction of steepest descent is easy. By the Cauchy–Schwarz Inequality,

$$(*) \qquad -\|\nabla f(\mathbf{x}^{(0)})\| = -\|\nabla f(\mathbf{x}^{(0)})\|\,\|\mathbf{u}\| \le \nabla f(\mathbf{x}^{(0)})\cdot\mathbf{u} = \nabla_{\mathbf{u}} f(\mathbf{x}^{(0)})$$
$$\le \|\nabla f(\mathbf{x}^{(0)})\|\,\|\mathbf{u}\| = \|\nabla f(\mathbf{x}^{(0)})\|.$$

Consequently, the directional derivative $\nabla_{\mathbf{u}} f(\mathbf{x}^{(0)})$ is as negative as possible when there is equality in the leftmost inequality in $(*)$. This happens when $\mathbf{u}$ is the unit vector

$$\mathbf{u} = -\frac{\nabla f(\mathbf{x}^{(0)})}{\|\nabla f(\mathbf{x}^{(0)})\|}.$$

Moreover, for this choice of $\mathbf{u}$, we see that the directional derivative of $f(\mathbf{x})$ at $\mathbf{x}^{(0)}$ in this direction is precisely $-\|\nabla f(\mathbf{x}^{(0)})\|$.

The Method of Steepest Descent can now be described very simply as follows: At each stage of the iteration, the method searches for the next point by minimizing the function in the direction of the negative gradient at the current point. More precisely, the method can be formulated as follows:

(3.2.1) The Method of Steepest Descent. Suppose that $f(\mathbf{x})$ is a function with continuous first partial derivatives on R^n and that $\mathbf{x}^{(0)} \in R^n$. Then the *Steepest Descent sequence* $\{\mathbf{x}^{(k)}\}$ *with initial point* $\mathbf{x}^{(0)}$ for minimizing $f(\mathbf{x})$ is defined by the following recurrence formula:

$$\mathbf{x}^{(k+1)} = \mathbf{x}^{(k)} - t_k \nabla f(\mathbf{x}^{(k)}),$$

where t_k is the value of $t \ge 0$ that minimizes the function

$$\varphi_k(t) = f(\mathbf{x}^{(k)} - t\nabla f(\mathbf{x}^{(k)})), \qquad t \ge 0.$$

(3.2.2) Example. Let us compute the first three terms of the Steepest Descent sequence for

$$f(x, y) = 4x^2 - 4xy + 2y^2,$$

with initial point $\mathbf{x}^{(0)} = (2, 3)$. Note that

$$\nabla f(x, y) = (8x - 4y, -4x + 4y).$$

Consequently, $\nabla f(2, 3) = (4, 4)$ and so

$$\varphi_0(t) = f(2 - 4t, 3 - 4t).$$

But then

$$\begin{aligned}\varphi_0'(t) &= -\nabla f(\mathbf{x}^{(0)} - t\nabla f(\mathbf{x}^{(0)}))\cdot\nabla f(\mathbf{x}^{(0)})\\ &= -\nabla f(2-4t, 3-4t)\cdot(4,4)\\ &= -([8(2-4t)-4(3-4t)]4 + [-4(2-4t)+4(3-4t)]4)\\ &= -16(2-4t),\end{aligned}$$

so that $\varphi_0(t)$ has but one critical point at $t = \frac{1}{2}$ and this critical point is a global minimizer because $\varphi_0''(\frac{1}{2}) = 64 > 0$. It follows that $t_0 = \frac{1}{2}$, and that $\mathbf{x}^{(1)} = \mathbf{x}^{(0)} - \frac{1}{2}\nabla f(\mathbf{x}^{(0)}) = (2, 3) - \frac{1}{2}(4, 4) = (0, 1)$.

We proceed to compute the second term $\mathbf{x}^{(2)}$ in the Steepest Descent sequence in a similar way. Because $\nabla f(\mathbf{x}^{(1)}) = (-4, 4)$, we see that

$$\begin{aligned}\varphi_1(t) &= f(\mathbf{x}^{(1)} - t\nabla f(\mathbf{x}^{(1)}))\\ &= f(4t, 1-4t).\end{aligned}$$

Consequently,

$$\begin{aligned}\varphi_1'(t) &= -\nabla f(\mathbf{x}^{(1)} - t\nabla f(\mathbf{x}^{(1)}))\cdot\nabla f(\mathbf{x}^{(1)})\\ &= -([8(4t)-4(1-4t)](-4) + [-4(4t)+4(1-4t)](4))\\ &= -16(2-20t).\end{aligned}$$

Thus, $t = \frac{1}{10}$ is the unique critical point of $\varphi_1(t)$ and this critical point is a minimizer because $\varphi_1''(\frac{1}{10}) = 320 > 0$. We conclude that $t_1 = \frac{1}{10}$, and that $\mathbf{x}^{(2)} = \mathbf{x}^{(1)} - \frac{1}{10}\nabla f(\mathbf{x}^{(1)}) = (0, 1) - \frac{1}{10}(-4, 4) = (\frac{4}{10}, \frac{6}{10})$. Another iteration of the above procedure shows that $\mathbf{x}^{(3)} = (0, \frac{2}{10})$.

Note that

$$f(x, y) = 2(x^2 + (x - y)^2),$$

so that $\mathbf{x}^* = (0, 0)$ is the global minimizer of $f(x, y)$. If we plot the progress of $\mathbf{x}^{(0)}, \mathbf{x}^{(1)}, \mathbf{x}^{(2)}, \mathbf{x}^{(3)}$ toward this minimizer, we see that the Method of Steepest Descent is following a zigzag path toward $\mathbf{x}^*$ with the right angles at each turn

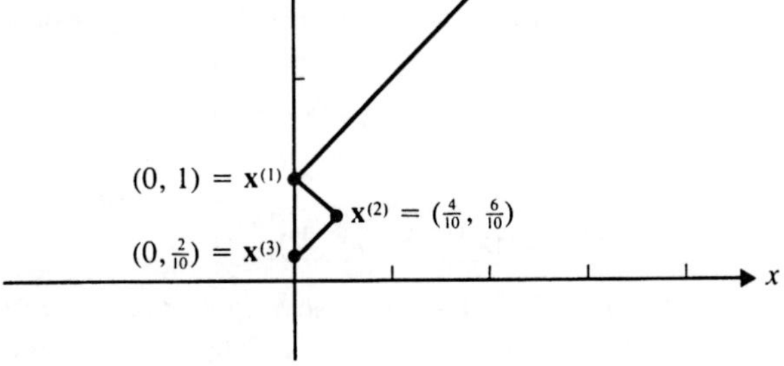

This feature is not an accident of this particular example, but rather it is a general characteristic of the Method of Steepest Descent as the following result shows.

(3.2.3) Theorem. *The Method of Steepest Descent moves in perpendicular steps. More precisely, if* $\{\mathbf{x}^{(k)}\}$ *is a Steepest Descent sequence for a function* $f(\mathbf{x})$, *then, for each* $k \in N$, *the vector joining* $\mathbf{x}^{(k)}$ *to* $\mathbf{x}^{(k+1)}$ *is orthogonal to the vector joining* $\mathbf{x}^{(k+1)}$ *to* $\mathbf{x}^{(k+2)}$.

PROOF. The recurrence formula in (3.2.1) for the Method of Steepest Descent shows that

$$(\mathbf{x}^{(k+1)} - \mathbf{x}^{(k)}) \cdot (\mathbf{x}^{(k+2)} - \mathbf{x}^{(k+1)}) = t_{k+1} t_k \nabla f(\mathbf{x}^{(k)}) \cdot \nabla f(\mathbf{x}^{(k+1)}).$$

Consequently, it suffices to show that $\nabla f(\mathbf{x}^{(k)}) \cdot \nabla f(\mathbf{x}^{(k+1)}) = 0$.

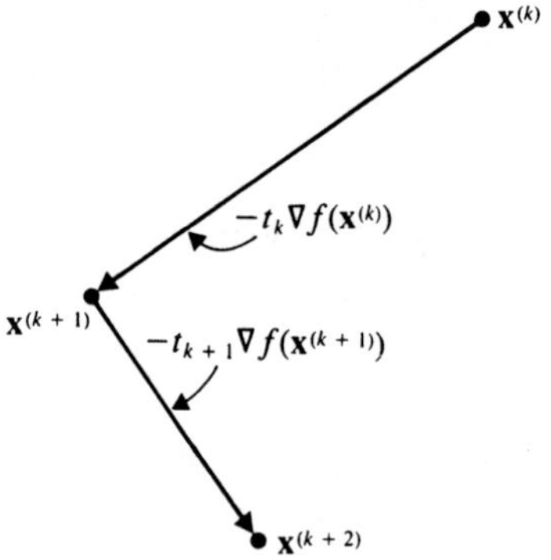

Since $\mathbf{x}^{(k+1)} = \mathbf{x}^{(k)} - t_k \nabla f(\mathbf{x}^{(k)})$ where t_k minimizes

$$\varphi_k(t) = f(\mathbf{x}^{(k)} - t\nabla f(\mathbf{x}^{(k)}))$$

for $t \geq 0$, it follows from the Chain Rule that

$$\begin{aligned} 0 = \varphi_k'(t_k) &= -\nabla f(\mathbf{x}^{(k)} - t_k \nabla f(\mathbf{x}^{(k)})) \cdot \nabla f(\mathbf{x}^{(k)}) \\ &= -\nabla f(\mathbf{x}^{(k+1)}) \cdot \nabla f(\mathbf{x}^{(k)}), \end{aligned}$$

which completes the proof.

The preceding theorem helps us to understand the geometry of the Method of Steepest Descent. Since the gradient vector $\nabla f(\mathbf{x}^{(k+1)})$ is perpendicular to the level surface

$$f(\mathbf{x}) = f(\mathbf{x}^{(k+1)})$$

at $\mathbf{x}^{(k+1)}$, and since

$$\nabla f(\mathbf{x}^{(k)}) \cdot \nabla f(\mathbf{x}^{(k+1)}) = 0$$

by (3.2.3), we see that $\nabla f(\mathbf{x}^{(k)})$ is parallel to the tangent plane to this level surface at $\mathbf{x}^{(k+1)}$. The diagram below describes this situation for functions of two variables

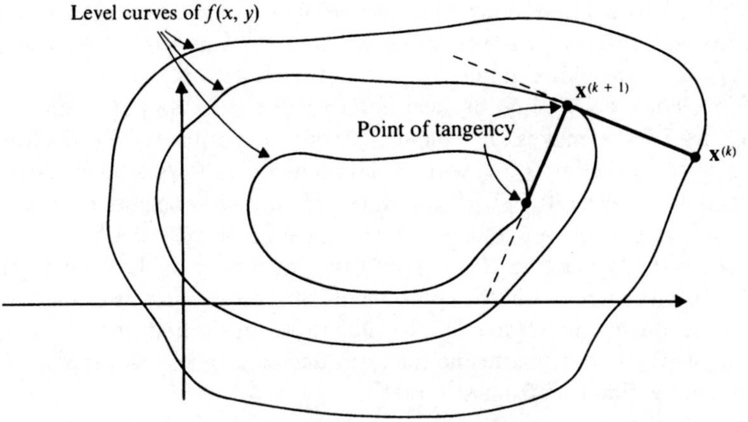

In spite of its appealing proof, the preceding theorem highlights the major drawback of the Method of Steepest Descent. In view of the remarks preceding (3.2.1), the direction of the negative gradient is the most promising direction of search for a minimizer of a function $f(\mathbf{x})$ at a given point, and that is precisely the direction prescribed by the Method of Steepest Descent at a given $\mathbf{x}^{(k)}$. However, as soon as we move in that direction, the direction ceases to be the best and continues to get worse until it is actually perpendicular to the best direction of search for a minimizer as (3.2.3) shows. This feature of the Method of Steepest Descent forces it to be inherently slow and hence computationally inefficient. To get a feeling for this problem, let's consider the following example.

(3.2.4) Example. Fix a constant $a > 1$ and let

$$f(x, y) = x^2 + a^2 y^2.$$

This function is strictly convex on R^2 and has a unique global minimizer at (0, 0). Note that

$$\nabla f(x, y) = (2x, 2a^2 y).$$

We will start the Method of Steepest Descent at a point $\mathbf{x}^{(0)}$ on the level curve $f(x, y) = 1$ and then estimate geometrically how fast the Steepest Descent sequence $\{\mathbf{x}^{(k)}\}$ proceeds toward the minimizer $\mathbf{x}^* = (0, 0)$. We will select the initial point $\mathbf{x}^{(0)}$ on the level curve $f(x, y) = 1$ so that $\nabla f(\mathbf{x}^{(0)})$ makes an angle of 45° with the x-axis. Then $x = a^2 y$ and a little algebra shows that

$$\mathbf{x}^{(0)} = \left(\frac{a}{\sqrt{1 + a^2}}, \frac{1}{a\sqrt{1 + a^2}}\right).$$

Note that $\|\mathbf{x}^{(0)}\| \leq 1$ (so that the initial point $\mathbf{x}^{(0)}$ is within one unit of the global minimizer $\mathbf{x}^* = (0, 0)$) and that

$$\nabla f(\mathbf{x}^{(0)}) = \left(\frac{2a}{\sqrt{1 + a^2}}, \frac{2a}{\sqrt{1 + a^2}}\right).$$

Let us find a lower bound for the number of iterations of the Steepest Descent algorithm that are required to move from $\mathbf{x}^{(0)}$ to within a distance of $2\sqrt{2}/a^2$ of the global minimizer at (0, 0).

First, since the Method of Steepest Descent proceeds in orthogonal steps and since $\nabla f(\mathbf{x}^{(0)})$ makes an angle of 45° with the positive x-axis, the Steepest Descent sequence for $f(x, y)$ with initial point $\mathbf{x}^{(0)}$ follows an ever-narrowing zigzag path toward (0, 0) with the slopes of successive segments of the path alternating between $+1$ and -1. The segment from $\mathbf{x}^{(k)}$ to $\mathbf{x}^{(k+1)}$ is perpendicular to the level curve $f(\mathbf{x}) = f(\mathbf{x}^{(k)})$ and tangent to the level curve $f(\mathbf{x}) = f(\mathbf{x}^{(k+1)})$. The decreases in the components of successive terms of the Steepest Descent sequence are always smaller than those for the first step. The diagram below helps us to estimate the (biggest) decrease in the x-component that occurs in the first step from $\mathbf{x}^{(0)}$ to $\mathbf{x}^{(1)}$.

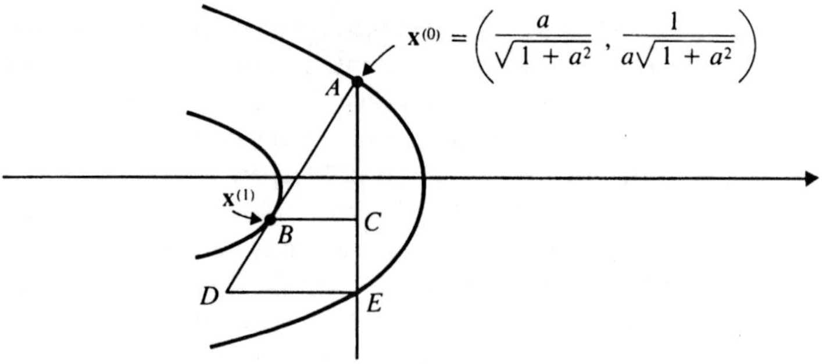

The decrease in the x-component between $\mathbf{x}^{(0)}$ and $\mathbf{x}^{(1)}$ is the length of the segment BC, which is certainly less than the length of

$$DE = AE = \frac{2}{a\sqrt{1+a^2}} < \frac{2}{a^2}.$$

Also, since $a > 1$, the x-component of $\mathbf{x}^{(0)}$ is

$$\frac{a}{\sqrt{1+a^2}} > \frac{a}{\sqrt{2a^2}} = \frac{1}{\sqrt{2}}.$$

Hence, since

$$\frac{\frac{1}{\sqrt{2}}}{\frac{2}{a^2}} = \frac{a^2}{2\sqrt{2}},$$

at least $a^2/2\sqrt{2}$ iterations are necessary to bring the Steepest Descent sequence to within a distance $\sqrt{2}(2/a^2)$ of (0, 0). For example, if $a = 20$, then at least $400/2\sqrt{2} = 141$ iterations are required to bring the Steepest Descent sequence to within a distance of $2\sqrt{2}/400 = 0.0071$ of the true minimum. (Actually, many more iterations are required for this accuracy because our estimation procedure is very rough.)

Thus, the Method of Steepest Descent pays a high penalty for moving in perpendicular increments. Actually, the main problem is that the steps it takes are *too long*, that is, there are points $\mathbf{y}^{(k)}$ between $\mathbf{x}^{(k)}$ and $\mathbf{x}^{(k+1)}$ where $-\nabla f(\mathbf{y}^{(k)})$ provides a better new search direction than $\nabla f(\mathbf{x}^{(k+1)})$. In a sense, the very greed implicit in the construction of the Steepest Descent sequence for $f(x, y)$ prevents it from converging quickly.

On a more positive note, the following theorem shows that the Method of Steepest Descent succeeds in reducing the value of the function with each iteration. This very desirable feature of a minimization method is often lacking in applications of Newton's Method.

(3.2.5) Theorem. *If $\{\mathbf{x}^{(k)}\}$ is the Steepest Descent sequence for $f(\mathbf{x})$ and if $\nabla f(\mathbf{x}^{(k)}) \neq \mathbf{0}$ for some k, then $f(\mathbf{x}^{(k+1)}) < f(\mathbf{x}^{(k)})$.*

PROOF. Recall that

$$\mathbf{x}^{(k+1)} = \mathbf{x}^{(k)} - t_k \nabla f(\mathbf{x}^{(k)}),$$

where t_k is the minimizer of

$$\varphi_k(t) = f(\mathbf{x}^{(k)} - t\nabla f(\mathbf{x}^{(k)}))$$

over all $t \geq 0$. Hence

$$f(\mathbf{x}^{(k+1)}) = \varphi_k(t_k) \leq \varphi_k(t)$$

for all $t \geq 0$. Also, by the Chain Rule,

$$\begin{aligned}\varphi_k'(0) &= -\nabla f(\mathbf{x}^{(k)} - 0 \cdot \nabla f(\mathbf{x}^{(k)})) \cdot \nabla f(\mathbf{x}^{(k)}) \\ &= -\nabla f(\mathbf{x}^{(k)}) \cdot \nabla f(\mathbf{x}^{(k)}) = -\|\nabla f(\mathbf{x}^{(k)})\|^2 < 0\end{aligned}$$

because $\nabla f(\mathbf{x}^{(k)}) \neq \mathbf{0}$ by hypothesis. But $\varphi_k'(0) < 0$ implies that there is a $\bar{t} > 0$ such that $\varphi_k(0) > \varphi_k(t)$ for $0 < t \leq \bar{t}$. Consequently,

$$f(\mathbf{x}^{(k+1)}) = \varphi_k(t_k) \leq \varphi_k(\bar{t}) < \varphi_k(0) = f(\mathbf{x}^{(k)}),$$

which is exactly what we wanted to prove.

Iterative minimization methods that produce a sequence $\{\mathbf{x}^{(k)}\}$ with the property described in Theorem (3.2.5) are called *descent methods.*

Another very appealing feature of the Method of Steepest Descent is that it has very strong convergence guarantees as the following theorem demonstrates.

(3.2.6) Theorem. *Suppose that $f(\mathbf{x})$ is a coercive function with continuous first partial derivatives on R^n. If $\mathbf{x}^{(0)}$ is any point in R^n and if $\{\mathbf{x}^{(k)}\}$ is the Steepest Descent sequence for $f(\mathbf{x})$ with initial point $\mathbf{x}^{(0)}$, then some subsequence of $\{\mathbf{x}^{(k)}\}$*

converges. The limit of any convergent subsequence of $\{\mathbf{x}^{(k)}\}$ *is a critical point of* $f(\mathbf{x})$.

Proof. Because $f(\mathbf{x})$ is continuous and coercive, Theorem (1.4.4) ensures that $f(\mathbf{x})$ has a global minimizer $\mathbf{x}^*$. Also, by Theorem (3.2.5) and the definition of the Steepest Descent sequence, we see that $\{f(\mathbf{x}^{(k)})\}$ is a decreasing sequence that is bounded below by $f(\mathbf{x}^*)$. It follows that $\{\mathbf{x}^{(k)}\}$ is a bounded sequence because $f(\mathbf{x})$ is coercive and therefore cannot be bounded on an unbounded set. The Bolzano–Weierstrass Property implies that $\{\mathbf{x}^{(k)}\}$ has at least one convergent subsequence, which completes the proof of the first assertion.

Now let $\{\mathbf{x}^{(k_p)}\}$ be a convergent subsequence of $\{\mathbf{x}^{(k)}\}$ and let $\mathbf{y}^*$ be its limit. Suppose, contrary to the second assertion of the theorem, that $\mathbf{y}^*$ is not a critical point of $f(\mathbf{x})$, that is, that $\nabla f(\mathbf{y}^*) \neq \mathbf{0}$. If we define $\varphi(t)$ for $t \geq 0$ by

$$\varphi(t) = f(\mathbf{y}^* - t\nabla f(\mathbf{y}^*)),$$

then we can see (as in the proof of (3.2.5)) that

$$\varphi'(0) = -\|\nabla f(\mathbf{y}^*)\|^2 < 0.$$

Also, if t^* is a global minimizer for $t \geq 0$ of $\varphi(t)$, then (again as in the proof of (3.2.5)) we can see that

$$f(\mathbf{y}^* - t^*\nabla f(\mathbf{y}^*)) = \varphi(t^*) < \varphi(0) = f(\mathbf{y}^*). \tag{1}$$

The continuity of $f(\mathbf{x})$ and its first partial derivatives implies that

$$\lim_p f(\mathbf{x}^{(k_p)} - t^*\nabla f(\mathbf{x}^{(k_p)})) = f(\mathbf{y}^* - t^*\nabla f(\mathbf{y}^*)). \tag{2}$$

By combining (1) and (2), we conclude that

$$f(\mathbf{x}^{(k_{\bar{p}})}) - t^*\nabla f(\mathbf{x}^{(k_{\bar{p}})})) < f(\mathbf{y}^*) \tag{3}$$

for a sufficiently large integer $\bar{p}$.

Because $\{f(\mathbf{x}^{(k)})\}$ is a decreasing sequence and because

$$\lim_p f(\mathbf{x}^{(k_p)}) = f(\mathbf{y}^*)$$

by the continuity of $f(\mathbf{x})$, it follows that

$$\lim_k f(\mathbf{x}^{(k)}) = f(\mathbf{y}^*).$$

Therefore,

$$f(\mathbf{y}^*) \leq f(\mathbf{x}^{(k_{\bar{p}}+1)}) = f(\mathbf{x}^{(k_{\bar{p}})}) - t_{k_{\bar{p}}}\nabla f(\mathbf{x}^{(k_{\bar{p}})})) \tag{4}$$

and the definition of $t_{k_{\bar{p}}}$ yields

$$f(\mathbf{x}^{(k_{\bar{p}})}) - t_{k_{\bar{p}}}\nabla f(\mathbf{x}^{(k_{\bar{p}})})) < f(\mathbf{x}^{(k_{\bar{p}})}) - t^*\nabla f(\mathbf{x}^{(k_{\bar{p}})})). \tag{5}$$

If we combine (3), (4), and (5), we obtain the contradiction $f(\mathbf{y}^*) < f(\mathbf{y}^*)$ so our supposition that $\nabla f(\mathbf{y}^*) \neq \mathbf{0}$ must be false. Therefore, $\mathbf{y}^*$ is a critical point of $f(\mathbf{x})$, which completes the proof of the theorem.

(3.2.7) Corollary. *If $f(\mathbf{x})$ is a strictly convex, coercive function with continuous first partial derivatives on R^n, then for any initial point $\mathbf{x}^{(0)}$, the Steepest Descent sequence with initial point $\mathbf{x}^{(0)}$ converges to the unique global minimizer of $f(\mathbf{x})$.*

PROOF. Because $f(\mathbf{x})$ is coercive and continuous, it has a global minimizer $\mathbf{x}^*$ by Theorem (1.4.4) and $\mathbf{x}^*$ is a critical point of $f(\mathbf{x})$ and the only global minimizer of $f(\mathbf{x})$ by (2.3.4).

Suppose that the Steepest Descent sequence $\{\mathbf{x}^{(k)}\}$ does not converge to $\mathbf{x}^*$. Then there is an $r > 0$ and a subsequence $\{\mathbf{x}^{(k_p)}\}$ of $\{\mathbf{x}^{(k)}\}$ such that

$$\|\mathbf{x}^{(k_p)} - \mathbf{x}^*\| \geq r \tag{6}$$

for all p. But the sequence $\{\mathbf{x}^{(k)}\}$ is bounded because $f(\mathbf{x})$ is coercive, so $\{\mathbf{x}^{(k_p)}\}$ is a bounded sequence in its own right. The Bolzano–Weierstrass Property implies that $\{\mathbf{x}^{(k_p)}\}$ has a convergent subsequence which is in turn a subsequence of $\{\mathbf{x}^{(k)}\}$. Theorem (3.2.6) asserts that the limit of this subsequence must be a critical point of $f(\mathbf{x})$. Since $\mathbf{x}^*$ is the unique critical point of $f(\mathbf{x})$, we have obtained a contradiction to (6). Therefore, the Steepest Descent sequence $\{\mathbf{x}^{(k)}\}$ converges to the global minimizer $\mathbf{x}^*$ of $f(\mathbf{x})$.

3.3. Beyond Steepest Descent

In the last section, we saw that the Method of Steepest Descent has strong convergence guarantees but that it may move laboriously slowly to a minimizer. The mathematical theorems dealing with this method are clean and appealing. Yet, as happens all too often in numerical work, solid mathematical theorems do not always translate into effective practical procedures. In fact, the Method of Steepest Descent has computational drawbacks so severe that it has fallen out of favor even though it was at one time a very popular technique. Let us look at this method with a critical eye.

One serious drawback is at the heart of the method. Recall that the constant t_k in the defining recurrence relation for the method

$$\mathbf{x}^{(k+1)} = \mathbf{x}^{(k)} - t_k \nabla f(\mathbf{x}^{(k)})$$

is set by minimizing

$$\varphi_k(t) = f(\mathbf{x}^{(k)} - t\nabla f(\mathbf{x}^{(k)}))$$

over all $t > 0$. Significant computational time may be needed to compute t_k. Finding t_k might take a lot of effort and is only one step of the real n-dimensional problem.

A second drawback to the Method of Steepest Descent is its movement by perpendicular steps (see Theorem (3.2.3) and Example (3.2.4)). The perpendicular steps force the method to be inherently slow in converging; hence it is not computationally efficient.

What can we do? We could fall back to Newton's Method but then, as we know from Section 3.1, real trouble can occur if the Hessian is not always positive definite. This seems to exhaust our options. However, as in most matters scientific, a better understanding of the problem can lead to better, more effective methods.

One approach to the development of a practical iterative method for unconstrained minimization that is free of the drawbacks of the Method of Steepest Descent and Newton's Method is to list criteria that a good method should satisfy to avoid these drawbacks, and then attempt to define an iterative method that meets these criteria. As we shall see, the criteria discussed below will do nicely.

For a given function $f(\mathbf{x})$ with continuous first partial derivatives on R^n and a given initial point $\mathbf{x}^{(0)}$, we want a method that produces a sequence $\mathbf{x}^{(k)}$ defined by a recurrence formula

$$\mathbf{x}^{(k+1)} = \mathbf{x}^{(k)} + t_k \mathbf{p}^{(k)}, \qquad t_k > 0.$$

For each k, we insist that the vector $\mathbf{p}^{(k)}$ and the parameter t_k are generated such that the following requirements are satisfied:

Criterion 1. $f(\mathbf{x}^{(k+1)}) < f(\mathbf{x}^{(k)})$ whenever $\nabla f(\mathbf{x}^{(k)}) \neq \mathbf{0}$.

Criterion 1 simply specifies that a "good" method should be a descent method. The Steepest Descent Method has this property while Newton's Method does not in general.

The following condition will assure that a "good" method must move in a promising direction for minimization at each step of the iteration process.

Criterion 2. $\mathbf{p}^{(k)} \cdot \nabla f(\mathbf{x}^{(k)}) < 0$.

If this condition is satisfied, then the restriction

$$\varphi_k(t) = f(\mathbf{x}^{(k)} + t\mathbf{p}^{(k)})$$

of $f(\mathbf{x})$ to the line through the current point $\mathbf{x}^{(k)}$ parallel to $\mathbf{p}^{(k)}$ has the following property

$$\varphi_k'(0) = \nabla f(\mathbf{x}^{(k)}) \cdot \mathbf{p}^{(k)} < 0.$$

Hence, for positive values of t that are sufficiently small, it is true that

$$f(\mathbf{x}^{(k)} + t\mathbf{p}^{(k)}) = \varphi_k(t) < \varphi(0) = f(\mathbf{x}^{(k)}).$$

It follows that if Criterion 2 is satisfied and if t_k is positive and sufficiently small, then $f(\mathbf{x}^{(k+1)}) = f(\mathbf{x}^{(k)} + t_k\mathbf{p}^{(k)}) < f(\mathbf{x}^{(k)})$, that is, Criterion 1 is also satisfied.

If we continue to take t_k to be very small and positive, then it may be impossible for the $\mathbf{x}^{(k)}$'s to move very rapidly toward the minimizer $\mathbf{x}^*$ because the successive steps are too small. To avoid excessively small steps, we also insist that a "good" iteration method satisfy:

Criterion 3. There is a β with $0 < \beta < 1$ such that

$$\mathbf{p}^{(k)} \cdot \nabla f(\mathbf{x}^{(k+1)}) > \beta \mathbf{p}^{(k)} \cdot \nabla f(\mathbf{x}^{(k)}).$$

Let us see how Criteria 2 and 3 in tandem prevent arbitrarily small choices for t_k. Assume that β is fixed in the range $0 < \beta < 1$ and that t_k is chosen to satisfy Criterion 3. Then

$$\mathbf{p}^{(k)} \cdot \nabla f(\mathbf{x}^{(k)} + t_k \mathbf{p}^{(k)}) > \beta \mathbf{p}^{(k)} \cdot \nabla f(\mathbf{x}^{(k)}) > \mathbf{p}^{(k)} \cdot \nabla f(\mathbf{x}^{(k)}),$$

since $\beta < 1$ and $\mathbf{p}^{(k)} \cdot \nabla f(\mathbf{x}^{(k)}) < 0$ by Criterion 2. Accordingly,

$$(*) \qquad \mathbf{p}^{(k)} \cdot \nabla f(\mathbf{x}^{(k)} + t_k \mathbf{p}^{(k)}) - \mathbf{p}^{(k)} \cdot \nabla f(\mathbf{x}^{(k)}) > (\beta - 1)\mathbf{p}^{(k)} \cdot \nabla f(\mathbf{x}^{(k)}) > 0,$$

and this means that t_k cannot be arbitrarily small. For if we let t_k approach 0 in (*), the left-hand side of the inequality approaches 0 while the right-hand side remains constant at $(\beta - 1)\mathbf{p}^{(k)} \cdot \nabla f(\mathbf{x}^{(k)}) > 0$, which is impossible. Thus, Criteria 2 and 3 together prevent arbitrarily small choices of t_k.

But what about excessively large t_k? A very large t_k will result in a large step from $\mathbf{x}^{(k)}$ to $\mathbf{x}^{(k+1)}$. However, as we saw in Example (3.2.4), one of the problems inherent in the Method of Steepest Descent is that it tends to set too large a value for t_k even when a much smaller value would achieve nearly the same reduction in the value of the objective function $f(\mathbf{x})$. We should be willing to make t_k large only if it results in a correspondingly large decrease in the value of $f(\mathbf{x})$. We quantify this requirement of a "good" iteration method as follows:

Criterion 4. There is an α with $0 < \alpha < \beta < 1$ such that

$$f(\mathbf{x}^{(k+1)}) \leq f(\mathbf{x}^{(k)}) + \alpha t_k \mathbf{p}^{(k)} \cdot \nabla f(\mathbf{x}^{(k)}).$$

To understand how this requirement achieves the desired result, note that the stated inequality can be written as

$$\frac{f(\mathbf{x}^{(k)}) - f(\mathbf{x}^{(k+1)})}{t_k} \geq \alpha[-\mathbf{p}^{(k)} \cdot \nabla f(\mathbf{x}^{(k)})].$$

The left-hand side of the preceding inequality represents the relative decrease in the value of $f(\mathbf{x})$ with respect to the increase in t-values between $\mathbf{x}^{(k)}$ and $\mathbf{x}^{(k+1)}$. The term

$$[-\mathbf{p}^{(k)} \cdot \nabla f(\mathbf{x}^{(k)})]$$

on the right-hand side, which is a positive number by Criterion 2, is a multiple of the magnitude of the rate of decrease in the direction $\mathbf{p}^{(k)}$ of $f(\mathbf{x})$ at $\mathbf{x}^{(k)}$. (See the discussion of Criterion 2.) Consequently, the meaning of Criterion 4 is that the decrease in the values of the objective function $f(\mathbf{x})$ relative to the size of t_k should exceed a preassigned fraction of the magnitude of the rate of decrease of $f(\mathbf{x})$ at $\mathbf{x}^{(k)}$ in the direction to $\mathbf{x}^{(k+1)}$.

A slightly different perspective on Criterion 4 is also enlightening. If we set

$$M = -\frac{\alpha \mathbf{p}^{(k)} \cdot \nabla f(\mathbf{x}^{(k)})}{\|\mathbf{p}^{(k)}\|},$$

then Criterion 4 yields the inequality

$$\frac{f(\mathbf{x}^{(k)}) - f(\mathbf{x}^{(k+1)})}{\|\mathbf{x}^{(k)} - \mathbf{x}^{(k+1)}\|} = \frac{f(\mathbf{x}^{(k)}) - f(\mathbf{x}^{(k+1)})}{t_k \|\mathbf{p}^{(k)}\|} \geq M.$$

We see from this that

$$(*) \qquad f(\mathbf{x}^{(k)}) - f(\mathbf{x}^{(k+1)}) \geq M \|\mathbf{x}^{(k)} - \mathbf{x}^{(k+1)}\|.$$

Thus, if the step from $\mathbf{x}^{(k)}$ to $\mathbf{x}^{(k+1)}$ is large (that is, $\|\mathbf{x}^{(k)} - \mathbf{x}^{(k+1)}\|$ is large), then the decrease in objective function value $f(\mathbf{x}^{(k)}) - f(\mathbf{x}^{(k+1)})$ between $\mathbf{x}^{(k)}$ and $\mathbf{x}^{(k+1)}$ must be proportionately large. Note that $(*)$ shows that Criterion 1 (that is, $f(\mathbf{x}^{(k+1)}) < f(\mathbf{x}^{(k)})$) is automatically satisfied when Criteria 2 and 4 are in force.

Our description of a "good" iteration method

$$\mathbf{x}^{(k+1)} = \mathbf{x}^{(k)} + t_k \mathbf{p}^{(k)}, \qquad k \geq 1,$$

for minimizing a function $f(\mathbf{x})$ can be recapitulated as follows: Given α, β with $0 < \alpha < \beta < 1$, we want to select $\mathbf{p}^{(k)}$ and $t_k > 0$ so that Criteria (1)–(4) are satisfied. Happily, it is always possible to make such a selection as the following theorem demonstrates.

(3.3.1) Theorem (Wolfe). *Suppose that $f(\mathbf{x})$ has continuous first partial derivatives and is bounded from below on R^n. Let α, β be fixed numbers with $0 < \alpha < \beta < 1$. If $\mathbf{p}^{(k)}$ and $\mathbf{x}^{(k)}$ are vectors in R^n satisfying*

$$\mathbf{p}^{(k)} \cdot \nabla f(\mathbf{x}^{(k)}) < 0,$$

then there are real numbers a_k and b_k such that $0 \leq a_k < b_k$, and

(i) *Criterion 4 is satisfied for any choice of t_k in $(0, b_k)$;*
(ii) *Criterion 3 is satisfied for any choice of t_k in (a_k, b_k).*

Consequently, both Criteria 3 and 4 are satisfied for any choice of t_k in (a_k, b_k).

Proof. Define for $t \geq 0$

$$\varphi_k(t) = f(\mathbf{x}^{(k)} + t\mathbf{p}^{(k)}).$$

As we have noted earlier,

$$0 > \mathbf{p}^{(k)} \cdot \nabla f(\mathbf{x}^{(k)}) = \varphi'(0) = \lim_{t \to 0} \frac{f(\mathbf{x}^{(k)} + t\mathbf{p}^{(k)}) - f(\mathbf{x}^{(k)})}{t}.$$

Also because $\mathbf{p}^{(k)} \cdot \nabla f(\mathbf{x}^{(k)}) < 0$ and $\alpha < 1$, we see

$$\alpha \mathbf{p}^{(k)} \cdot \nabla f(\mathbf{x}^{(k)}) > \mathbf{p}^{(k)} \cdot \nabla f(\mathbf{x}^{(k)}) = \lim_{t \to 0} \frac{f(\mathbf{x}^{(k)} + t\mathbf{p}^{(k)}) - f(\mathbf{x}^{(k)})}{t}.$$

Consequently, there is an $\varepsilon > 0$ such that

$$\alpha \mathbf{p}^{(k)} \cdot \nabla f(\mathbf{x}^{(k)}) > \frac{f(\mathbf{x}^{(k)} + t\mathbf{p}^{(k)}) - f(\mathbf{x}^{(k)})}{t}$$

for $0 < t < \varepsilon$. Hence, for $0 < t < \varepsilon$, we have

$$\text{(A)} \qquad f(\mathbf{x}^{(k)} + t\mathbf{p}^{(k)}) < f(\mathbf{x}^{(k)}) + t\alpha \mathbf{p}^{(k)} \cdot \nabla f(\mathbf{x}^{(k)}).$$

Let t^* be the largest such ε. Notice that there will be a largest such ε, since otherwise

$$f(\mathbf{x}^{(k)} + t\mathbf{p}^{(k)}) < f(\mathbf{x}^{(k)}) + t\alpha \mathbf{p}^{(k)} \cdot \nabla f(\mathbf{x}^{(k)})$$

for all $t > 0$. This is impossible because $f(\mathbf{x})$ is bounded from below and $f(\mathbf{x}^{(k)}) + t\alpha \mathbf{p}^{(k)} \cdot \nabla f(\mathbf{x}^{(k)})$ tends to $-\infty$ as t tends to $+\infty$. This proves t^* is a finite real number and shows that Criterion 4 is in force for any choice of t in $(0, t^*)$.

Next, note that

$$\text{(B)} \qquad f(\mathbf{x}^{(k)} + t^*\mathbf{p}^{(k)}) = f(\mathbf{x}^{(k)}) + t^*\alpha \mathbf{p}^{(k)} \cdot \nabla f(\mathbf{x}^{(k)})$$

because, if not, then by continuity there is a $\delta > 0$ such that

$$f(\mathbf{x}^{(k)} + t\mathbf{p}^{(k)}) < f(\mathbf{x}^{(k)}) + t\alpha \mathbf{p}^{(k)} \cdot \nabla f(\mathbf{x}^{(k)})$$

for $t^* - \delta < t < t^* + \delta$. It follows that $\varepsilon = t^* + \delta > t^*$ works in (A), which contradicts the fact that t^* is the largest ε for which (A) works.

Rewrite (B) as follows:

$$\text{(C)} \qquad f(\mathbf{x}^{(k)} + t^*\mathbf{p}^{(k)}) - f(\mathbf{x}^{(k)}) = t^*\alpha \mathbf{p}^{(k)} \cdot \nabla f(\mathbf{x}^{(k)}),$$

and apply the Mean Value Theorem to the function

$$\varphi(t) = f(\mathbf{x}^{(k)} + t\mathbf{p}^{(k)})$$

on the interval $[0, t^*]$ to learn

$$\text{(D)} \qquad f(\mathbf{x}^{(k)} + t^*\mathbf{p}^{(k)}) - f(\mathbf{x}^{(k)}) = \varphi(t^*) - \varphi(0) = \varphi'(t^{**})(t^* - 0)$$

for some t^{**} in $(0, t^*)$. Now

$$\varphi'(t^{**}) = \nabla f(\mathbf{x}^{(k)} + t^{**}\mathbf{p}^{(k)}) \cdot \mathbf{p}^{(k)}.$$

Hence, (D) becomes

$$\text{(E)} \qquad f(\mathbf{x}^{(k)} + t^*\mathbf{p}^{(k)}) - f(\mathbf{x}^{(k)}) = t^*\mathbf{p}^{(k)} \cdot \nabla f(\mathbf{x}^{(k)} + t^{**}\mathbf{p}^{(k)}).$$

Combining (C) and (E) gives

$$\alpha t^*\mathbf{p}^{(k)} \cdot \nabla f(\mathbf{x}^{(k)}) = t^*\nabla f(\mathbf{x}^{(k)} + t^{**}\mathbf{p}^{(k)}) \cdot \mathbf{p}^{(k)}$$

and because $0 < \alpha < \beta$, it follows that

$$\text{(F)} \qquad \beta \mathbf{p}^{(k)} \cdot \nabla f(\mathbf{x}^{(k)}) < \nabla f(\mathbf{x}^{(k)} + t^{**}\mathbf{p}^{(k)}) \cdot \mathbf{p}^{(k)}.$$

Recall now that because $0 < t^{**} < t^*$, we have from (A) that

(G) $$f(\mathbf{x}^{(k)} + t^{**}\mathbf{p}^{(k)}) < f(\mathbf{x}^{(k)}) + \alpha t^{**}\mathbf{p}^{(k)} \cdot \nabla f(\mathbf{x}^{(k)}).$$

By the continuity of the functions involved, there is an interval (a_k, b_k) with $0 < a_k < t^{**} < b_k < t^*$ such that

$$f(\mathbf{x}^{(k)} + t\mathbf{p}^{(k)}) < f(\mathbf{x}^{(k)}) + \alpha t\mathbf{p}^{(k)} \cdot \nabla f(\mathbf{x}^{(k)}),$$

and

$$\beta\mathbf{p}^{(k)} \cdot \nabla f(\mathbf{x}^{(k)}) < \nabla f(\mathbf{x}^{(k)} + t\mathbf{p}^{(k)}) \cdot \mathbf{p}^{(k)}$$

for all t in (a_k, b_k). Therefore, any choice of t_k in (a_k, b_k) satisfies Criteria 3 and 4. This proves (ii) of the theorem. To prove (i) notice that $0 < b_k < t^*$ and Criterion 4 is in force for any t_k in $(0, t^*)$; therefore, it is in force for any t_k in $(0, b_k)$. This completes the proof.

Wolfe's Theorem guarantees that it is always possible to set t_k such that Criteria 1, 3, and 4 are in force provided $\mathbf{p}^{(k)}$ satisfies Criterion 2. The science of mathematics via the last proof has made this indisputable. But in practice, how do we find t_k?

Finding a numerically efficient, reliable method for setting t_k in practice involves a substantial amount of numerical art as well as science. In most routines, we begin by setting $t = 1$ and then we apply a "backtracking" procedure to reduce t until we reach a value for which Criteria 3 and 4 are satisfied. One prescription for doing this is to check for failure of Criterion 4; in other words, determine whether

$$f(\mathbf{x}^{(k)} + t\mathbf{p}^{(k)}) > f(\mathbf{x}^{(k)}) + \alpha t\mathbf{p}^{(k)} \cdot \nabla f(\mathbf{x}^{(k)}),$$

and if this inequality holds, we replace t by st for some appropriate $s < 1$. Making an appropriate choice for s is an interesting problem in its own right. For more detail on this subject, see Section 6.3.2 in *Numerical Methods for Unconstrained Optimization and Nonlinear Equations* by J. E. Dennis and R. B. Schnabel (Prentice-Hall, Englewood Cliffs, NJ, 1983).

Finally, we examine the possibilities for the choice of the search direction $\mathbf{p}^{(k)}$. As we have already noted, we want

(*) $$\mathbf{p}^{(k)} \cdot \nabla f(\mathbf{x}^{(k)}) < 0$$

so that the values of $f(\mathbf{x})$ will decrease, at least at first, as we leave $\mathbf{x}^{(k)}$ in the direction $\mathbf{p}^{(k)}$. Of course, the easiest choice of $\mathbf{p}^{(k)}$ to accomplish this is $\mathbf{p}^{(k)} = -\nabla f(\mathbf{x}^{(k)})$, the search direction of the Method of Steepest Descent, but we know that choice may result in very slow convergence of the resulting iteration method. However, note that if $Q(\mathbf{x})$ is any function on R^n whose values are positive definite matrices, then the choice

$$\mathbf{p}^{(k)} = -Q(\mathbf{x}^{(k)}) \cdot \nabla f(\mathbf{x}^{(k)})$$

still forces (*). This allows considerable latitude in the choice of $\mathbf{p}^{(k)}$. In particular, it includes the search directions for the Method of Steepest Descent (where $Q(x)$ is the identity matrix) and, if $Hf(\mathbf{x})$ is positive definite, for Newton's Method (where $Q(\mathbf{x}) = Hf(\mathbf{x})^{-1}$). We have seen in Example (3.1.7) that the latter choice is not very useful if $Hf(\mathbf{x})$ is not positive definite. However, even when $Hf(\mathbf{x})$ is not positive definite, it is possible to perturb $Hf(\mathbf{x})$ by a positive multiple μ_k of the identity matrix I so that the resulting matrix

$$Q(\mathbf{x}) = Hf(\mathbf{x}) + \mu_k I$$

is positive definite, and so that the corresponding choice of $\mathbf{p}^{(k)}$ produces an acceptable iteration method.

To determine the size of μ_k, observe that for any positive number μ

$$\mathbf{x} \cdot (Hf(\mathbf{x}) + \mu I)\mathbf{x} = \mathbf{x} \cdot Hf(\mathbf{x})\mathbf{x} + \mu \|\mathbf{x}\|^2.$$

Define

$$\bar{\mu}_k = \max_{\|\mathbf{x}\|=1} |\mathbf{x} \cdot Hf(\mathbf{x}^{(k)})\mathbf{x}|.$$

(It can be shown (see Exercise 13) that $\bar{\mu}_k$ is the absolute value of the eigenvalue of largest absolute value of $Hf(\mathbf{x})$; means are available for computing an upper bound for $\bar{\mu}_k$.) Note that if $\mu_k > \bar{\mu}_k$, and $\mathbf{x} \neq \mathbf{0}$

$$\begin{aligned} \mathbf{x} \cdot (Hf(\mathbf{x}^{(k)}) + \mu_k I)\mathbf{x} &= \mathbf{x} \cdot Hf(\mathbf{x}^{(k)})\mathbf{x} + \mu_k \|\mathbf{x}\|^2 \\ &= \|\mathbf{x}\|^2 \left[\frac{\mathbf{x}}{\|\mathbf{x}\|} \cdot Hf(\mathbf{x}^{(k)}) \frac{\mathbf{x}}{\|\mathbf{x}\|} + \mu_k \right] > \|\mathbf{x}\|^2(-\bar{\mu}_k + \mu_k) > 0, \end{aligned}$$

so $Hf(\mathbf{x}^{(k)}) + \mu_k I$ is positive definite.

In view of the conclusion of the preceding paragraph, we can proceed as follows even if the Hessian $Hf(x)$ is not positive definite at all points of R^n:

(1) Given $\mathbf{x}^{(k)}$.

(2) Compute μ_k such that

$$Hf(\mathbf{x}^{(k)}) + \mu_k I$$

is positive definite.

(3) Solve for $\mathbf{p}^{(k)}$

$$(Hf(\mathbf{x}^{(k)}) + \mu_k I)\mathbf{p}^{(k)} = -\nabla f(\mathbf{x}^{(k)}).$$

(4) Set t_k by backtracking.

(5) Update

$$\mathbf{x}^{(k+1)} = \mathbf{x}^{(k)} + t_k \mathbf{p}^{(k)}.$$

(6) Iterate.

For most well-behaved functions, the preceding procedure works fine. However, if $\mathbf{x}^{(k)}$ is in a region for which the $\mathbf{p}^{(k)}$ computed from Step 3 is too large to be numerically helpful, then this algorithm must be modified. We will discuss this modification in Section 5.4.

The features of "good" iterative methods that we have discussed in this section are incorporated in much of the professionally written optimization software. These packaged routines are carefully written to be reliable and robust and therefore are to be preferred over most "homemade" programs that a user might write. Although it is certainly instructive to write simple programs to implement the minimization techniques discussed in this text, there is no doubt that the reader interested in applying optimization methods to the solution of practical problems will find it far more useful to learn to use and evaluate a variety of the commercially prepared programs that are available.

3.4. Broyden's Method

Every computation has an associated cost. The main computational costs of the iterative methods we have studied so far derive from repeated evaluations of functions, gradients, Hessians, and Jacobians. Of these, the most costly and error-prone are the calculations of Jacobian and Hessian matrices. This section is devoted to a discussion of Broyden's Method, which like Newton's Method is a zero-finder but is designed specifically to avoid the computation of Jacobian matrices.

To get a feeling for Broyden's Method, it is helpful to recall Newton's Method for solving a system

$$\mathbf{g}(\mathbf{x}) = \mathbf{0} \tag{1}$$

of n (usually nonlinear) equations in n unknowns where the component functions of $\mathbf{g}$: $R^n \to R^n$ are assumed to have continuous first partial derivatives. At the current point $\mathbf{x}^{(k)}$ in the Newton iteration, we evaluate the Jacobian

$$\nabla\mathbf{g}(\mathbf{x}^{(k)}) = \left(\frac{\partial g_i}{\partial x_j}: i, j = 1, \ldots, n\right). \tag{2}$$

Then we construct the linear approximation

$$L_k(\mathbf{x}) = \mathbf{g}(\mathbf{x}^{(k)}) + \nabla\mathbf{g}(\mathbf{x}^{(k)})(\mathbf{x} - \mathbf{x}^{(k)}) \tag{3}$$

to $\mathbf{g}(\mathbf{x})$ at $\mathbf{x}^{(k)}$ and take $\mathbf{x}^{(k+1)}$ to be the solution of $L_k(\mathbf{x}) = \mathbf{0}$. The fact that it is necessary to compute the n^2 partial derivatives involved in the Jacobian $\nabla\mathbf{g}(\mathbf{x}^{(k)})$ at each iteration of Newton's Method forces the computational cost of this method to be too high to be of practical value. Essentially, the same situation exists when Newton's Method is applied to function minimization. In that case, the function $\mathbf{g}(\mathbf{x})$ in (1) is taken to be the gradient $\nabla f(\mathbf{x})$ of the function $f(\mathbf{x})$ to be minimized while the Jacobian matrix becomes the Hessian matrix $Hf(\mathbf{x})$, and this latter matrix must again be computed at each iteration. Note that the modification of Newton's Method discussed in the last section also shares this drawback.

Broyden's Method is a direct relative of Newton's Method but it avoids repeated evaluations of the Jacobian or Hessian matrices. To motivate the construction of Broyden's Method for solving systems (1) of equations, it is helpful to recall the construction of Newton's Method for solving $g(x) = 0$ for a differentiable function $g(x)$ of one variable as indicated in the figure below:

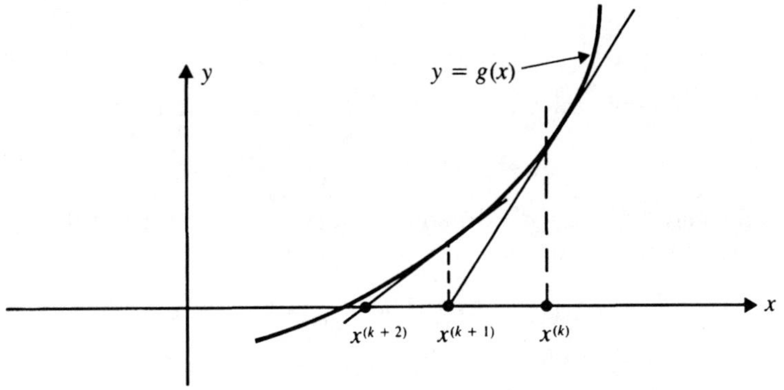

In the single variable case, the Jacobian matrix of g is just (dg/dx) and so

$$x^{(k+1)} = x^{(k)} - [\nabla g(x^{(k)})]^{-1} g(x^{(k)}) \tag{4}$$

simply identifies the zero $x^{(k+1)}$ of the tangent line to the graph of $g(x)$ at $(x^{(k)}, g(x^{(k)}))$.

Broyden's Method retains the idea of linear approximation but replaces the potentially complicated and computationally expensive Jacobian matrix with a simpler choice. More precisely, we define a linear function

$$l_k(\mathbf{x}) = \mathbf{g}(\mathbf{x}^{(k)}) + D_k(\mathbf{x} - \mathbf{x}^{(k)}),$$

where D_k is an $n \times n$-matrix possibly different from $\nabla g(\mathbf{x}^{(k)})$. Observe that

$$l_k(\mathbf{x}^{(k)}) = \mathbf{g}(\mathbf{x}^{(k)}),$$

and we can compute $\mathbf{x}^{(k+1)}$ by solving the linear system $l_k(\mathbf{x}) = 0$ to obtain

$$\mathbf{x}^{(k+1)} = \mathbf{x}^{(k)} - D_k^{-1}\mathbf{g}(\mathbf{x}^{(k)}). \tag{5}$$

But how do we choose the matrix D_k so that the linear approximation $l_k(\mathbf{x})$ is reasonable (that is, fairly close to the Newton approximation $L_k(\mathbf{x})$) and yet relatively simple to compute? And then how do we choose the next "update" matrix D_{k+1} needed to continue the iteration?

A return to the case of functions of one variable can help us to answer these questions. There, an alternative to Newton's Method for solving $g(x) = 0$ is the Secant Method. The geometry of that method is pictured as follows:

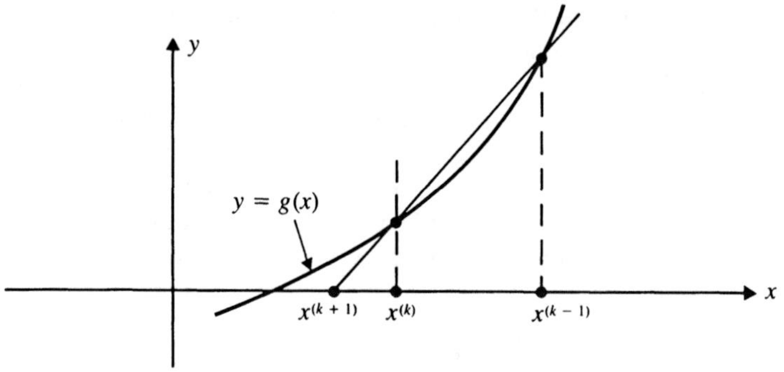

Here, D_k is a number selected so that the linear approximation

$$l_k(x) = g(x^{(k)}) + D_k(x - x^{(k)})$$

passes through $(x^{(k-1)}, g(x^{(k-1)}))$ and $(x^{(k)}, g(x^{(k)}))$, and we solve $0 = l_k(x)$ for $x^{(k+1)}$ and update to a new linear approximation

$$l_{k+1}(x) = g(x^{(k+1)}) + D_{k+1}(x - x^{(k+1)}),$$

where D_{k+1} is selected so that this linear approximation passes through $(x^{(k)}, g(x^{(k)}))$ and $(x^{(k+1)}, g(x^{(k+1)}))$. Then we iterate.

For Broyden's Method for functions $\mathbf{g}: R^n \to R^n$, we retain this basic feature of the Secant Method in one dimension: We insist that the matrix D_{k+1} in linear approximation

$$l_{k+1}(\mathbf{x}) = \mathbf{g}(\mathbf{x}^{(k+1)}) + D_{k+1}(\mathbf{x} - \mathbf{x}^{(k+1)})$$

is selected so that $l_{k+1}(\mathbf{x}^{(k)}) = \mathbf{g}(\mathbf{x}^{(k)})$ (and that $l_{k+1}(\mathbf{x}^{(k+1)}) = \mathbf{g}(\mathbf{x}^{(k+1)})$, but this automatically follows from the definition of l_{k+1}). Thus, we insist that D_{k+1} is selected so that the *secant condition*

$$D_{k+1}(\mathbf{x}^{(k+1)} - \mathbf{x}^{(k)}) = \mathbf{g}(\mathbf{x}^{(k+1)}) - \mathbf{g}(\mathbf{x}^{(k)}) \tag{6}$$

is satisfied. Note that equation (6) alone allows considerable latitude in the choice of D_{k+1} since (6) prescribes only one image vector for the matrix transformation D_{k+1} on R^n, so we have $n - 1$ "degrees of freedom" to use to meet our other objective for D_{k+1}—that D_{k+1} should be easy to compute.

It is useful to think of the process of going from the matrix D_k to the matrix D_{k+1} as follows. Write

$$D_{k+1} = D_k + (D_{k+1} - D_k)$$

and call $(D_{k+1} - D_k) = U_k$ the kth *update matrix*. Our objective will be to make the kth update matrix U_k as simple to compute as possible.

One way to achieve the desired simplicity in the computation of U_k is to define this matrix in such a way that it is completely defined by prescribing two vectors in R^n. Let us take a close look at matrices of this type before we go on with our discussion of Broyden's Method.

Suppose that $\mathbf{a}$ and $\mathbf{b}$ are two nonzero vectors in R^n. The *outer product or tensor product* $\mathbf{a} \otimes \mathbf{b}$ is the $n \times n$-matrix $U = (u_{ij})$ defined by

$$u_{ij} = a_i b_j \quad \text{for } i, j = 1, 2, \ldots, n.$$

For example, if $\mathbf{a} = (2, 3, -1)$ and $\mathbf{b} = (-1, 4, -5)$ then

$$\mathbf{a} \otimes \mathbf{b} = \begin{pmatrix} 2 \\ 3 \\ -1 \end{pmatrix} (-1, 4, -5) = \begin{pmatrix} -2 & 8 & -10 \\ -3 & 12 & -15 \\ 1 & -4 & 5 \end{pmatrix}.$$

The action of $\mathbf{a} \otimes \mathbf{b}$ on vectors $\mathbf{x}$ in R^n is easy to describe. The outer product $\mathbf{a} \otimes \mathbf{b}$ is just the ordinary matrix product of $\mathbf{a}$ and $\mathbf{b}^T$ when $\mathbf{a}$, $\mathbf{b}$ are regarded as column vectors. If $\mathbf{x}$ is a (column) vector in R^n, then

$$(\mathbf{a} \otimes \mathbf{b})\mathbf{x} = (\mathbf{b} \cdot \mathbf{x})\mathbf{a}.$$

Thus, the range of $\mathbf{a} \otimes \mathbf{b}$ is the one-dimensional subspace of R^n spanned by $\mathbf{a}$, and the rank of $\mathbf{a} \otimes \mathbf{b}$ is one. If $\mathbf{a} \cdot \mathbf{b} \neq 0$, then the matrix $\mathbf{a} \otimes \mathbf{b}$ has only one nonzero eigenvalue $\mathbf{b} \cdot \mathbf{a}$ and 0 is an eigenvalue of multiplicity $n - 1$. Also, note that $(\mathbf{a} \otimes \mathbf{b})^T = \mathbf{b} \otimes \mathbf{a}$; in particular, for any nonzero vector $\mathbf{a}$ in R^n, $\mathbf{a} \otimes \mathbf{a}$ is a symmetric positive semidefinite $n \times n$-matrix.

Now let us return to our discussion of Broyden's Method. For the sake of computational simplicity, we will choose the update matrices

$$U_k = D_{k+1} - D_k$$

to be rank-one matrices of the form

$$U_k = \mathbf{a}^{(k)} \otimes \mathbf{b}^{(k)}$$

for appropriate choices of $\mathbf{a}^{(k)}$ and $\mathbf{b}^{(k)}$ in R^n. But how do we choose $\mathbf{a}^{(k)}$ and $\mathbf{b}^{(k)}$? Recall that we have already imposed the secant condition on D_{k+1}:

$$D_{k+1}(\mathbf{x}^{(k+1)} - \mathbf{x}^{(k)}) = \mathbf{g}(\mathbf{x}^{(k+1)}) - \mathbf{g}(\mathbf{x}^{(k)}). \tag{6}$$

Since $D_{k+1} = D_k + \mathbf{a}^{(k)} \otimes b^{(k)}$, this means that

$$D_k(\mathbf{x}^{(k+1)} - \mathbf{x}^{(k)}) + (\mathbf{a}^{(k)} \otimes b^{(k)})(\mathbf{x}^{(k+1)} - \mathbf{x}^{(k)}) = \mathbf{g}(\mathbf{x}^{(k+1)}) - \mathbf{g}(\mathbf{x}^{(k)}).$$

Rewrite this equation as

$$[\mathbf{b}^{(k)} \cdot (\mathbf{x}^{(k+1)} - \mathbf{x}^{(k)})]\mathbf{a}^{(k)} = \mathbf{g}(\mathbf{x}^{(k+1)}) - \mathbf{g}(\mathbf{x}^{(k)}) - D_k(\mathbf{x}^{(k+1)} - \mathbf{x}^{(k)}).$$

It follows that once $\mathbf{b}^{(k)}$ has been chosen, then $\mathbf{a}^{(k)}$ is determined by the equation

$$\mathbf{a}^{(k)} = \frac{\mathbf{g}(\mathbf{x}^{(k+1)}) - \mathbf{g}(\mathbf{x}^{(k)}) - D_k(\mathbf{x}^{(k+1)} - \mathbf{x}^{(k)})}{\mathbf{b}^{(k)} \cdot (\mathbf{x}^{(k+1)} - \mathbf{x}^{(k)})}. \tag{7}$$

To help us to understand the significance of the choice of $\mathbf{b}^{(k)}$, let us do a little algebra with the linear approximations $l_k(\mathbf{x})$ and $l_{k+1}(\mathbf{x})$. Note that

$$\begin{aligned} l_{k+1}(\mathbf{x}) - l_k(\mathbf{x}) &= \mathbf{g}(\mathbf{x}^{(k+1)}) + D_{k+1}(\mathbf{x} - \mathbf{x}^{(k+1)}) - \mathbf{g}(\mathbf{x}^{(k)}) - D_k(\mathbf{x} - \mathbf{x}^{(k)}) \\ &= \mathbf{g}(\mathbf{x}^{(k+1)}) - \mathbf{g}(\mathbf{x}^{(k)}) + D_{k+1}[(\mathbf{x} - \mathbf{x}^{(k)}) + (\mathbf{x}^{(k)} - \mathbf{x}^{(k+1)})] \\ &\quad - D_k(\mathbf{x} - \mathbf{x}^{(k)}) \\ &= \mathbf{g}(\mathbf{x}^{(k+1)}) - \mathbf{g}(\mathbf{x}^{(k)}) - D_{k+1}(\mathbf{x}^{(k+1)} - \mathbf{x}^{(k)}) + U_k(\mathbf{x} - \mathbf{x}^{(k)}), \end{aligned}$$

where $U_k = D_{k+1} - D_k$ is the kth update matrix $\mathbf{a}^{(k)} \otimes \mathbf{b}^{(k)}$. But the secant condition (6) asserts that

$$\mathbf{g}(\mathbf{x}^{(k+1)}) - \mathbf{g}(\mathbf{x}^{(k)}) - D_{k+1}(\mathbf{x}^{(k+1)} - \mathbf{x}^{(k)}) = 0,$$

so

$$l_{k+1}(\mathbf{x}) - l_k(\mathbf{x}) = [\mathbf{b}^{(k)} \cdot (\mathbf{x} - \mathbf{x}^{(k)})]\mathbf{a}^{(k)}. \tag{8}$$

Because $\mathbf{a}^{(k)}$ is determined from $\mathbf{b}^{(k)}$ by equation (7), we see from equation (8) that the difference between the linear approximations $l_{k+1}(\mathbf{x})$ and $l_k(\mathbf{x})$ is completely determined by $\mathbf{b}^{(k)}$.

Broyden's Method fixes $\mathbf{b}^{(k)}$ by requiring that no "white noise" be allowed to have any influence. After all, we have information only at $\mathbf{x}^{(k)}$ and $\mathbf{x}^{(k+1)}$. Hence, it is not possible to obtain better information about how $\mathbf{g}(\mathbf{x})$ is changing in directions very different from that of $\mathbf{x}^{(k)}$ to $\mathbf{x}^{(k+1)}$. Thus, we insist that if $\mathbf{x}$ is a vector in R^n such that $\mathbf{x} - \mathbf{x}^{(k)}$ is orthogonal to $\mathbf{x}^{(k+1)} - \mathbf{x}^{(k)}$, then $l_{k+1}(\mathbf{x}) = l_k(\mathbf{x})$ that is, the change in $\mathbf{g}(\mathbf{x})$ predicted by D_{k+1} should be the same as that predicted by D_k in directions orthogonal to $\mathbf{x}^{(k+1)} - \mathbf{x}^{(k)}$. According to equation (8) this requirement amounts to

$$\mathbf{0} = l_{k+1}(\mathbf{x}) - l_k(\mathbf{x}) = [\mathbf{b}^{(k)} \cdot (\mathbf{x} - \mathbf{x}^{(k)})]\mathbf{a}^{(k)},$$

whenever $(\mathbf{x} - \mathbf{x}^{(k)}) \cdot (\mathbf{x}^{(k+1)} - \mathbf{x}^{(k)}) = 0$. We see from this that the choice $\mathbf{b}^{(k)} = \mathbf{x}^{(k+1)} - \mathbf{x}^{(k)}$ will do. Thus, the kth update matrix $U_k = \mathbf{a}^{(k)} \otimes \mathbf{b}^{(k)}$ is completely determined. We have shown that if we define

$$D_{k+1} = D_k + \mathbf{a}^{(k)} \otimes \mathbf{b}^{(k)},$$

where

$$\mathbf{b}^{(k)} = \mathbf{x}^{(k+1)} - \mathbf{x}^{(k)},$$

and

$$\mathbf{a}^{(k)} = \frac{\mathbf{g}(\mathbf{x}^{(k+1)}) - \mathbf{g}(\mathbf{x}^{(k)}) - D_k\mathbf{b}^{(k)}}{\mathbf{b}^{(k)} \cdot \mathbf{b}^{(k)}}$$

then, in the presence of the secant condition

$$D_{k+1}\mathbf{b}^{(k)} = \mathbf{g}(\mathbf{x}^{(k+1)}) - \mathbf{g}(\mathbf{x}^{(k)}),$$

the white noise condition

$$l_{k+1}(\mathbf{x}) = l_k(\mathbf{x}) \quad \text{whenever} \quad (\mathbf{x} - \mathbf{x}^{(k)}) \cdot \mathbf{b}^{(k)} = 0$$

is satisfied. This completely defines the famous *Broyden rank-one update*.

We can now summarize the entire preceding discussion with a statement of the resulting method.

(3.4.1) Broyden's Method for Solving Systems of Equations. Suppose that the function $\mathbf{g}: R^n \to R^n$ has component functions with continuous first partial derivatives, that $\mathbf{x}^{(0)} \in R^n$ and that D_0 is an $n \times n$-matrix. Then the *Broyden's Method sequence* $\{\mathbf{x}^{(k)}\}$ *(with initial point* $\mathbf{x}^{(0)}$ *for solving the system* $\mathbf{g}(\mathbf{x}) = \mathbf{0}$) is defined by the following recurrence formula:

$$\mathbf{x}^{(k+1)} = \mathbf{x}^{(k)} - D_k^{-1}\mathbf{g}(\mathbf{x}^{(k)}),$$

where we

(1) Solve $D_k(\mathbf{x} - \mathbf{x}^{(k)}) = -\mathbf{g}(\mathbf{x}^{(k)})$ for $\mathbf{x}$ and set $\mathbf{x}^{(k+1)} = \mathbf{x}$.
(2) Set $\mathbf{d}^{(k)} = \mathbf{x}^{(k+1)} - \mathbf{x}^{(k)}$ and $\mathbf{y}^{(k)} = \mathbf{g}(\mathbf{x}^{(k+1)}) - \mathbf{g}(\mathbf{x}^{(k)})$.
(3) Update

$$D_{k+1} = D_k + \frac{(\mathbf{y}^{(k)} - D_k(\mathbf{d}^{(k)})) \otimes \mathbf{d}^{(k)}}{\mathbf{d}^{(k)} \cdot \mathbf{d}^{(k)}}.$$

There are many possibilities for choosing the matrix D_0 needed to start Broyden's Method. One obvious possibility is to allow a single evaluation of the Jacobian matrix $\nabla\mathbf{g}(\mathbf{x}^{(0)})$ and use this for D_0. Other strategies can be found in Chapter 8 of *Numerical Methods for Unconstrained Optimization and Nonlinear Equations* by J. E. Dennis and R. B. Schnabel (Prentice-Hall, Englewood Cliffs, NJ, 1983).

To develop a feel for the mechanics of Broyden's Method, let us return to a system of equations considered earlier in conjunction with Newton's Method.

(3.4.2) Example. As we observed in (3.1.2), the system of equations

$$(*) \qquad \begin{aligned} x^2 + y^2 + z^2 &= 3, \\ x^2 + y^2 - z &= 1, \\ x + y + z &= 3, \end{aligned}$$

has the unique solution (1, 1, 1). If we take the initial point $\mathbf{x}^{(0)}$ to be (1, 0, 1) and the initial matrix D_0 for Broyden's Method to be the Jacobian matrix $\nabla\mathbf{g}(1, 0, 1)$ of the function $\mathbf{g}: R^3 \to R^3$ corresponding to $(*)$

$$\mathbf{g}(x, y, z) = (x^2 + y^2 + z^2 - 3, x^2 + y^2 - z - 1, x + y + z - 3),$$

then the first step of Broyden's Method is identical to the first step of Newton's Method in (3.1.2) so $\mathbf{x}^{(1)} = (\frac{3}{2}, \frac{1}{2}, 1)$.

To prepare for the second iteration, we compute the update D_1 of D_0 as follows:

$$\mathbf{d}^{(0)} = \mathbf{x}^{(1)} - \mathbf{x}^{(0)} = (\tfrac{1}{2}, \tfrac{1}{2}, 0),$$

$$\mathbf{y}^{(0)} = \mathbf{g}(\mathbf{x}^{(1)}) - \mathbf{g}(\mathbf{x}^{(0)}) = (\tfrac{1}{2}, \tfrac{1}{2}, 0) - (-1, -1, -1) = (\tfrac{3}{2}, \tfrac{3}{2}, 1).$$

Then

$$\mathbf{y}^{(0)} - D_0\mathbf{d}^{(0)} = \begin{pmatrix} \frac{3}{2} \\ \frac{3}{2} \\ 1 \end{pmatrix} - \begin{pmatrix} 2 & 0 & 2 \\ 2 & 0 & -1 \\ 1 & 1 & 1 \end{pmatrix} \begin{pmatrix} \frac{1}{2} \\ \frac{1}{2} \\ 0 \end{pmatrix} = \begin{pmatrix} \frac{1}{2} \\ \frac{1}{2} \\ 0 \end{pmatrix},$$

so

$$D_1 = D_0 + \frac{(\mathbf{y}^{(0)} - D_0\mathbf{d}^{(0)}) \otimes \mathbf{d}^{(0)}}{\mathbf{d}^{(0)} \cdot \mathbf{d}^{(0)}}$$

$$= \begin{pmatrix} 2 & 0 & 2 \\ 2 & 0 & -1 \\ 1 & 1 & 1 \end{pmatrix} + 2 \begin{pmatrix} \frac{1}{2} \\ \frac{1}{2} \\ 0 \end{pmatrix} (\tfrac{1}{2}, \tfrac{1}{2}, 0) = \begin{pmatrix} \frac{5}{2} & \frac{1}{2} & 2 \\ \frac{5}{2} & \frac{1}{2} & -1 \\ 1 & 1 & 1 \end{pmatrix}.$$

We can now compute $\mathbf{x}^{(2)}$ by solving the system $D_1(\mathbf{x} - \mathbf{x}^{(1)}) = -\mathbf{g}(\mathbf{x}^{(1)})$, that is,

$$\begin{pmatrix} \frac{5}{2} & \frac{1}{2} & 2 \\ \frac{5}{2} & \frac{1}{2} & -1 \\ 1 & 1 & 1 \end{pmatrix} \begin{pmatrix} x - \frac{3}{2} \\ y - \frac{1}{2} \\ z - 1 \end{pmatrix} = - \begin{pmatrix} \frac{1}{2} \\ \frac{1}{2} \\ 0 \end{pmatrix}.$$

The solution is readily checked to be $\mathbf{x}^{(2)} = (\tfrac{5}{4}, \tfrac{3}{4}, 1)$.

The primary objective in the development of Broyden's Method was to avoid the expensive computation of the Jacobian matrix (required by Newton's Method) by employing a matrix D_k that was relatively inexpensive to update at each iteration. We have already seen that there is at least one sense in which the matrix D_{k+1} is a simple perturbation of D_k when we constructed the update matrix U_k as a matrix of rank one. There is another very interesting sense in which D_{k+1} is a small perturbation of D_k. We will show that the matrix D_{k+1} is the "closest" matrix to D_k that satisfies the secant condition (6). However, before we can state and prove this result, we need to make the meaning of "closest" precise for matrices, that is, we need to develop a measure of distance between matrices.

(3.4.3) Definition. If A, B are $n \times n$-matrices, the distance $d(A, B)$ between A and B is

$$d(A, B) = \max\{\|A\mathbf{x} - B\mathbf{x}\| : \|\mathbf{x}\| \leq 1\}.$$

It is often convenient to write $\|A - B\|$ in place of $d(A, B)$.

(3.4.4) Examples

(a) If I is the $n \times n$-identity matrix and 0 is the $n \times n$-zero matrix, then

$$d(I, 0) = \max\{\|\mathbf{x} - \mathbf{0}\|: \|\mathbf{x}\| \le 1\} = 1$$

so $d(I, 0) = \|I\| = 1$.

(b) If $\mathbf{a}$ and $\mathbf{b}$ are nonzero vectors in R^n and if $\|\mathbf{x}\| \le 1$, then by the Cauchy–Schwarz Inequality,

$$|\mathbf{b}\cdot\mathbf{x}| \le \|\mathbf{b}\|$$

with equality holding for $\|\mathbf{x}\| = 1$ when $\mathbf{x}$ is a multiple of $\mathbf{b}$. Therefore, the distance between the outer product $\mathbf{a} \otimes \mathbf{b}$ and the zero matrix 0 is given by

$$d(\mathbf{a} \otimes \mathbf{b}, 0) = \max\{\|(\mathbf{b}\cdot\mathbf{x})\mathbf{a}\|: \|\mathbf{x}\| \le 1\} = \|\mathbf{a}\|\,\|\mathbf{b}\|.$$

that is, $\|\mathbf{a} \otimes \mathbf{b}\| = \|\mathbf{a}\|\,\|\mathbf{b}\|$.

The distance $d(A, B) = \|A - B\|$ between two matrices A, B has a number of properties that are similar to those of the usual distance $d(\mathbf{x}, \mathbf{y}) = \|\mathbf{x} - \mathbf{y}\|$ between vectors in R^n. For example,

(1) $d(A, B) = d(B, A)$ for all $n \times n$-matrices A and B.
(2) $d(A, B) = 0$ if and only if $A = B$.
(3) $d(A, B) \le d(A, C) + d(C, B)$ for all $n \times n$-matrices A, B, C (the Triangle Inequality).
(4) $\|AB\| \le \|A\|\,\|B\|$ for all $n \times n$-matrices A and B.

The proofs of these results are straightforward.

Now that we have made precise the concept of distance between matrices, we are ready to establish the following description of the Broyden Method update matrices. Note that the secant condition (6) assumes the form

$$D_{k+1}\mathbf{d}^{(k)} = \mathbf{y}^{(k)}$$

in the notation established in (3.4.1).

(3.4.5) Theorem. *Suppose that D is an $n \times n$-matrix that satisfies the secant condition $D\mathbf{d}^{(k)} = \mathbf{y}^{(k)}$. Then*

$$d(D_{k+1}, D_k) \le d(D, D_k),$$

that is, D_{k+1} is as close to D_k as any $n \times n$-matrix that satisfies the secant condition.

Proof. If D is an $n \times n$-matrix satisfying the secant condition $D\mathbf{d}^{(k)} = \mathbf{y}^{(k)}$, then by (3.4.1) we see that

$$d(D_{k+1}, D_k) = \|D_{k+1} - D_k\| = \left\|\frac{(\mathbf{y}^{(k)} - D_k\mathbf{d}^{(k)}) \otimes \mathbf{d}^{(k)}}{\mathbf{d}^{(k)}\cdot\mathbf{d}^{(k)}}\right\|$$

$$= \left\|\frac{(D\mathbf{d}^{(k)} - D_k\mathbf{d}^{(k)}) \otimes \mathbf{d}^{(k)}}{\mathbf{d}^{(k)}\cdot\mathbf{d}^{(k)}}\right\| = \left\|(D - D_k)\left[\frac{\mathbf{d}^{(k)} \otimes \mathbf{d}^{(k)}}{\|\mathbf{d}^{(k)}\|^2}\right]\right\|$$

$$\le \|D - D_k\| \left\|\left[\frac{\mathbf{d}^{(k)}}{\|\mathbf{d}^{(k)}\|}\right] \otimes \left[\frac{\mathbf{d}^{(k)}}{\|\mathbf{d}^{(k)}\|}\right]\right\| = d(D, D_k),$$

where the inequality follows from Property (4) of the matrix distance $d(A, B) = \|A - B\|$ and

$$\left\| \frac{\mathbf{d}^{(k)}}{\|\mathbf{d}^{(k)}\|} \otimes \frac{\mathbf{d}^{(k)}}{\|\mathbf{d}^{(k)}\|} \right\| = 1$$

by Example (3.4.4)(b). This completes the proof of the theorem.

The preceding theorem reveals the following striking feature of Broyden's Method. Even though the matrix D_{k+1} is easily computed from the matrix D_k, the theorem states that no other matrix D (no matter how complicated its relation to D_k) with the required secant property is any closer to D_k than D_{k+1}.

Some important questions remain concerning Broyden's Method for solving systems $\mathbf{g}(\mathbf{x}) = \mathbf{0}$ for $\mathbf{g}: R^n \to R^n$. For example:

(1) Does Broyden's Method actually work? In other words, if we make a reasonably good choice of an initial point $\mathbf{x}^{(0)}$, does the Broyden Method sequence $\{\mathbf{x}^{(k)}\}$ converge to a point $\mathbf{x}^*$ such that $\mathbf{g}(\mathbf{x}^*) = \mathbf{0}$?
(2) What happens to the matrices D_k in the long run? Does D_k eventually approximate the Jacobian $\nabla \mathbf{g}(\mathbf{x}^{(k)})$? If so, in what sense?

The answers to these questions are interconnected. Under reasonable hypotheses, it can be shown that the Broyden steps $\mathbf{d}^{(k)} = \mathbf{x}^{(k+1)} - \mathbf{x}^{(k)} = -D_k^{-1}\mathbf{g}(\mathbf{x}^{(k)})$ act a lot like the Newton steps—$[\nabla \mathbf{g}(\mathbf{x}^{(k)})]^{-1}\mathbf{g}(\mathbf{x}^{(k)})$. In fact, if $\mathbf{x}^*$ is a solution of the system $\mathbf{g}(\mathbf{x}) = \mathbf{0}$ and if $\mathbf{x}^{(0)}$ is not too far from $\mathbf{x}^*$, then it can be shown that

$$\lim_k \frac{\|(D_k - \nabla \mathbf{g}(\mathbf{x}^{(k)}))(\mathbf{x}^{(k)} - \mathbf{x}^*)\|}{\|\mathbf{x}^{(k)} - \mathbf{x}^*\|} = 0. \tag{9}$$

This connection between the Broyden's Method steps and the Newton's Method steps is the chief theoretical fact needed to prove that the method works in the sense explained in Question 1 above. Detailed proofs of (9) and the resulting convergence theorem can be found in the Dennis and Schnabel text cited after (3.4.1).

The answer to Question 2 posed above is provided in part by (9) since that result guarantees that, in the long run, $D_k(\mathbf{x}^{(k)} - \mathbf{x}^*)$ acts like $\nabla \mathbf{g}(\mathbf{x}^*)(\mathbf{x}^{(k)} - \mathbf{x}^*)$. This does not mean that the matrices $\{D_k\}$ necessarily converge to the Jacobian $\nabla \mathbf{g}(\mathbf{x}^*)$. In fact, they might not. It is one of the beauties of this method that even though the matrices $\{D_k\}$ might not converge to $\nabla \mathbf{g}(\mathbf{x}^*)$, the action of D_k on the "important" vectors $\mathbf{x}^{(k)} - \mathbf{x}^*$ *does approximate the action of* $\nabla \mathbf{g}(\mathbf{x}^*)$ on these vectors. Thus, Broyden's Method has the essential reliability of Newton's Method once $\mathbf{x}^{(k)}$ is close enough to $\mathbf{x}^*$.

Just as with Newton's Method we can adapt Broyden's Method to minimize a function $f: R^n \to R$. The technique is simple: We simply take

$$\mathbf{g}(\mathbf{x}) = \nabla f(\mathbf{x})$$

and use Broyden's Method to find a critical point of $f(\mathbf{x})$.

Also, just as with Newton's Method, we can modify Broyden's Method by introducing a parameter t_k in the formula for the kth iterate

$$\mathbf{x}^{(k+1)} = \mathbf{x}^{(k)} - t_k D_k^{-1} \nabla f(\mathbf{x}^{(k)}),$$

and use this parameter to speed convergence of the method. One problem with this approach is that the search directions prescribed by Broyden's Method are not always descent directions. However, the Broyden search directions are usually descent directions for $f(\mathbf{x})$ and in this case line search methods such as those discussed in Section 3.3 can identify values of t_k that will speed the search for solutions of $\nabla f(\mathbf{x}) = \mathbf{0}$. Modern computer codes often incorporate line searches for precisely this purpose.

3.5. Secant Methods for Minimization

This section will present two very effective minimization methods, the Broyden–Fletcher–Goldfarb–Shanno (BFGS) Method and the Davidon–Fletcher–Powell (DFP) Method. Both methods require no evaluation of the Hessian matrix and retain the secant feature of Broyden's Method.

At the end of the last section, we mentioned that Broyden's zero-finding method could be used to minimize a function $f(\mathbf{x})$ on R^n by applying the method to find a zero of the gradient $\nabla f(\mathbf{x})$. If we follow this approach, we encounter the problem that the Broyden search direction $-D_k^{-1}(\nabla f(\mathbf{x}^{(k)}))$ at $\mathbf{x}^{(k)}$ need not be a descent direction for $f(\mathbf{x})$ because there is no guarantee that D_k is positive definite. Our objective in this section is to develop a Broyden-like method for which the D_k's are all positive definite. This will allow us to apply Criteria 1–4 of Section 3.3 to prescribe an iterative sequence $\{\mathbf{x}^{(k)}\}$ that will head toward a minimizer of $f(\mathbf{x})$ under reasonably general conditions on $f(\mathbf{x})$.

Here is a list of features that we want for our iterative method

$$\mathbf{x}^{(k+1)} = \mathbf{x}^{(k)} - t_k D_k^{-1} \nabla f(\mathbf{x}^{(k)})$$

for minimizing $f(\mathbf{x})$ on R^n:

(1) Each D_k should be positive definite.
(2) The Secant Condition relative to $\nabla f(\mathbf{x})$ should be satisfied

$$D_{k+1}(\mathbf{x}^{(k+1)} - \mathbf{x}^{(k)}) = \nabla f(\mathbf{x}^{(k+1)}) - \nabla f(\mathbf{x}^{(k)}),$$

that is, $D_{k+1}(\mathbf{d}^{(k)}) = \mathbf{y}^{(k)}$ where $\mathbf{d}^{(k)} = \mathbf{x}^{(k+1)} - \mathbf{x}^{(k)}$ and $\mathbf{y}^{(k)} = \nabla f(\mathbf{x}^{(k+1)}) - \nabla f(\mathbf{x}^{(k)})$.
(3) The update from D_k to D_{k+1} should be as simple and computationally inexpensive as possible.

As we shall see, all of this can be accomplished.

The underlying theoretical tool is the following fact from linear algebra.

(3.5.1) Theorem. *Let* **a**, **b** *be vectors in* R^n *such that* $\mathbf{a}\cdot\mathbf{b} > 0$. *Then there is a positive definite matrix* A *such that* $A\mathbf{a} = \mathbf{b}$.

PROOF. Since any two vectors in R^n lie in a two-dimensional subspace M of R^n, we can visualize these vectors in R^2 as follows:

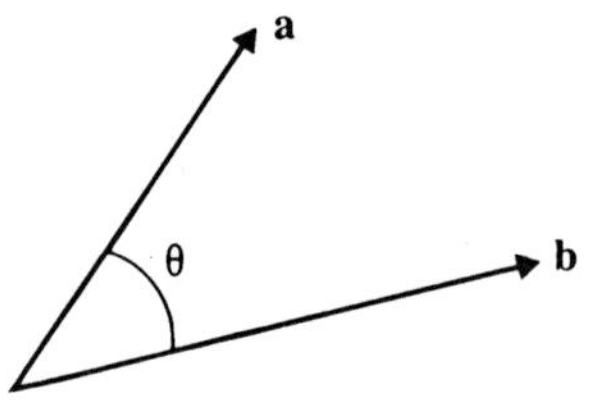

where the angle θ between **a** and **b** satisfies $0 \le \theta < \pi/2$ since $0 < \mathbf{a}\cdot\mathbf{b} = \|\mathbf{a}\|\,\|\mathbf{b}\|\cos\theta$. We shall consider two cases:

Case 1. $0 < \theta < \pi/2$.

In this case, choose orthogonal vectors $\mathbf{a}^{(1)}$ and $\mathbf{b}^{(1)}$ in M such that the following picture is realized:

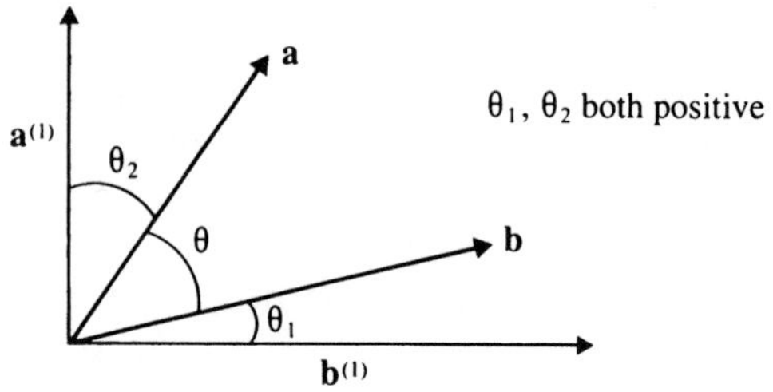

It is evident from this picture that there exist positive numbers s_1, s_2, t_1, t_2 such that

$$\mathbf{a} = s_1\mathbf{a}^{(1)} + t_1\mathbf{b}^{(1)},$$
$$\mathbf{b} = s_2\mathbf{a}^{(1)} + t_2\mathbf{b}^{(1)}.$$

Then the matrix

$$A_1 = \begin{pmatrix} \dfrac{s_2}{s_1} & 0 \\ 0 & \dfrac{t_2}{t_1} \end{pmatrix}$$

defines a linear transformation of M relative to the basis $\mathbf{a}^{(1)}, \mathbf{b}^{(1)}$ by

$$A_1(s\mathbf{a}^{(1)} + t\mathbf{b}^{(1)}) = s\left(\frac{s_2}{s_1}\right)\mathbf{a}^{(1)} + t\left(\frac{t_2}{t_1}\right)\mathbf{b}^{(1)}$$

and $A_1(\mathbf{a}) = \mathbf{b}$. Pick vectors $\mathbf{c}^{(1)}, \ldots, \mathbf{c}^{(n-2)}$ such that $\{\mathbf{a}^{(1)}, \mathbf{b}^{(1)}, \mathbf{c}^{(1)}, \ldots, \mathbf{c}^{(n-2)}\}$ is an orthogonal basis of R^n. The matrix

$$A = \begin{pmatrix} \frac{s_2}{s_1} & 0 & & \cdots & 0 \\ 0 & \frac{t_2}{t_1} & 0 & \cdots & 0 \\ & 0 & 1 & & \\ \vdots & \vdots & & & 0 \\ 0 & 0 & & 0 & 1 \end{pmatrix}$$

defines a transformation of R^n relative to this basis. Note that A is positive definite and that $A(\mathbf{a}) = \mathbf{b}$ as required.

Case 2. $\theta = 0$.

In this case, $\mathbf{b} = \lambda\mathbf{a}$ where $\lambda = (\mathbf{a}\cdot\mathbf{b})/\|\mathbf{a}\| > 0$. The construction of the required positive definite matrix A such that $A\mathbf{a} = \mathbf{b}$ is left to the reader as an exercise. This completes the proof.

Suppose that D_k is positive definite, that $\mathbf{p}^{(k)} = -D_k^{-1}\nabla f(\mathbf{x}^{(k)})$ and that t_k has been selected so that Criteria 1–4 of Section 3.3 are satisfied. Then, if $\mathbf{d}^{(k)} = \mathbf{x}^{(k+1)} - \mathbf{x}^{(k)}$ and $\mathbf{y}^{(k)} = \nabla f(\mathbf{x}^{(k+1)}) - \nabla f(\mathbf{x}^{(k)})$,

$$\begin{aligned}\mathbf{d}^{(k)}\cdot\mathbf{y}^{(k)} &= t_k\mathbf{p}^{(k)}\cdot\mathbf{y}^{(k)} = t_k\mathbf{p}^{(k)}\cdot(\nabla f(\mathbf{x}^{(k+1)}) - \nabla f(\mathbf{x}^{(k)}))\\ &= t_k(\mathbf{p}^{(k)}\cdot\nabla f(\mathbf{x}^{(k+1)}) - \mathbf{p}^{(k)}\cdot\nabla f(\mathbf{x}^{(k)}))\\ &> t_k(\beta\mathbf{p}^{(k)}\cdot\nabla f(\mathbf{x}^{(k)}) - \mathbf{p}^{(k)}\cdot\nabla f(\mathbf{x}^{(k)})) \qquad \text{(by Criterion 3)}\\ &= t_k(\beta - 1)\mathbf{p}^{(k)}\cdot\nabla f(\mathbf{x}^{(k)})\\ &= t_k(\beta - 1)(-D_k^{-1}\nabla f(\mathbf{x}^{(k)}))\cdot\nabla f(\mathbf{x}^{(k)}) > 0,\end{aligned}$$

since D_k is positive definite, $\beta < 1$ and $t_k > 0$. Because $\mathbf{d}^{(k)}\cdot\mathbf{y}^{(k)} > 0$, there is a positive definite matrix A_k such that

$$A_k\mathbf{d}^{(k)} = \mathbf{y}^{(k)}$$

by Theorem (3.5.1). This tells us that if we take $D_{k+1} = A_k$, then features (1) and (2) prescribed in the remarks preceding (3.5.1) for a suitable iterative minimization method are present. If we can now find a simple and economical way to compute a positive definite matrix A_k such that $A_k\mathbf{d}^{(k)} = \mathbf{y}^{(k)}$, then the choice $D_{k+1} = A_k$ will also have the remaining feature (3) that we have prescribed for our method.

As we saw in Section 3.4, the Secant Condition, which reduces to

$$D_{k+1}(\mathbf{d}^{(k)}) = A_k\mathbf{d}^{(k)} = \mathbf{y}^{(k)}$$

in our present setting, does not determine $A_k = D_{k+1}$ uniquely and, in fact, allows considerable flexibility in the choice of this matrix. To find a simple and economical update from D_k to D_{k+1}, we will mimic the approach that we took with the Broyden zero-finding method in Section 3.4.

For Broyden's Method, we set

$$D_{k+1} = D_k + \mathbf{a}^{(k)} \otimes \mathbf{b}^{(k)}$$

and then we chose $\mathbf{a}^{(k)}$ and $\mathbf{b}^{(k)}$ so that the Secant Condition was satisfied. This will not work if we also want D_{k+1} to be positive definite and hence symmetric. For the rank-one update $\mathbf{a}^{(k)} \otimes \mathbf{b}^{(k)}$ is not symmetric unless $\mathbf{a}^{(k)}$, $\mathbf{b}^{(k)}$ are multiples of one another, and $\mathbf{a}^{(k)} \otimes \mathbf{b}^{(k)}$ is indefinite unless $\mathbf{a}^{(k)}$, $\mathbf{b}^{(k)}$ are positive multiples of one another. If we must choose $\mathbf{a}^{(k)} = \lambda_k \mathbf{b}^{(k)}$ for $\lambda_k > 0$, we no longer have enough flexibility to obtain the Secant Condition. To get around this difficulty, we set

$$D_{k+1} = D_k + \alpha_k(\mathbf{a}^{(k)} \otimes \mathbf{a}^{(k)}) + \beta_k(\mathbf{b}^{(k)} \otimes \mathbf{b}^{(k)}),$$

where the real numbers α_k, β_k and the vectors $\mathbf{a}^{(k)}$, $\mathbf{b}^{(k)}$ are yet to be determined. The Secant Condition

$$D_{k+1}(\mathbf{d}^{(k)}) = \mathbf{y}^{(k)}$$

forces

$$\mathbf{y}^{(k)} = D_k(\mathbf{d}^{(k)}) + \alpha_k(\mathbf{a}^{(k)} \cdot \mathbf{d}^{(k)})\mathbf{a}^{(k)} + \beta_k(\mathbf{b}^{(k)} \cdot \mathbf{d}^{(k)})\mathbf{b}^{(k)}.$$

If we set $\mathbf{a}^{(k)} = \mathbf{y}^{(k)}$ and $\mathbf{b}^{(k)} = D_k(\mathbf{d}^{(k)})$, then the preceding equation yields

$$\mathbf{y}^{(k)} - D_k(\mathbf{d}^{(k)}) = \alpha_k(\mathbf{y}^{(k)} \cdot \mathbf{d}^{(k)})\mathbf{y}^{(k)} + \beta_k(D_k\mathbf{d}^{(k)} \cdot \mathbf{d}^{(k)})D_k\mathbf{d}^{(k)}.$$

This equation is satisfied if we set

$$\alpha_k(\mathbf{y}^{(k)} \cdot \mathbf{d}^{(k)}) = 1, \qquad \beta_k(D_k\mathbf{d}^{(k)} \cdot \mathbf{d}^{(k)}) = -1;$$

that is,

$$\alpha_k = \frac{1}{\mathbf{y}^{(k)} \cdot \mathbf{d}^{(k)}}, \qquad \beta_k = -\frac{1}{D_k\mathbf{d}^{(k)} \cdot \mathbf{d}^{(k)}}.$$

Thus our proposed update is

$$D_{k+1} = D_k + \frac{\mathbf{y}^{(k)} \otimes \mathbf{y}^{(k)}}{\mathbf{y}^{(k)} \cdot \mathbf{d}^{(k)}} - \frac{(D_k\mathbf{d}^{(k)}) \otimes (D_k\mathbf{d}^{(k)})}{(D_k\mathbf{d}^{(k)} \cdot \mathbf{d}^{(k)})}. \qquad (*)$$

This update is simple. The new matrix D_{k+1} automatically satisfies the Secant Condition and, as we will now see, D_{k+1} fulfills our remaining expectations.

(3.5.2) Theorem. *Suppose that D_k is positive definite and that $\mathbf{x}^{(k)}$ has been set. If $t_k > 0$ is selected so that*

$$\mathbf{x}^{(k+1)} = \mathbf{x}^{(k)} - t_k D_k^{-1}(\nabla f(\mathbf{x}^{(k)}))$$

satisfies Criterion 3 of Section 3.3, then the matrix D_{k+1} defined by $(*)$ is positive definite.

PROOF. We have already seen in the computation following (3.5.1) that if Criterion 3 is satisfied then $\mathbf{d}^{(k)} \cdot \mathbf{y}^{(k)} > 0$. But then, by Theorem (3.5.1), there is a positive definite matrix A_k such that $A_k(\mathbf{d}^{(k)}) = \mathbf{y}^{(k)}$. Thus

$$D_{k+1} = D_k + \frac{A_k\mathbf{d}^{(k)} \otimes A_k\mathbf{d}^{(k)}}{\mathbf{d}^{(k)} \cdot A_k\mathbf{d}^{(k)}} - \frac{D_k\mathbf{d}^{(k)} \otimes D_k\mathbf{d}^{(k)}}{\mathbf{d}^{(k)} \cdot D_k\mathbf{d}^{(k)}}.$$

For $\mathbf{x} \neq \mathbf{0}$, the preceding formula gives

$$\mathbf{x} \cdot D_{k+1}\mathbf{x} = \mathbf{x} \cdot D_k\mathbf{x} + \frac{(A_k\mathbf{d}^{(k)} \cdot \mathbf{x})^2}{\mathbf{d}^{(k)} \cdot A_k\mathbf{d}^{(k)}} - \frac{(D_k\mathbf{d}^{(k)} \cdot \mathbf{x})^2}{\mathbf{d}^{(k)} \cdot D_k\mathbf{d}^{(k)}}.$$

Since D_k is a positive definite matrix, it has a positive definite square root $D_k^{1/2}$ (see Exercise 7). Therefore

$$\mathbf{x} \cdot D_{k+1}\mathbf{x} = \mathbf{x} \cdot D_k^{1/2} D_k^{1/2}\mathbf{x} + \frac{(A_k\mathbf{d}^{(k)} \cdot \mathbf{x})^2}{\mathbf{d}^{(k)} \cdot A_k\mathbf{d}^{(k)}} - \frac{(D_k^{1/2}\mathbf{d}^{(k)} \cdot D_k^{1/2}\mathbf{x})^2}{D_k^{1/2}\mathbf{d}^{(k)} \cdot D_k^{1/2}\mathbf{d}^{(k)}}$$

$$= \frac{1}{\|D_k^{1/2}\mathbf{d}^{(k)}\|^2}[\|D_k^{1/2}\mathbf{d}^{(k)}\|^2 \|D_k^{1/2}\mathbf{x}\|^2 - (D_k^{1/2}\mathbf{d}^{(k)} \cdot D_k^{1/2}\mathbf{x})^2] + \frac{(A_k\mathbf{d}^{(k)} \cdot \mathbf{x})^2}{\mathbf{d}^{(k)} \cdot A_k\mathbf{d}^{(k)}}.$$

The Cauchy–Schwarz Inequality guarantees that the bracketed expression is nonnegative, and the second term is nonnegative since A_k is positive definite, so

$$\mathbf{x} \cdot D_{k+1}\mathbf{x} \geq 0$$

whenever $\mathbf{x} \neq \mathbf{0}$. Moreover, if $\mathbf{x} \cdot D_{k+1}\mathbf{x} = 0$, then we must have

$$D_k^{1/2}\mathbf{d}^{(k)} \cdot D_k^{1/2}\mathbf{x} = \|D_k^{1/2}\mathbf{d}^{(k)}\|^2 \|D_k^{1/2}\mathbf{x}\|^2,$$

and

$$A_k\mathbf{d}^{(k)} \cdot \mathbf{x} = 0.$$

But if the first of these equations holds (that is, equality holds in the Cauchy–Schwarz Inequality), there must be a $\lambda \neq 0$ such that

$$D_k^{1/2}\mathbf{d}^{(k)} = \lambda D_k^{1/2}\mathbf{x}.$$

This yields $\mathbf{d}^{(k)} = \lambda\mathbf{x}$ since $D_k^{1/2}$ is invertible and so

$$0 = A_k\mathbf{d}^{(k)} \cdot \mathbf{x} = A_k(\lambda\mathbf{x}) \cdot \mathbf{x} = \lambda\mathbf{x} \cdot A_k\mathbf{x} \neq 0,$$

a contradiction. Therefore, $\mathbf{x} \cdot D_{k+1}\mathbf{x} > 0$ if $\mathbf{x} \neq \mathbf{0}$, which completes the proof.

The method that we have just arrived at is the famous *Broyden–Fletcher–Goldfarb–Shanno* (BFGS) *Method*. Here is a recapitulation.

(3.5.3) The BFGS Method. To minimize $f(\mathbf{x})$ on R^n, select an initial point $\mathbf{x}^{(0)}$ and an initial positive definite matrix D_0. If $\mathbf{x}^{(k)}$ and D_k have been computed, then:

(1) Set $t_k > 0$ such that

$$\mathbf{x}^{(k+1)} = \mathbf{x}^{(k)} - t_k D_k^{-1}(\nabla f(\mathbf{x}^{(k)}))$$

satisfies Criteria 1–4 of Section 3.3.

(2) Update

$$D_{k+1} = D_k + \frac{\mathbf{y}^{(k)} \otimes \mathbf{y}^{(k)}}{\mathbf{d}^{(k)} \cdot \mathbf{y}^{(k)}} - \frac{D_k\mathbf{d}^{(k)} \otimes D_k\mathbf{d}^{(k)}}{\mathbf{d}^{(k)} \cdot D_k\mathbf{d}^{(k)}},$$

where $\mathbf{d}^{(k)} = \mathbf{x}^{(k+1)} - \mathbf{x}^{(k)}$, $\mathbf{y}^{(k)} = \nabla f(\mathbf{x}^{(k+1)}) - \nabla f(\mathbf{x}^{(k)})$.

Criteria 1–4 and Theorem (3.5.2) show that the BFGS Method has a number of very desirable theoretical characteristics. However, the real test of a minimization method is its performance on concrete problems. The BFGS Method has proven to be remarkably reliable and efficient. In fact, J. E. Dennis and R. B. Schnabel flatly state (*Numerical Methods for Unconstrained Optimization and Nonlinear Equations* (Prentice-Hall, Englewood Cliffs, NJ, 1983)) that the BFGS Method is the "best" Hessian update currently known. The reader should consult this text for a wealth of additional information on the BFGS Method. Of particular interest is their derivation of the BFGS update which reveals that if

$$D_{k+1} = L_{k+1}L_{k+1}^{\mathrm{T}}, \qquad D_k = L_k L_k^{\mathrm{T}}$$

are the Cholesky factorizations of D_{k+1} and D_k then L_{k+1} is the lower triangular matrix satisfying

$$\|L_{k+1} - L_k\| \le \|L - L_k\|$$

for any lower triangular matrix L such that $LL^{\mathrm{T}}\mathbf{d}^{(k)} = \mathbf{y}^{(k)}$. This illustrates another way in which the BFGS Method is a close relative of Broyden's Method (cf. (3.4.5)).

Another famous method, which has been used for nearly twenty years, is the Davidon–Fletcher–Powell (DFP) Method. Instead of updating the matrices D_k in

$$\mathbf{x}^{(k+1)} = \mathbf{x}^{(k)} - t_k D_k^{-1}(\nabla f(\mathbf{x}^{(k)}))$$

as in the BFGS Method, the DFP Method updates their inverses and yet retains the features of a secant method. More precisely, the DFP Method

$$\mathbf{x}^{(k+1)} = \mathbf{x}^{(k)} - t_k D_k(\nabla f(\mathbf{x}^{(k)}))$$

is defined to achieve the following objectives:

(1) The parameter t_k is set so that Criteria 1–4 of Section 3.3 are satisfied.
(2) Each D_k is positive definite.
(3) The *Inverse Secant Condition*

$$D_{k+1}(\mathbf{y}^{(k)}) = \mathbf{d}^{(k)}$$

is satisfied where $\mathbf{y}^{(k)} = \nabla f(\mathbf{x}^{(k+1)}) - \nabla f(\mathbf{x}^{(k)})$ and $\mathbf{d}^{(k)} = \mathbf{x}^{(k+1)} - \mathbf{x}^{(k)}$.

(4) The update from D_k to D_{k+1} is simple and computationally economical.

To arrive at the method, we will use the same strategy as we did with the BFGS Method. We write

$$D_{k+1} = D_k + \alpha_k \mathbf{a}^{(k)} \otimes \mathbf{a}^{(k)} + \beta_k \mathbf{b}^{(k)} \otimes \mathbf{b}^{(k)},$$

where the numbers α_k, β_k and the vectors $\mathbf{a}^{(k)}$, $\mathbf{b}^{(k)}$ are to be determined. The Inverse Secant Condition prescribes that

$$\mathbf{d}^{(k)} = D_k \mathbf{y}^{(k)} + \alpha_k(\mathbf{a}^{(k)} \cdot \mathbf{y}^{(k)})\mathbf{a}^{(k)} + \beta_k(\mathbf{b}^{(k)} \cdot \mathbf{y}^{(k)})\mathbf{b}^{(k)}.$$

This equation holds if we set $\mathbf{a}^{(k)} = \mathbf{d}^{(k)}$ and $\mathbf{b}^{(k)} = D_k \mathbf{y}^{(k)}$ provided that we set

$$\alpha_k = \frac{1}{\mathbf{d}^{(k)} \cdot \mathbf{y}^{(k)}}, \qquad \beta_k = \frac{-1}{\mathbf{y}^{(k)} \cdot D_k \mathbf{y}^{(k)}}.$$

With this choice, the update becomes

$$D_{k+1} = D_k + \frac{\mathbf{d}^{(k)} \otimes \mathbf{d}^{(k)}}{\mathbf{y}^{(k)} \cdot \mathbf{d}^{(k)}} - \frac{(D_k \mathbf{y}^{(k)}) \otimes (D_k \mathbf{y}^{(k)})}{\mathbf{y}^{(k)} \cdot D_k \mathbf{y}^{(k)}}.$$

This is the well-known Davidon–Fletcher–Powell (DFP) update. It provides the basis for the following:

(3.5.4) The DFP Method. To minimize $f(\mathbf{x})$ on R^n, select an initial point $\mathbf{x}^{(0)}$ and an initial positive definite matrix D_0. If $\mathbf{x}^{(k)}$ and D_k have been computed, then:

(1) Set $t_k > 0$ such that

$$\mathbf{x}^{(k+1)} = \mathbf{x}^{(k)} - t_k D_k(\nabla f(\mathbf{x}^{(k)}))$$

satisfies Criteria 1–4 of Section 3.3.

(2) Update

$$D_{k+1} = D_k + \frac{\mathbf{d}^{(k)} \otimes \mathbf{d}^{(k)}}{\mathbf{y}^{(k)} \cdot \mathbf{d}^{(k)}} - \frac{D_k \mathbf{y}^{(k)} \otimes D_k \mathbf{y}^{(k)}}{\mathbf{y}^{(k)} \cdot D_k \mathbf{y}^{(k)}},$$

where $\mathbf{d}^{(k)} = \mathbf{x}^{(k+1)} - \mathbf{x}^{(k)}$, $\mathbf{y}^{(k)} = \nabla f(\mathbf{x}^{(k+1)}) - \nabla f(\mathbf{x}^{(k)})$.

Using techniques very similar to those employed for the BFGS updates in Theorem (3.5.2), one can show that the DFP updates are positive definite. (See Exercise 19.)

A comparison of the BFGS update formula in (3.5.3) and the DFP update formula in (3.5.4) shows a duality in the roles of $\mathbf{y}^{(k)}$ and $\mathbf{d}^{(k)}$. This duality reflects the fact that the BFGS Method satisfies the Secant Condition $D_{k+1}(\mathbf{d}^{(k)}) = \mathbf{y}^{(k)}$, while the DFP Method satisfies the Inverse Secant Condition $D_{k+1}(\mathbf{y}^{(k)}) = \mathbf{d}^{(k)}$. One apparent advantage of the DFP Method over the BFGS Method is that the BFGS search direction $\mathbf{p}^{(k)} = D_k^{-1}(\nabla f(\mathbf{x}^{(k)}))$ for the latter must be computed by solving the system $D_k \mathbf{p}^{(k)} = \nabla f(\mathbf{x}^{(k)})$ while the search direction $\mathbf{p}^{(k)} = D_k(\nabla f(\mathbf{x}^{(k)}))$ can be computed directly.

It turns out that this advantage is offset by some computational advantages of the BFGS Method over the DFP Method. For example, although the D_k's produced by both methods are positive definite in theory, the DFP Method has a tendency to produce D_k's that are not positive definite because of computer round-off error, while the BFGS Method does not seem to share this defect. Thus, although the DFP Method was applied very successfully to many problems for many years using professionally written programs, the method has been displaced in recent years by BFGS routines.

The reader can learn more about the BFGS, DFP, and other unconstrained minimization methods by consulting, for example, *Numerical Methods for Unconstrained Optimization and Nonlinear Equations* by J. E. Dennis and R. B. Schnabel (Prentice-Hall, Englewood Cliffs, NJ, 1983) and *Introduction to Linear and Non-Linear Programming* by D. G. Luenberger (Addison-Wesley, Reading, MA, 1973).

Exercises

1. Suppose that $g(x)$ is a twice differentiable real-valued function that changes sign on the interval $[a, b]$, that is, $g(a)g(b) < 0$. Suppose further that there exist positive constants m and M such that

$$|g'(x)| \geq m, \qquad |g''(x)| \leq M$$

for all $x \in [a, b]$.

(a) Prove that the equation $g(x) = 0$ has one and only one solution r in $[a, b]$. (Hint: Apply the Intermediate Value Theorem and Rolle's Theorem.)

(b) If $x^{(k)}$ belongs to $[a, b]$ and if $x^{(k+1)}$ is defined in terms of $x^{(k)}$ by Newton's Formula

$$x^{(k+1)} = x^{(k)} - \frac{g(x^{(k)})}{g'(x^{(k)})},$$

show that

$$|x^{(k+1)} - r| \leq \frac{M}{2m}|x^{(k)} - r|^2,$$

where r is the unique solution to $g(x) = 0$ in $[a, b]$. (Hint: Apply Taylor's Formula for $g(x)$ at the point $x = x^{(k)}$.)

(c) Use the inequality obtained in part (b) to identify a subinterval $(r - c, r + c)$ of $[a, b]$ with the property that the sequence $\{\mathbf{x}^{(k)}\}$ produced by Newton's Method always converges to r provided that $x^{(0)}$ belongs to $(r - c, r + c)$.

(d) Show that if $x^{(0)} \neq 0$, the sequence $\{x^{(k)}\}$ produced by Newton's Method does not converge to the unique solution $r = 0$ of the equation

$$x^{1/3} = 0.$$

2. Show that the function $f(x)$ defined on R^1 by

$$f(x) = x^{4/3}$$

has a unique global minimizer at $x^* = 0$ but that, for any nonzero initial point $x^{(0)}$, the Newton's Method sequence $\{x^{(k)}\}$ with initial point $x^{(0)}$ for minimizing $f(x)$ diverges.

3. (a) Compute the quadratic approximation $q(\mathbf{x})$ for

$$f(x_1, x_2) = 8x_1^2 + 8x_2^2 - x_1^4 - x_2^4 - 1$$

at the point $(\frac{1}{2}, \frac{1}{2})$.

(b) Compute the minimum $\mathbf{x}^*$ of the quadratic approximation

$$q(\mathbf{x}) \quad \text{at} \quad (\tfrac{1}{2}, \tfrac{1}{2}).$$

4. Compute the first two terms $\mathbf{x}^{(1)}, \mathbf{x}^{(2)}$ of the Newton's Method sequence $\{\mathbf{x}^{(k)}\}$ for minimizing the function

$$f(x_1, x_2) = 2x_1^4 + x_2^2 - 4x_1x_2 + 5x_2$$

with initial point $\mathbf{x}^{(0)} = (0, 0)$.

5. Prove that if A is a positive definite matrix, then A^{-1} exists and is positive definite.

6. (a) Find a factorization $A = LU$ of the matrix

$$A = \begin{pmatrix} 1 & 1 & 2 \\ 2 & 4 & 1 \\ -1 & 1 & 3 \end{pmatrix},$$

where L is a lower triangular matrix, U is an upper triangular matrix and the diagonal elements of L and U are nonzero. (Hint: Use row reduction to compute U from A.)

(b) If a square matrix A has a factorization $A = LU$ of the sort described in part (a), show that A also has a factorization of the form

$$A = L_1DU_1,$$

where L_1 is a lower triangular matrix, U_1 is an upper triangular matrix, L_1 and U_1 have diagonal entries equal to 1, and D is a diagonal matrix with nonzero diagonal elements. (Hint: If you get stuck, try to find the required factorization for the 3×3-matrix in part (a).)

(c) Show that if a square matrix A has a factorization of the sort described in (b), then it has only one such factorization. (Hint: If $L_1D_1U_1 = L_2D_2U_2$, show that L_i^{-1} and U_i^{-1} exist and have the same triangularity properties as L_i and U_i, respectively, for $i = 1, 2$; then reduce the equation $L_1D_1U_1 = L_2D_2U_2$ to the form: upper triangular matrix = lower triangular matrix.

(d) Show that if A is a symmetric matrix that has a factorization $A = L_1DU_1$ of the sort described in part (b), then $U_1 = L_1^T$ and so

$$A = L_1DL_1^T,$$

where D is a diagonal matrix with nonzero diagonal entries and L_1 is a lower triangular matrix with ones on the main diagonal.

7. (a) Show that if A is a positive definite $n \times n$-matrix, then there is a unique positive definite matrix S such that $S^2 = A$. The matrix S is called the *square root* of A and is denoted by $A^{1/2}$. (Hint: Diagonalize A with an orthogonal matrix P and observe that a diagonal matrix with positive diagonal entries has an obvious square root.)

(b) Construct the square root of the matrix

$$A = \begin{pmatrix} 4 & 2 & 2 \\ 2 & 4 & 2 \\ 2 & 2 & 4 \end{pmatrix}.$$

8. Compute the first two terms $\mathbf{x}^{(1)}$, $\mathbf{x}^{(2)}$ of the Steepest Descent sequence $\{\mathbf{x}^{(k)}\}$ for the function $f(x_1, x_2)$ and the initial point $\mathbf{x}^{(0)}$ in Exercise 4.

9. In Section 3.1, we specialized Newton's Method for solving the system $\mathbf{g}(\mathbf{x}) = \mathbf{0}$ for $\mathbf{g}\colon R^n \to R^n$ to the problem of minimizing a function $f(\mathbf{x})$ on R^n by taking $\mathbf{g}(\mathbf{x}) = \nabla f(\mathbf{x})$. It is possible to reverse matters and seek a solution of the system $\mathbf{g}(\mathbf{x}) = \mathbf{0}$ by minimizing the function

$$G(\mathbf{x}) = \tfrac{1}{2}\|\mathbf{g}(\mathbf{x})\|^2 = \tfrac{1}{2}\mathbf{g}(\mathbf{x})\cdot\mathbf{g}(\mathbf{x}).$$

(a) Compute the function $G(\mathbf{x})$ for the system in Example (3.1.2).
(b) Show that the Hessian of $G(\mathbf{x})$ is given by

$$HG(\mathbf{x}) = [\nabla \mathbf{g}(\mathbf{x})]^{\mathrm{T}}\mathbf{g}(\mathbf{x}).$$

(c) Show that the Newton direction

$$\mathbf{x}^{(k+1)} - \mathbf{x}^{(k)} = -\nabla \mathbf{g}(\mathbf{x}^{(k)})\cdot \mathbf{g}(\mathbf{x}^{(k)})$$

for minimizing $G(\mathbf{x})$ is a direction of decreasing function values.

10. Compute the first two iterates $\mathbf{x}^{(1)}$, $\mathbf{x}^{(2)}$ of the Newton's Method sequence and the modified Newton's Method sequence (that is, the function is minimized in the direction of the Newton step $Hf(\mathbf{x}^{(k)})^{-1}\nabla f(\mathbf{x}^{(k)})$ at $\mathbf{x}^{(k)}$) for

$$f(x_1, x_2) = \frac{x_1^4}{4} + x_2^2$$

with $\mathbf{x}^{(0)} = (1, 0)$.

11. Compute the first two iterates $\mathbf{x}^{(1)}$, $\mathbf{x}^{(2)}$ in the Steepest Descent sequence $\{\mathbf{x}^{(k)}\}$ for

$$f(x_1, x_2) = 2x_1^2 + x_2^2 - x_1x_2$$

starting with the initial point $\mathbf{x}^{(0)} = (1, 4)$.

12. Suppose that $f(\mathbf{x})$ is a quadratic function of n variables

$$f(\mathbf{x}) = a + \mathbf{b}\cdot\mathbf{x} + \tfrac{1}{2}\mathbf{x}\cdot A\mathbf{x},$$

where $a \in R$, $\mathbf{b} \in R^n$, and A is a positive definite $n \times n$-matrix.
(a) Show that $f(\mathbf{x})$ has a unique global minimizer.
(b) Show that if the initial point $\mathbf{x}^{(0)}$ is selected so that $\mathbf{x}^{(0)} - \mathbf{x}^*$ is an eigenvector of A, then the Steepest Descent sequence $\{\mathbf{x}^{(k)}\}$ with initial point $\mathbf{x}^{(0)}$ reaches $\mathbf{x}^*$ in one step, that is, $\mathbf{x}^{(1)} = \mathbf{x}^*$.

13. (a) Show that if A is a symmetric matrix, then

$$\lambda = \max\{\mathbf{x}\cdot A\mathbf{x}\colon \mathbf{x} \in R^n,\ \|\mathbf{x}\| = 1\},$$

where λ is the largest eigenvalue of A. (Hint: Diagonalize A with an orthogonal matrix.)

(b) Show that

$$\max\{|\mathbf{x}\cdot A\mathbf{x}|: \mathbf{x}\in R^n, \|\mathbf{x}\| = 1\}$$

is equal to the absolute value of the eigenvalue of A with the largest absolute value.

(c) Show that if μ is a number greater than the absolute value of all eigenvalues of a symmetric matrix A, then

$$A + \mu I$$

is positive definite.

(d) For each of the following symmetric matrices A find a positive number μ such that $A + \mu I$ is positive definite:

$$\text{(i) } A = \begin{pmatrix} 3 & 7 & 0 \\ 7 & 5 & 0 \\ 0 & 0 & 1 \end{pmatrix} \qquad \text{(ii) } A = \begin{pmatrix} -1 & 0 & 0 \\ 0 & -3 & 0 \\ 0 & 0 & -4 \end{pmatrix}.$$

14. Compute the first two iterates $\mathbf{x}^{(1)}, \mathbf{x}^{(2)}$ using the Broyden's Method for the initial point $\mathbf{x}^{(0)}$ and function in Exercise 11 with
(a) $D_0 = I$.
(b) $D_0 = Hf(\mathbf{x}^{(0)})$.

15. Show that in Broyden's Method (3.4.1), the vector $\mathbf{y}^{(k)} - D_k(\mathbf{p}^{(k)})$ is just $g(\mathbf{x}^{(k+1)})$. Explain why this means that it is not necessary to evaluate $\mathbf{y}^{(k)}$ explicitly.

16. Show that if the variables in the function $f(x, y) = x^2 + a^2y^2$ in Example (3.2.4) are "rescaled" by the change of variables

$$\begin{cases} x' = x, \\ y' = ay, \end{cases}$$

then the convergence rate of the Method of Steepest Descent improves dramatically. More precisely, show that for any initial point $\mathbf{x}^{(0)}$, the Steepest Descent sequence for $f(x', y')$ converges in one step, that is, $\mathbf{x}^{(1)} = \mathbf{0}$.

17. Suppose that A is a positive definite matrix. Modify the recurrence formula defining the Method of Steepest Descent as follows:

$$\mathbf{x}^{(k+1)} = \mathbf{x}^{(k)} - t_k A\nabla f(\mathbf{x}^{(k)}),$$

where t_k is the value of t that minimizes

$$\bar{\varphi}_k(t) = f(\mathbf{x}^{(k)} - tA\nabla f(\mathbf{x}^{(k)})); \qquad t \geq 0.$$

(a) Show that if $\nabla f(\mathbf{x}^{(k)}) \neq \mathbf{0}$, then $f(\mathbf{x}^{(k+1)}) < f(\mathbf{x}^{(k)})$ so that this modified Method of Steepest Descent is a descent method.
(b) Show that a proper choice of A can force this modified Method of Steepest Descent to converge in a single step to the minimizer for the function and initial point in Example (3.2.4). (Hint: See Exercise 16.)
(c) Prove the analogs of Theorem 3.2.6 and Corollary 3.2.7 for this method.

18. (a) Show that if D_{k+1} is the BFGS update from D_k (see (3.5.3)) then

$$D_{k+1} = D_k + \frac{(D_{k+1}\mathbf{d}^{(k)}) \otimes (D_{k+1}\mathbf{d}^{(k)})}{\mathbf{d}^{(k)} \cdot D_{k+1}\mathbf{d}^{(k)}} - \frac{(D_k\mathbf{d}^{(k)}) \otimes (D_k\mathbf{d}^{(k)})}{\mathbf{d}^{(k)} \cdot D_k\mathbf{d}^{(k)}}.$$

(b) Show that if D_{k+1} is the DFP update from D_k (see (3.5.4)), then

$$D_{k+1} = D_k + \frac{(D_{k+1}\mathbf{y}^{(k)}) \otimes (D_{k+1}\mathbf{y}^{(k)})}{\mathbf{y}^{(k)} \cdot D_{k+1}\mathbf{y}^{(k)}} - \frac{(D_k\mathbf{y}^{(k)}) \otimes (D_k\mathbf{y}^{(k)})}{\mathbf{y}^{(k)} \cdot D_k\mathbf{y}^{(k)}}.$$

19. Prove that the DFP updates D_k are positive definite under the same hypotheses as Theorem (3.5.2) for the BFGS updates.

20. Compute the first two terms of the BFGS sequence and the first two terms of the DFP sequence for minimizing the function

$$f(x_1, x_2) = x_1^2 - x_1x_2 + \tfrac{3}{2}x_2^2$$

starting with the initial point $\mathbf{x}^{(0)} = (1, 2)$ and $D_0 = I$. For each case, choose $t_k > 0$ to be the exact minimizer of $f(\mathbf{x})$ in the search direction from $\mathbf{x}^{(k)}$.

21. Discuss why it is possible to pick t_k in the BFGS and DFP Methods so that Criteria 1–4 are satisfied.

CHAPTER 4

Least Squares Optimization

The techniques of least squares optimization have their origins in problems of curve fitting, and of finding the best possible solution for a system of linear equations with infinitely many solutions. Curve fitting problems begin with data points $(t_1, s_1), \ldots, (t_n, s_n)$ and a given class of functions (for example, linear functions, polynomial functions, exponential functions), and seek to identify the function $s = f(t)$ that "best fits" the data points. On the other hand, such problems as finding the minimum distance in geometric contexts or minimum variance in statistical contexts can often be solved by finding the solution of minimum norm for an underdetermined linear system of equations. We will consider the least squares technique that derives from curve fitting in Section 4.1, and then proceed to develop minimum norm methods in Sections 4.2 and 4.3.

4.1. Least Squares Fit

Suppose that in a certain experiment or study, we record a series of observed values $(t_1, s_1), (t_2, s_2), \ldots, (t_n, s_n)$ of two variables s, t that we have reason to believe are related by a function $s = f(t)$ of a certain type. For example, we might know that s and t are related by a polynomial function

$$p(t) = x_0 + x_1 t + \cdots + x_k t^k$$

of degree $\leq k$, where k is prescribed in advance, but we do not know the specific values of the coefficients $x_0, x_1, \ldots, x_k$ of $p(t)$. We are interested in choosing the values of these coefficients so that the deviations

$$|s_i - p(t_i)|,$$

between the observed value s_i at t_i and the value $p(t_i)$ of $p(t)$ at t_i, are all as small as possible.

One reasonable approach to this problem is to minimize the function

$$\begin{aligned}\varphi(x_0, x_1, \ldots, x_k) &= \sum_{i=1}^{n} (s_i - p(t_i))^2 \\ &= \sum_{i=1}^{n} \left(s_i - \sum_{j=0}^{k} x_j t_i^j\right)^2\end{aligned}$$

over all $(x_0, x_1, \ldots, x_k)$ in R^{k+1}. Although the use of the "square deviation" $(s_i - p(t_i))^2$ in place of the raw deviation $|s_i - p(t_i)|$ can be justified purely in terms of the convenience afforded by the resulting differentiability of φ, this choice has some theoretical advantages that will soon become evident.

Recall that the customary approach to the minimization of the function $\varphi(x_0, x_1, \ldots, x_k)$ is to set the gradient of φ equal to zero and solve the resulting system for $x_0, x_1, \ldots, x_k$ (cf. Exercise 18 of Chapter 1). This produces the minimizers of φ because φ is a convex function of $x_0, x_1, \ldots, x_k$ (Why?) and so any critical point of φ is a global minimizer. Our approach to this minimization problem is similar but somewhat more refined.

We first observe that the function $\varphi(x_0, x_1, \ldots, x_k)$ can be expressed conveniently in terms of the norm on R^{k+1}. Specifically, if we set

$$A = \begin{pmatrix} 1 & t_1 & t_1^2 & \cdots & t_1^k \\ 1 & t_2 & t_2^2 & & t_2^k \\ \vdots & \vdots & \vdots & & \vdots \\ 1 & t_n & t_n^2 & & t_n^k \end{pmatrix}; \qquad \mathbf{b} = \begin{pmatrix} s_1 \\ s_2 \\ \vdots \\ s_n \end{pmatrix}; \qquad \mathbf{x} = \begin{pmatrix} x_0 \\ x_1 \\ \vdots \\ x_k \end{pmatrix},$$

then

$$\begin{aligned}\varphi(x_0, x_1, \ldots, x_k) &= \|\mathbf{b} - A\mathbf{x}\|^2 = (\mathbf{b} - A\mathbf{x})\cdot(\mathbf{b} - A\mathbf{x}) \\ &= \mathbf{b}\cdot\mathbf{b} - 2\mathbf{b}\cdot A\mathbf{x} + A\mathbf{x}\cdot A\mathbf{x} \\ &= \mathbf{b}\cdot\mathbf{b} - 2A^{\mathrm{T}}\mathbf{b}\cdot\mathbf{x} + \mathbf{x}\cdot A^{\mathrm{T}}A\mathbf{x}.\end{aligned}$$

Therefore, the gradient and Hessian of φ are given by

$$\nabla\varphi(\mathbf{x}) = -2A^{\mathrm{T}}\mathbf{b} + 2A^{\mathrm{T}}A\mathbf{x}; \qquad H\varphi(\mathbf{x}) = 2A^{\mathrm{T}}A.$$

Now here is the pertinent observation: Since the numbers $t_1, t_2, \ldots, t_n$ are *distinct* values of the independent variable t, the columns of the matrix A are linearly independent. This means that if $A\mathbf{x} = \mathbf{0}$ then $\mathbf{x} = \mathbf{0}$ since $A\mathbf{x}$ is simply a linear combination of the column vectors of A. But then, because

$$\mathbf{x}\cdot A^{\mathrm{T}}A\mathbf{x} = A\mathbf{x}\cdot A\mathbf{x} = \|A\mathbf{x}\|^2,$$

we see that $H\varphi(\mathbf{x})$ is positive definite. It follows from (2.3.7) that $\varphi(\mathbf{x})$ is strictly convex on R^{k+1} and so $\varphi(\mathbf{x})$ has unique global minimizer at the point $\mathbf{x}^*$ for which $\nabla\varphi(\mathbf{x}^*) = 0$. Since $\nabla\varphi(\mathbf{x}) = -2A^{\mathrm{T}}\mathbf{b} + 2A^{\mathrm{T}}A\mathbf{x}$, we see that the minimizer

$\mathbf{x}^*$ of φ is characterized by the so-called *normal equation*

$$A^{\mathrm{T}}A\mathbf{x}^* = A^{\mathrm{T}}\mathbf{b}. \tag{1}$$

The matrix $A^{\mathrm{T}}A$ is invertible because it is positive definite, so $\mathbf{x}^*$ is also given by

$$\mathbf{x}^* = (A^{\mathrm{T}}A)^{-1}A^{\mathrm{T}}\mathbf{b}. \tag{2}$$

If $\mathbf{x}^* = (x_0^*, x_1^*, \ldots, x_k^*)$, then the polynomial

$$p(t) = x_0^* + x_1^*t + \cdots + x_k^*t^k$$

is called the *best least squares* (*k*th *degree polynomial*) *fit* for the given data.

In the special case when $k = 1$, the least squares procedure fits a line

$$p(t) = x_0^* + x_1^*t$$

to the given data points $(t_1, s_1), (t_2, s_2), \ldots, (t_n, s_n)$. This line is called the *linear regression line* for the given data.

(4.1.1) Example. Convenient formulas that enable one to compute the coefficients $\mathbf{x}_0^*$, $\mathbf{x}_1^*$ for the linear regression line from the given data are easy to obtain. Since

$$A = \begin{pmatrix} 1 & t_1 \\ 1 & t_2 \\ \vdots & \vdots \\ 1 & t_n \end{pmatrix}; \qquad \mathbf{b} = \begin{pmatrix} s_1 \\ s_2 \\ \vdots \\ s_n \end{pmatrix};$$

it follows that

$$A^{\mathrm{T}}A = \begin{pmatrix} n & \sum t_i \\ \sum t_i & \sum t_i^2 \end{pmatrix}; \qquad A^{\mathrm{T}}\mathbf{b} = \begin{pmatrix} \sum s_i \\ \sum s_i t_i \end{pmatrix},$$

where each summation extends from $i = 1$ to $i = n$. If we solve the normal equation $A^{\mathrm{T}}A\mathbf{x} = A^{\mathrm{T}}\mathbf{b}$ by Cramer's Rule for x_1^*, we obtain

$$x_1^* = \frac{n\sum s_i t_i - (\sum s_i)(\sum t_i)}{n\sum t_i^2 - (\sum t_i)^2}.$$

If we set $\bar{t}$, $\bar{s}$ to be the means of the component data, that is,

$$\bar{t} = \frac{1}{n}\sum_{i=1}^{n} t_i, \qquad \bar{s} = \frac{1}{n}\sum_{i=1}^{n} s_i,$$

then x_1^* can be expressed in the convenient form

$$x_1^* = \frac{\frac{1}{n}\sum s_i t_i - \bar{t}\bar{s}}{\frac{1}{n}\sum t_i^2 - (\bar{t})^2}.$$

The coefficient $\mathbf{x}_0^*$ is then given by

$$\mathbf{x}_0^* = \bar{s} - \mathbf{x}_1^* \bar{t}$$

as can be seen from the first equation in the system $A^{\mathrm{T}}A\mathbf{x} = A^{\mathrm{T}}\mathbf{b}$. These formulas allow one to compute the coefficients $\mathbf{x}_0^*$, $\mathbf{x}_1^*$ of the linear regression line directly from the data points (t_1, s_1), (t_2, s_2), $\ldots$, (t_n, s_n) without constructing the matrix A. In fact, many hand-held calculators do this computation with a single keystroke after the data points have been entered.

Our development of least squares curve fitting actually made no use of the specific entries of the matrix A. The only fact about A that was important to our derivation of the normal equation (1) and the corresponding equation (2) for $\mathbf{x}^*$ was that the columns of A must be linearly independent. Consequently, the same analysis proves the following theorem.

(4.1.2) Theorem. *Suppose that A is a $m \times n$-matrix whose columns are linearly independent and that $\mathbf{b} \in R^m$. Then the vector $\mathbf{x}^*$ given by*

$$\mathbf{x}^* = (A^{\mathrm{T}}A)^{-1}A^{\mathrm{T}}\mathbf{b}$$

satisfies

$$\|A\mathbf{x}^* - \mathbf{b}\| \leq \|A\mathbf{x} - \mathbf{b}\|$$

for all $\mathbf{x} \in R^n$.

The vector $\mathbf{x}^* = (A^{\mathrm{T}}A)^{-1}A^{\mathrm{T}}\mathbf{b}$ is called the *best least squares solution* of the (possibly inconsistent) system

$$A\mathbf{x} = \mathbf{b}.$$

The matrix $(A^{\mathrm{T}}A)^{-1}A^{\mathrm{T}}$ is called the *generalized inverse* of the matrix A and is denoted by $A^{\dagger}$. This choice of terminology is reasonable because if A is an invertible square matrix then

$$A^{\dagger} = (A^{\mathrm{T}}A)^{-1}A^{\mathrm{T}} = A^{-1}(A^{\mathrm{T}})^{-1}A^{\mathrm{T}} = A^{-1}.$$

Although the best least squares solution $\mathbf{x}^*$ of the system $A\mathbf{x} = \mathbf{b}$ is most conveniently described by $\mathbf{x}^* = A^{\dagger}\mathbf{b}$, the vector $\mathbf{x}^*$ is computed by solving the normal equation

$$A^{\mathrm{T}}A\mathbf{x} = A^{\mathrm{T}}\mathbf{b}$$

because the calculation of $A^{\dagger}$ requires the inversion of the matrix $A^{\mathrm{T}}A$, a relatively expensive computational procedure.

The following example presents a simple illustration of the general least squares procedure.

(4.1.3) Example. Find the best least squares solution to the inconsistent linear system

$$x + y = 2,$$
$$x + 2y = 3,$$
$$x + 3y = 3.$$

In this case, the coefficient matrix A and the vector $\mathbf{b}$ are given by

$$A = \begin{pmatrix} 1 & 1 \\ 1 & 2 \\ 1 & 3 \end{pmatrix}; \qquad \mathbf{b} = \begin{pmatrix} 2 \\ 3 \\ 3 \end{pmatrix},$$

so the best least squares solution $\mathbf{x}^*$ is the solution of the system

$$A^{\mathrm{T}}A\mathbf{x} = \begin{pmatrix} 3 & 6 \\ 6 & 14 \end{pmatrix}\begin{pmatrix} x \\ y \end{pmatrix} = \begin{pmatrix} 8 \\ 17 \end{pmatrix} = A^{\mathrm{T}}\mathbf{b}.$$

From this we see that

$$\mathbf{x}^* = (\tfrac{5}{3}, \tfrac{1}{2}).$$

It would appear from the presentation up to this point that least squares computations are simple and problem-free. All that we need to do to find $\mathbf{x}^*$ is to solve the system of linear equations

$$A^{\mathrm{T}}A\mathbf{x} = A^{\mathrm{T}}\mathbf{b}.$$

Unfortunately, the solution of this system is often fraught with numerical difficulties. The basic problem from the mathematical point of view is that although $A^{\mathrm{T}}A$ is always positive definite in theory, the matrix that is actually computed may not be positive definite because of round-off error. The following example, which is found in *Introduction to Matrix Computations* by G. W. Stewart (Academic Press, New York, 1973), illustrates this point.

(4.1.4) Example. If

$$A = \begin{pmatrix} 1.000 & 1.020 \\ 1.000 & 1.000 \\ 1.000 & 1.000 \end{pmatrix}$$

then $A^{\mathrm{T}}A$ is given by

$$A^{\mathrm{T}}A = \begin{pmatrix} 3.000 & 3.020 \\ 3.020 & 3.040 \end{pmatrix}$$

to three decimal places. Note that $A^{\mathrm{T}}A$ as computed is indefinite since

$$\det(A^{\mathrm{T}}A) < 0.$$

This example points out a difference that often exists between mathematical theory and actual computation where round-off error can eliminate the relevancy of the computation to the problem at hand. Numerical experience has shown that, even with the use of double precision, the numerical solution of the system

$$A^T A\mathbf{x} = A^T\mathbf{b}$$

may not be numerically reliable.

Fortunately, there is a way around this difficulty. The key idea depends on a little geometry. Let us begin by recalling the Gram–Schmidt Process from linear algebra. Given m linearly independent vectors $\mathbf{a}^{(1)}, \ldots, \mathbf{a}^{(m)}$ in R^n, we define m vectors $\mathbf{u}^{(1)}, \ldots, \mathbf{u}^{(m)}$ as follows:

(1) $\mathbf{e}^{(1)} = \mathbf{a}^{(1)}$,
$\mathbf{u}^{(1)} = \mathbf{e}^{(1)}/\|\mathbf{e}^{(1)}\|$;
(2) $\mathbf{e}^{(2)} = \mathbf{a}^{(2)} - (\mathbf{u}^{(1)}\cdot\mathbf{a}^{(2)})\mathbf{u}^{(1)}$,
$\mathbf{u}^{(2)} = \mathbf{e}^{(2)}/\|\mathbf{e}^{(2)}\|$;
⋮
(m) $\mathbf{e}^{(m)} = \mathbf{a}^{(m)} - (\mathbf{u}^{(1)}\cdot\mathbf{a}^{(m)})\mathbf{u}^{(1)} - \cdots - (\mathbf{u}^{(m-1)}\cdot\mathbf{a}^{(m)})\mathbf{u}^{(m-1)}$,
$\mathbf{u}^{(m)} = \mathbf{e}^{(m)}/\|\mathbf{e}^{(m)}\|$.

At the $(k+1)$st step, this process constructs a unit vector $\mathbf{u}^{(k+1)}$ perpendicular to the previously constructed mutually perpendicular vectors $\mathbf{u}^{(1)}, \ldots, \mathbf{u}^{(k)}$. Consequently, the subspace spanned by $\mathbf{u}^{(1)}, \ldots, \mathbf{u}^{(k)}$ is identical to the subspace spanned by $\mathbf{a}^{(1)}, \ldots, \mathbf{a}^{(k)}$ at each stage. The resulting set of vectors $\{\mathbf{u}^{(1)}, \ldots, \mathbf{u}^{(m)}\}$ is *orthonormal*, that is,

(a) $\|\mathbf{u}^{(i)}\| = 1$ for $i = 1, 2, \ldots, m$;
(b) $\mathbf{u}^{(i)}\cdot\mathbf{u}^{(j)} = 0$ for $i \neq j$, $i, j = 1, \ldots, m$.

Any orthonormal set is automatically linearly independent.

Now suppose that A is an $m \times n$-matrix whose column vectors $\mathbf{a}^{(1)}, \mathbf{a}^{(2)}, \ldots, \mathbf{a}^{(n)}$ are linearly independent. Let $\mathbf{u}^{(1)}, \mathbf{u}^{(2)}, \ldots, \mathbf{u}^{(n)}$ be the corresponding orthonormal set of vectors in R^m obtained by applying the Gram–Schmidt Process to $\mathbf{a}^{(1)}, \mathbf{a}^{(2)}, \ldots, \mathbf{a}^{(n)}$ and let Q be the $m \times n$-matrix with $\mathbf{u}^{(i)}$ as the ith column for $i = 1, \ldots, n$. Because the columns of Q are orthonormal, we see that $Q^T Q = I$. Moreover, the Gram–Schmidt Process can be viewed as a "column reduction" process that transforms A into Q by subtracting multiples of columns to the left of the pivot column. Consequently, we can obtain Q from A by right multiplication by an $n \times n$-upper triangular matrix L, that is,

$$Q = AL.$$

The matrix L is invertible since Q has linearly independent columns, so

$$A = QR,$$

where $R = L^{-1}$ is an upper triangular $n \times n$-matrix.

This proves the following very useful theorem:

(4.1.5) Theorem. *If A is an $m \times n$-matrix whose columns are linearly independent, then there exists an $m \times n$-matrix Q whose columns are orthonormal and an $n \times n$-invertible upper triangular matrix R such that*

$$(*) \qquad A = QR.$$

We refer to $(*)$ as the *QR-factorization* of the matrix A.

Now return to our possibly inconsistent linear system

$$A\mathbf{x} = \mathbf{b}$$

for which we assume that A has linearly independent columns. Then the best least squares solution of this system is

$$\begin{aligned}
\mathbf{x}^* &= (A^{\mathrm{T}}A)^{-1}A^{\mathrm{T}}\mathbf{b} \\
&= ((QR)^{\mathrm{T}}QR)^{-1}(QR)^{\mathrm{T}}\mathbf{b} \\
&= (R^{\mathrm{T}}Q^{\mathrm{T}}QR)^{-1}R^{\mathrm{T}}Q^{\mathrm{T}}\mathbf{b} \\
&= R^{-1}(Q^{\mathrm{T}}Q)^{-1}(R^{\mathrm{T}})^{-1}R^{\mathrm{T}}Q^{\mathrm{T}}\mathbf{b} \\
&= R^{-1}I \cdot IQ^{\mathrm{T}}\mathbf{b} \qquad (\text{since } Q^{\mathrm{T}}Q = I) \\
&= R^{-1}Q^{\mathrm{T}}\mathbf{b}.
\end{aligned}$$

This computation of the best least squares least squares answer $\mathbf{x}^*$ involves only the multiplication $Q^{\mathrm{T}}\mathbf{b}$ followed by a quick and easy back-substitution to solve the triangular system

$$R\mathbf{x}^* = Q^{\mathrm{T}}\mathbf{b}.$$

(4.1.6) Example. Let us find the best least squares solution to the inconsistent linear system

$$\begin{pmatrix} 1 & 0 \\ 0 & 1 \\ 1 & 1 \end{pmatrix}\begin{pmatrix} x \\ y \end{pmatrix} = \begin{pmatrix} 1 \\ 1 \\ 1 \end{pmatrix}.$$

Here $\mathbf{a}^{(1)} = (1, 0, 1)$ and $\mathbf{a}^{(2)} = (0, 1, 1)$ so

$$\mathbf{u}^{(1)} = \mathbf{a}^{(1)}/\sqrt{2} = \left(\frac{1}{\sqrt{2}}, 0, \frac{1}{\sqrt{2}}\right)$$

$$\begin{aligned}
\mathbf{e}^{(2)} &= \mathbf{a}^{(2)} - (\mathbf{a}^{(2)} \cdot \mathbf{u}^{(1)})\mathbf{u}^{(1)} \\
&= (0, 1, 1) - \frac{1}{\sqrt{2}}\left(\frac{1}{\sqrt{2}}, 0, \frac{1}{\sqrt{2}}\right) \\
&= (0, 1, 1) - (\tfrac{1}{2}, 0, \tfrac{1}{2}) \\
&= (-\tfrac{1}{2}, 1, \tfrac{1}{2}),
\end{aligned}$$

and

$$\mathbf{u}^{(2)} = \frac{(-\frac{1}{2}, 1, \frac{1}{2})}{\sqrt{\frac{1}{4}+1+\frac{1}{4}}} = \frac{(-\frac{1}{2}, 1, \frac{1}{2})}{(\sqrt{3/2})}$$
$$= \left(-\frac{1}{\sqrt{6}}, \frac{\sqrt{2}}{\sqrt{3}}, \frac{1}{\sqrt{6}}\right).$$

Now

$$QR = A,$$

or

$$\begin{pmatrix} \frac{1}{\sqrt{2}} & -\frac{1}{\sqrt{6}} \\ 0 & \frac{\sqrt{2}}{\sqrt{3}} \\ \frac{1}{\sqrt{2}} & \frac{1}{\sqrt{6}} \end{pmatrix} \begin{pmatrix} r_{11} & r_{12} \\ 0 & r_{22} \end{pmatrix} = \begin{pmatrix} 1 & 0 \\ 0 & 1 \\ 1 & 1 \end{pmatrix},$$

which gives

$$\frac{r_{11}}{\sqrt{2}} = 1,$$

$$\frac{r_{12}}{\sqrt{2}} - \frac{r_{22}}{\sqrt{6}} = 0,$$

$$\frac{\sqrt{2}}{\sqrt{3}} r_{22} = 1.$$

Hence $r_{11} = \sqrt{2}$, $r_{22} = \sqrt{3}/\sqrt{2}$, and $r_{12} = (1/\sqrt{3})r_{22} = 1/\sqrt{2}$. Thus, the QR-factorization of A is

$$A = \begin{pmatrix} 1 & 0 \\ 0 & 1 \\ 1 & 1 \end{pmatrix} = \begin{pmatrix} \frac{1}{\sqrt{2}} & -\frac{1}{\sqrt{6}} \\ 0 & \frac{\sqrt{2}}{\sqrt{3}} \\ \frac{1}{\sqrt{2}} & \frac{1}{\sqrt{6}} \end{pmatrix} \begin{pmatrix} \sqrt{2} & \frac{1}{\sqrt{2}} \\ 0 & \frac{\sqrt{3}}{\sqrt{2}} \end{pmatrix} = QR.$$

The best least squares solution (x^*, y^*) is given by

$$R\begin{pmatrix} x^* \\ y^* \end{pmatrix} = Q^{\mathrm{T}}\begin{pmatrix} 1 \\ 1 \\ 1 \end{pmatrix}.$$

that is,

$$\begin{pmatrix} \sqrt{2} & \dfrac{1}{\sqrt{2}} \\ 0 & \dfrac{\sqrt{3}}{\sqrt{2}} \end{pmatrix} \begin{pmatrix} x^* \\ y^* \end{pmatrix} = \begin{pmatrix} \dfrac{1}{\sqrt{2}} & 0 & \dfrac{1}{\sqrt{2}} \\ -\dfrac{1}{\sqrt{6}} & \dfrac{\sqrt{2}}{\sqrt{3}} & \dfrac{1}{\sqrt{6}} \end{pmatrix} \begin{pmatrix} 1 \\ 1 \\ 1 \end{pmatrix}.$$

By applying back-substitution, we see that

$$y^* = \tfrac{3}{4}, \qquad x^* = \tfrac{5}{8}.$$

For small problems, the computational procedure used in the preceding example is practical and probably easier than using Gaussian elimination to solve $A^TA\mathbf{x}^* = A^T\mathbf{b}$. However, if A has four or more columns, then round-off error can create trouble for either approach. Since most nontextbook problems are not small, a more sophisticated procedure is needed. Thankfully there is a numerically stable way called "orthonornal triangulation," used for finding the factorization $A = QR$ which is available in most software libraries. This numerical routine is based on an idea called Householder transformations. For more on this, see Chapter 5 of *Introduction to Matrix Computations* by G. W. Stewart (Academic Press, New York, 1973).

4.2. Subspaces and Projections

This section is devoted to the problem of finding the best approximation of a vector $\mathbf{x}$ in R^m by a member $\mathbf{m}$ of a subspace of R^m. Specifically, let M be a subspace of R^m and let $\mathbf{x}$ be a vector not in M. We are going to find a way to select a vector $\mathbf{m}^*$ in M such that

$$\|\mathbf{x} - \mathbf{m}^*\| \leq \|\mathbf{x} - \mathbf{m}\|$$

for all other $\mathbf{m}$ in M. Let us recall some terminology and results from linear algebra.

(4.2.1) Definition. A subset M of n-dimensional space R^m is a *subspace* if $\mathbf{y}^{(1)} + \mathbf{y}^{(2)} \in M$ and $\lambda\mathbf{y}^{(1)} \in M$ whenever $\mathbf{y}^{(1)}, \mathbf{y}^{(2)} \in M$ and λ is any real number.

Less formally, a subset M of R^m is a subspace if it contains the sum of any two of its members and if it contains all scalar multiples of any of its members. In particular, since $\mathbf{0} = 0 \cdot \mathbf{x}$ for any $\mathbf{x} \in M$, any subspace of R^m contains $\mathbf{0}$. It follows readily that if M is a subspace of R^m, then M contains all linear combinations of its members, that is, if $\mathbf{x}^{(1)}, \ldots, \mathbf{x}^{(p)}$ belong to M and if $\lambda_1, \ldots,$

λ_p are scalars, then the linear combination

$$\lambda_1 \mathbf{x}^{(1)} + \cdots + \lambda_p \mathbf{x}^{(p)}$$

belongs to M.

The prime example of a subspace of R^m is the range of an $m \times n$-matrix.

(4.2.2) Definition. Let A be an $m \times n$-matrix. Then *the range of* A is the subspace $\mathscr{R}(A)$ of R^m defined by

$$\mathscr{R}(A) = \{\mathbf{y} \in R^m: \mathbf{y} = A\mathbf{x} \text{ for some } \mathbf{x} \in R^n\}.$$

Actually, *any* subspace of R^m can be regarded as the range of an appropriate $m \times n$-matrix as the following theorem demonstrates.

(4.2.3) Theorem. *If M is a subspace of R^m of dimension n, then there is an $m \times n$-matrix A whose columns are linearly independent such that*

$$M = \mathscr{R}(A).$$

PROOF. Since M is a subspace of R^m, M has a basis $\mathbf{a}^{(1)}, \ldots, \mathbf{a}^{(n)}$, that is, there is a linearly independent set $\{\mathbf{a}^{(1)}, \ldots, \mathbf{a}^{(n)}\}$ such that each $\mathbf{m} \in M$ is a linear combination of vectors in this set

$$\mathbf{m} = \sum_{j=1}^{n} x_j \mathbf{a}^{(j)}.$$

If we write $\mathbf{x} = (x_1, \ldots, x_n)$, we see that another way of saying this is that $\mathbf{m} \in M$ if and only if there is $\mathbf{x} = (x_1, \ldots, x_n)$ in R^n such that

$$\mathbf{m} = A\mathbf{x} = \sum_{j=1}^{n} x_j \mathbf{a}^{(j)},$$

where A is the matrix whose jth column is $\mathbf{a}^{(j)}$. This shows $M = \mathscr{R}(A)$ which completes the proof.

Given a subspace M of R^m, the *orthogonal complement* $M^\perp$ of M is defined by

$$M^\perp = \{\mathbf{y} \in R^m: \mathbf{x} \cdot \mathbf{y} = 0 \text{ for all } \mathbf{x} \in M\}.$$

It is readily seen that the orthogonal complement $M^\perp$ of a subspace M of R^m is itself a subspace of R^m.

This next theorem is the basic result of this section.

(4.2.4) Theorem. *If M is a subspace of R^m and if $\mathbf{x} \in R^m$, then there is a unique $\mathbf{m}^* \in M$ such that*

$$\|\mathbf{x} - \mathbf{m}^*\| \le \|\mathbf{x} - \mathbf{m}\|$$

for all $\mathbf{m} \in M$. The point $\mathbf{m}^*$ of M is characterized by $\mathbf{x} - \mathbf{m}^* \in M^\perp$.

The content of this theorem is summarized by the following diagram in R^2:

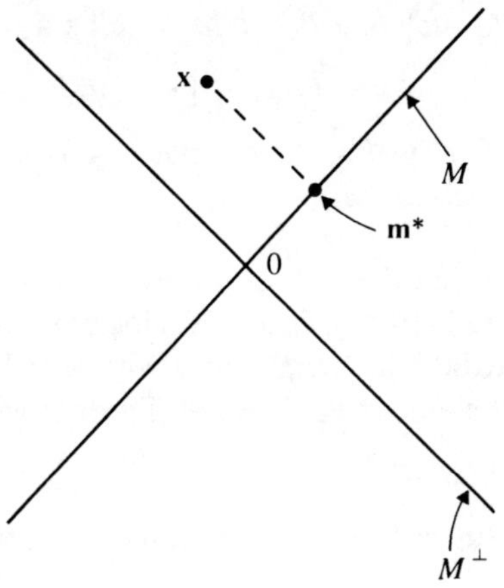

PROOF OF THEOREM (4.2.4). If the dimension of M is n, then Theorem (4.2.3) assets that there is an $m \times n$-matrix A such that $M = \mathscr{R}(A)$. Observe that $\|\mathbf{x} - \mathbf{m}^*\| \le \|\mathbf{x} - \mathbf{m}\|$ for all $\mathbf{m} \in M$ if and only if $\|\mathbf{x} - A\mathbf{y}^*\| \le \|\mathbf{x} - A\mathbf{y}\|$ for all $\mathbf{y} \in R^n$. According to Theorem (4.1.2), the vector $\mathbf{y}^* \in R^n$ that satisfies the latter inequality is given by

$$\mathbf{y}^* = A^\dagger \mathbf{x} = (A^T A)^{-1} A^T \mathbf{x},$$

so $\mathbf{m}^* = A\mathbf{y}^* = AA^\dagger \mathbf{x}^*$ is the required vector.

Note that $\mathbf{x} - \mathbf{n} \in M^\perp$ if and only if

$$(*) \qquad (\mathbf{x} - \mathbf{n}) \cdot \mathbf{m} = 0$$

for all $\mathbf{m} \in M$. Equation $(*)$ holds if and only if

$$(\mathbf{x} - \mathbf{n}) \cdot A\mathbf{y} = 0$$

for all $\mathbf{y} \in R^n$. But $\mathbf{n} = A\mathbf{z}$ for some $\mathbf{z} \in R^n$, so $(*)$ is satisfied if and only if

$$\begin{aligned} 0 &= (\mathbf{x} - A\mathbf{z}) \cdot A\mathbf{y} = A^T(\mathbf{x} - A\mathbf{z}) \cdot \mathbf{y} \\ &= (A^T\mathbf{x} - A^T A\mathbf{z}) \cdot \mathbf{y} \end{aligned}$$

for all $\mathbf{y} \in R^n$, that is, if and only if

$$A^T\mathbf{x} - A^T A\mathbf{z} = 0.$$

Thus, we arrive at the conclusion that $\mathbf{x} - \mathbf{n} \in M^\perp$ if and only if $\mathbf{n} = A\mathbf{z}$ where $\mathbf{z}$ is the unique solution of the normal equation

$$A^T A\mathbf{z} = A^T\mathbf{x}.$$

Therefore $\mathbf{x} - \mathbf{n} \in M^\perp$ if and only if $\mathbf{n} = \mathbf{m}^*$ where $\mathbf{m}^* = A\mathbf{y}^* = AA^\dagger\mathbf{x}$.

The following result is a corollary of the proof of the preceding theorem.

(4.2.5) Theorem. *Suppose that M is a subspace of R^m of dimension n. Then there is an $m \times m$-matrix P_M such that $P_M\mathbf{x} \in M$ for all $\mathbf{x} \in R^m$ and*

$$\|\mathbf{x} - P_M\mathbf{x}\| \leq \|\mathbf{x} - \mathbf{m}\|$$

for all $\mathbf{m} \in M$. Specifically, $P_M = AA^\dagger$ where A is an $m \times n$-matrix whose columns constitute a basis for M.

The matrix P_M is called the *orthogonal projection of R^m onto M*. For each $\mathbf{x} \in R^m$, then, according to (4.2.5), $P_M\mathbf{x}$ is the closest vector to $\mathbf{x}$ in the subspace M. Although the matrix A in (4.2.5) varies with the choice and order of the basis vectors for M, the matrix P_M does not. The following example illustrates this point.

(4.2.6) Example. Let us construct the orthogonal projection P_M of R^3 onto the subspace

$$M = \{\mathbf{m} \in R^3 : m_1 + m_2 + m_3 = 0\}.$$

In this case, a basis for M is provided by the vectors

$$\mathbf{u}^{(1)} = \begin{pmatrix} -1 \\ 1 \\ 0 \end{pmatrix}, \qquad \mathbf{u}^{(2)} = \begin{pmatrix} -1 \\ 0 \\ 1 \end{pmatrix},$$

so the associated matrix A is

$$A = \begin{pmatrix} -1 & -1 \\ 1 & 0 \\ 0 & 1 \end{pmatrix}$$

and

$$P_M = A(A^\mathrm{T}A)^{-1}A^\mathrm{T} = \begin{pmatrix} \frac{2}{3} & -\frac{1}{3} & -\frac{1}{3} \\ -\frac{1}{3} & \frac{2}{3} & -\frac{1}{3} \\ -\frac{1}{3} & -\frac{1}{3} & \frac{2}{3} \end{pmatrix}.$$

If we use the basis

$$\mathbf{v}^{(1)} = \begin{pmatrix} -1 \\ 1 \\ 0 \end{pmatrix}, \qquad \mathbf{v}^{(2)} = \begin{pmatrix} -2 \\ 1 \\ 1 \end{pmatrix}$$

for M, then

$$A = \begin{pmatrix} -1 & -2 \\ 1 & 1 \\ 0 & 1 \end{pmatrix},$$

but the matrix $P_M = A(A^TA)^{-1}A^T$ is identical to the one constructed above for the basis $\{\mathbf{u}^{(1)}, \mathbf{u}^{(2)}\}$ for M.

(4.2.7) Theorem. *If M is a subspace of R^m, then $(M^\perp)^\perp = M$.*

PROOF. If $\mathbf{x} \in M$, then $\mathbf{x}\cdot\mathbf{y} = 0$ for all $\mathbf{y} \in M^\perp$, and so $\mathbf{x} \in (M^\perp)^\perp$ by definition of $(M^\perp)^\perp$. Therefore, $M \subset (M^\perp)^\perp$. On the other hand, if $\mathbf{x} \in (M^\perp)^\perp$, then

$$0 = \mathbf{x}\cdot(\mathbf{x} - P_M\mathbf{x}),$$

since $\mathbf{x} - P_M\mathbf{x} \in M^\perp$ by (4.2.4) and (4.2.5). Also

$$0 = P_M\mathbf{x}\cdot(\mathbf{x} - P_M\mathbf{x}),$$

because $\mathbf{x} - P_M\mathbf{x} \in M^\perp$ and $P_M\mathbf{x} \in M$. Consequently,

$$\begin{aligned} 0 &= \mathbf{x}\cdot(\mathbf{x} - P_M\mathbf{x}) - P_M\mathbf{x}\cdot(\mathbf{x} - P_M\mathbf{x}) \\ &= (\mathbf{x} - P_M\mathbf{x})\cdot(\mathbf{x} - P_M\mathbf{x}) = \|\mathbf{x} - P_M\mathbf{x}\|^2, \end{aligned}$$

so $\mathbf{x} = P_M\mathbf{x} \in M$. We conclude that $M = (M^\perp)^\perp$ which completes the proof.

4.3. Minimum Norm Solutions of Underdetermined Linear Systems

Suppose that A is an $m \times n$-matrix and that $\mathbf{b} \in R^m$. The system

$$A\mathbf{x} = \mathbf{b}$$

of m linear equations in n unknowns is *underdetermined* if it has more than one solution. The goal of this section is to find an effective procedure for computing the solution of an underdetermined system that has minimum norm.

The following theorem from linear algebra will be basic to our considerations.

(4.3.1) Theorem. *Suppose that*

$$A\mathbf{x} = \mathbf{b} \tag{1}$$

is an underdetermined system of m linear equations in n unknowns and that $\mathbf{x}^{(0)}$ is a fixed solution of this system. Then $\mathbf{x}$ is a solution of (1) if and only if

$$\mathbf{x} = \mathbf{x}^{(0)} - \mathbf{y},$$

where $\mathbf{y}$ is a solution of the corresponding homogeneous system

$$A\mathbf{y} = \mathbf{0}. \tag{2}$$

PROOF. If $\mathbf{x} = \mathbf{x}^{(0)} - \mathbf{y}$ where $\mathbf{y}$ is a solution of (2), then

$$A\mathbf{x} = A(\mathbf{x}^{(0)} - \mathbf{y}) = A\mathbf{x}^{(0)} - A\mathbf{y} = \mathbf{b} - \mathbf{0} = \mathbf{b}.$$

Therefore, $\mathbf{x}$ is a solution of (1). On the other hand, if $\mathbf{x}$ is any solution of (1), then

$$A(\mathbf{x}^{(0)} - \mathbf{x}) = A\mathbf{x}^{(0)} - A\mathbf{x} = \mathbf{b} - \mathbf{b} = \mathbf{0}.$$

Consequently, $\mathbf{y} = \mathbf{x}^{(0)} - \mathbf{x}$ is a solution of (2).

With this result in mind, let us begin our search for the solution $\mathbf{x}^*$ of minimum norm for a given underdetermined system $A\mathbf{x} = \mathbf{b}$ of m linear equations in n unknowns. If $\mathbf{x}^{(0)}$ is a fixed solution of $A\mathbf{x} = \mathbf{b}$, then by (4.3.1) $\mathbf{x}$ is a solution of $A\mathbf{x} = \mathbf{b}$ if and only if

$$\mathbf{x} = \mathbf{x}^{(0)} - \mathbf{y},$$

where $\mathbf{y}$ is a solution of the associated homogeneous system $A\mathbf{y} = \mathbf{0}$. Thus, the minimum norm solution $\mathbf{x}^*$ of $A\mathbf{x} = \mathbf{b}$ can be written as

$$\mathbf{x}^* = \mathbf{x}^{(0)} - \mathbf{y}^*,$$

where $\mathbf{y}^*$ is the solution of $A\mathbf{x} = \mathbf{0}$ that satisfies

$$\|\mathbf{x}^{(0)} - \mathbf{y}^*\| \le \|\mathbf{x}^{(0)} - \mathbf{y}\|$$

for all solution $\mathbf{y}$ of $A\mathbf{y} = \mathbf{0}$.

By definition of matrix multiplication, the set

$$M = \{\mathbf{y} \in R^n \colon A\mathbf{y} = \mathbf{0}\}$$

of all solutions of the associated homogeneous system (2) has the following alternate description:

$$M = \{\mathbf{y} \in R^n \colon \mathbf{a}_{(i)} \cdot \mathbf{y} = 0 \text{ for } i = 1, \ldots, m\},$$

where $\mathbf{a}_{(i)}$ is the ith row vector of the matrix A. But $\mathbf{a}_{(i)}$ is the ith column vector of the transpose A^{T} of A, so we conclude that

$$M = \mathscr{R}(A^{\mathrm{T}})^{\perp},$$

that is, the set of all solutions of $A\mathbf{y} = \mathbf{0}$ is simply the orthogonal complement of the range $\mathscr{R}(A^{\mathrm{T}})$ of A^{T}.

Suppose that the row vectors $\mathbf{a}_{(1)}, \ldots, \mathbf{a}_{(m)}$ of A are linearly independent. (This is not a severe restriction on the system of linear equations $A\mathbf{x} = \mathbf{b}$; it simply requires that there are no redundant equations in the system.) Then the column vectors of A^{T} are linearly independent. Consequently, since the minimum norm solution

$$\mathbf{x}^* = \mathbf{x}^{(0)} - \mathbf{y}^*$$

is characterized by the fact that

$$\|\mathbf{x}^*\| = \|\mathbf{x}^{(0)} - \mathbf{y}^*\| \le \|\mathbf{x}^{(0)} - \mathbf{y}\| \tag{3}$$

for all $\mathbf{y} \in M$, it follows from (4.2.4) that $\mathbf{x}^*$ is the minimum norm solution of $A\mathbf{x} = \mathbf{b}$ if and only if $\mathbf{x}^* = \mathbf{x}^{(0)} - \mathbf{y}^* \in M^{\perp}$. But we observed above that

$M = \mathscr{R}(A^T)^\perp$, and $M^\perp = (\mathscr{R}(A^T)^\perp)^\perp = \mathscr{R}(A^T)$ by (4.2.7). Consequently, we have established the following result.

(4.3.2) Theorem. *Suppose that A is an $m \times n$-matrix with linearly independent row vectors, that $\mathbf{b} \in R^m$ and that the system $A\mathbf{x} = \mathbf{b}$ is underdetermined. Then a solution $\mathbf{x}^*$ of $A\mathbf{x} = \mathbf{b}$ is the minimum norm solution of this system if and only if $\mathbf{x}^* \in \mathscr{R}(A^T)$.*

Although (4.3.2) provides a clear description of the solution $\mathbf{x}^*$ of minimum norm for an underdetermined linear system $A\mathbf{x} = \mathbf{b}$, we need to do a bit more work to obtain an effective computational procedure for $\mathbf{x}^*$. First, we note that since $\mathbf{x}^* \in \mathscr{R}(A^T)$ and the columns of A^T are the (linearly independent) rows $\mathbf{a}_{(1)}, \ldots, \mathbf{a}_{(m)}$ of A, there is a unique vector $\mathbf{w} = (w_1, w_2, \ldots, w_m)$ in R^m such that

$$\mathbf{x}^* = w_1 \mathbf{a}_{(1)} + w_2 \mathbf{a}_{(2)} + \cdots + w_m \mathbf{a}_{(m)} = A^T\mathbf{w}.$$

Second, we observe that since $\mathbf{x}^*$ is a solution to the system $A\mathbf{x} = \mathbf{b}$, we must have

$$\mathbf{b} = A\mathbf{x}^* = A(A^T\mathbf{w}) = (AA^T)\mathbf{w}.$$

Thus, we have established the following computational procedure for $\mathbf{x}^*$.

(4.3.3) The minimum norm solution $\mathbf{x}^*$ of the underdetermined system of m linear equations in n unknowns can be computed as follows:

Step 1. Solve the $m \times m$-system

$$AA^T\mathbf{w} = \mathbf{b}$$

for its unique solution $\mathbf{w}^* = (w_1^*, w_2^*, \ldots, w_m^*)$.

Step 2. Let

$$\mathbf{x}^* = w_1^*\mathbf{a}_{(1)} + w_2^*\mathbf{a}_{(2)} + \cdots + w_m^*\mathbf{a}_{(m)},$$

where $\mathbf{a}_{(i)}$ is the ith row vector of A for $i = 1, 2, \ldots, m$.

The matrix $G = AA^T$ is often called the *Gram matrix* of the row vectors of A. The Gram matrix is symmetric and, since the row vectors of A are linearly independent, G is positive definite. (See Exercise 12.)

Let us illustrate the minimum norm procedure (4.3.3) by solving the following simple geometry problem.

(4.3.4) Example. Find the point on the line L of intersection of the two planes

$$x + y + z = 1, \qquad -x - y + z = 0,$$

that is nearest the origin.

Solution. A point (x, y, z) is on L if and only if it is a solution of the underdetermined system

$$x + y + z = 1,$$
$$-x - y + z = 0.$$

Consequently, the point on L that is nearest the origin is the solution of this system of minimum norm. In the notation of (4.3.3), we see in this case that

$$A = \begin{pmatrix} 1 & 1 & 1 \\ -1 & -1 & 1 \end{pmatrix}; \qquad \mathbf{b} = \begin{pmatrix} 1 \\ 0 \end{pmatrix},$$

and so the Gram matrix G is given by

$$G = AA^{\mathrm{T}} = \begin{pmatrix} 3 & -1 \\ -1 & 3 \end{pmatrix}.$$

The system $AA^{\mathrm{T}}\mathbf{w} = \mathbf{b}$ has the unique solution

$$w_1 = \tfrac{3}{8}, \qquad w_2 = \tfrac{1}{8},$$

so the minimum norm solution (x^*, y^*, z^*) of the system $A\mathbf{x} = \mathbf{b}$ is given by

$$\begin{pmatrix} x^* \\ y^* \\ z^* \end{pmatrix} = \tfrac{3}{8}\begin{pmatrix} 1 \\ 1 \\ 1 \end{pmatrix} + \tfrac{1}{8}\begin{pmatrix} -1 \\ -1 \\ 1 \end{pmatrix} = \begin{pmatrix} \frac{1}{4} \\ \frac{1}{4} \\ \frac{1}{2} \end{pmatrix}.$$

Thus, the nearest point to the origin on the line L of intersection of the two planes is $(\frac{1}{4}, \frac{1}{4}, \frac{1}{2})$.

4.4. Generalized Inner Products and Norms; The Portfolio Problem

Although the results of the preceding three sections were stated and proved for R^m equipped with the usual (Euclidean) norm

$$\|\mathbf{x}\| = \left(\sum_{i=1}^{m} x_i^2\right)^{1/2}$$

and inner product

$$\mathbf{x} \cdot \mathbf{y} = \sum_{i=1}^{m} x_i y_i,$$

these results are actually valid for any norm and inner product on R^m that is associated with an $m \times m$-positive definite matrix. More precisely, if H is an $m \times m$-positive definite matrix and if we define:

(1) the *H-inner product* $\mathbf{x} \cdot_H \mathbf{y}$ of $\mathbf{x}$ and $\mathbf{y}$ in R^m by

$$\mathbf{x} \cdot_H \mathbf{y} = \mathbf{x} \cdot H\mathbf{y};$$

(2) the *H-norm* $\|\mathbf{x}\|_H$ of $\mathbf{x} \in R^m$ by

$$\|\mathbf{x}\|_H = [\mathbf{x} \cdot_H \mathbf{x}]^{1/2} = [\mathbf{x} \cdot H\mathbf{x}]^{1/2};$$

(3) $\mathbf{x}$ and $\mathbf{y}$ to be *H-orthogonal vectors* in R^m (written $\mathbf{x} \perp_H \mathbf{y}$) if

$$0 = \mathbf{x} \cdot_H \mathbf{y} = \mathbf{x} \cdot H\mathbf{y};$$

(4) the *H-orthogonal complement* $(M)_H^\perp$ of a subspace M of R^m as

$$(M)_H^\perp = \{\mathbf{x} \in R^m : \mathbf{x} \cdot_H \mathbf{m} = 0 \text{ for all } \mathbf{m} \in M\};$$

then (4.1.2), (4.2.4), (4.3.1), and (4.3.2) remain valid when the usual norm, inner product, and orthogonal complement are replaced with the H-norm, H-inner product, and H-orthogonal complement associated with an $m \times m$-matrix H which is positive definite. This is due to the fact that these results do not depend on the defining formulas for the usual norm and inner product, but rather on the fact that the usual inner product is linear and symmetric in its two variables and that the usual norm and inner product have the following properties:

(1) $\|\mathbf{x}\| \geq 0$ for all $\mathbf{x} \in R^m$;
(2) $\|\mathbf{x}\| = 0$ if and only if $\mathbf{x} = \mathbf{0}$;
(3) $\|\alpha\mathbf{x}\| = |\alpha| \, \|\mathbf{x}\|$ for all $\mathbf{x} \in R^m$ and all real numbers α;
(4) $\|\mathbf{x} + \mathbf{y}\| \leq \|\mathbf{x}\| + \|\mathbf{y}\|$ for all $\mathbf{x}$, $\mathbf{y}$ in R^m;
(5) $|\mathbf{x} \cdot \mathbf{y}| \leq \|\mathbf{x}\| \, \|\mathbf{y}\|$ for all $\mathbf{x}$, $\mathbf{y}$ in R^m.

These properties are valid for any H-norm. For example, (1) and (2) are immediate consequence of the fact that H is positive definite while (3) follows from the computation

$$\|\alpha\mathbf{x}\|_H^2 = (\alpha\mathbf{x}) \cdot H(\alpha\mathbf{x}) = \alpha^2 \mathbf{x} \cdot H\mathbf{x} = \alpha^2 \|\mathbf{x}\|_H^2.$$

A good exercise is to verify that properties (4) and (5) hold for any H-inner product and the associated H-norm (see Exercise 18).

Given that (4.1.2) remains valid when the Euclidean inner product is replaced with the H-norm and H-inner product associated with a positive definite matrix H, it is an easy matter to formulate the least squares criterion corresponding to (4.2.4) and the corresponding generalized inverse formula.

(4.4.1) Theorem. *Suppose that H is an $m \times m$-matrix that is positive definite, that A is an $m \times n$-matrix with linearly independent columns and that $\mathbf{b} \in R^m$. Then $\mathbf{x}^* \in R^n$ minimizes $\|A\mathbf{x} - \mathbf{b}\|_H$ over all $\mathbf{x} \in R^n$ if and only if*

$$A^T H A \mathbf{x}^* = A^T H \mathbf{b}.$$

Also,

$$\mathbf{x}^* = (A^\dagger)_H H \mathbf{b},$$

where $(A^\dagger)_H$ *is the H-generalized inverse of A defined by*

$$(A^\dagger)_H = (A^T H A)^{-1} A^T,$$

Theorem (4.3.2) remains valid when the orthogonal complement is replaced by the H-orthogonal complement. It follows that the minimum norm criterion (4.3.2) has the following generalization to H-norms.

(4.4.2) Theorem. *Suppose that H is an $n \times n$-matrix that is positive definite, that A is an $m \times n$-matrix and that $\mathbf{b}$ is a vector in R^m for which $A\mathbf{x} = \mathbf{b}$ is an underdetermined system. Then the minimum H-norm solution $\mathbf{x}^*$ of $A\mathbf{x} = \mathbf{b}$ is the solution of $A\mathbf{x} = \mathbf{b}$ that lies in the range $\mathcal{R}(H^{-1}A^T)$ of $H^{-1}A^T$.*

The procedure (4.3.3) for computing the minimum norm solution of $A\mathbf{x} = \mathbf{b}$ can now be generalized to H-norms on the basis of (4.4.2).

(4.4.3) Suppose that H is an $n \times n$-matrix that is positive definite, that A is an $m \times n$-matrix with linearly independent row vectors $\mathbf{a}_{(1)}, \ldots, \mathbf{a}_{(m)}$, and that $\mathbf{b}$ is a vector in R^m for which $A\mathbf{x} = \mathbf{b}$ is an infinitely many solutions. Then the solution $\mathbf{x}^*$ of $A\mathbf{x} = \mathbf{b}$ of minimum H-norm is

$$\mathbf{x}^* = H^{-1}(w_1 \mathbf{a}_{(1)} + \cdots + w_m \mathbf{a}_{(m)}),$$

where $\mathbf{w} = (w_1, \ldots, w_m)$ is the unique solution of the system

$$\mathbf{b} = AH^{-1}A^T\mathbf{w}.$$

The following example provides a simple illustration of the procedure described in (4.4.3).

(4.4.4) Example. Minimize

$$f(x, y) = 5x^2 + 4xy + y^2,$$

subject to

$$3x + 2y = 5.$$

SOLUTION. In this case, the function $f(x, y)$ is the square of the H-norm associated with the positive definite matrix

$$H = \begin{pmatrix} 5 & 2 \\ 2 & 1 \end{pmatrix},$$

and $A = [3, 2]$, $\mathbf{b} = (5)$. The matrix $AH^{-1}A^T$ reduces to the scalar (5) so $\mathbf{w} = (1)$ is the unique solution of the system

$$\mathbf{b} = AH^{-1}A^T\mathbf{w}.$$

Since

$$H^{-1} = \begin{pmatrix} 1 & -2 \\ -2 & 5 \end{pmatrix}$$

it follows that

$$\mathbf{x}^* = H^{-1}A^{\mathrm{T}}\mathbf{w} = \begin{pmatrix} 1 & -2 \\ -2 & 5 \end{pmatrix}\begin{pmatrix} 3 \\ 2 \end{pmatrix} = \begin{pmatrix} -1 \\ 4 \end{pmatrix}$$

is the minimum H-norm solution of $A\mathbf{x} = \mathbf{b}$ and hence $x = -1, y = 4$ provides the desired minimum of $f(x, y)$.

The following example provides a more substantial illustration of the application of minimum H-norm solutions of underdetermined linear systems.

(4.4.5) Example (Portfolio Management). A young urban professional, Jodi Hardy, has C dollars to invest in the stock market. After some consultation with a stockbroker, she selects n stocks $S_1, \ldots, S_n$ for investment consideration. On the basis of historical data for these stocks, she estimates the expected annual return r_i and the variance s_{ii} on each dollar invested in stock S_i as well as the covariance s_{ij} of the returns r_i, r_j. Given this information, she could maximize the expected return by simply investing all of the money in the stock with the largest rate of return. However, this strategy would disregard market risk. A more conservative investment strategy would be to invest in several stocks with the objective of achieving a total return of at least R dollars per year while minimizing the total variance of her entire stock portfolio. We will now develop a mathematical model to implememt this latter investment strategy.

Suppose that x_i is the number of dollars Jodi invests in the stock S_i for $i = 1, \ldots, n$. Then since she intends to invest a total of C dollars in her entire stock portfolio, it follows that

$$x_1 + x_2 + \cdots + x_n = C.$$

The return on x_i dollars invested in stock S_i is $x_i r_i$ so that the requirement that the total return on the entire stock portfolio should be at least R dollars is expressed by the inequality

$$x_1 r_1 + x_2 r_2 + \cdots + x_n r_n \geq R.$$

The variance V of the total return on the stock portfolio can be expressed in terms of the variances s_{ii} and covariances s_{ij} of the rates of return by

$$V = \sum_{i=1}^{n} \sum_{j=1}^{n} s_{ij} x_i x_j.$$

Consequently, the conservative portfolio management strategy described above can be formulated as follows:

$$\text{Minimize} \quad V = \sum_{i=1}^{n} \sum_{j=1}^{n} s_{ij} x_i x_j,$$

subject to the constraints

$$\sum_{i=1}^{n} x_i = C, \tag{1}$$

$$\sum_{i=1}^{n} x_i r_i \geq R, \tag{2}$$

$$x_i \geq 0 \quad \text{for } i = 1, 2, \ldots, n. \tag{3}$$

Stated in this form, we see that Jodi's portfolio management problem reduces to the minimization of a quadratic function of n variables subject to linear equality and inequality constraints and nonnegativity constraints on the variables. As such, the problem is ideally suited to the application of a special "quadratic programming" technique that we will discuss in detail in Chapter 7.

The technique developed in this section does permit us to solve the portfolio management problem in the special case when the desired total return is specified to be some attainable value R. In this case, the problem assumes the form

$$\begin{aligned} &\text{Minimize} \quad V = \mathbf{x} \cdot H\mathbf{x}, \\ &\text{subject to} \quad A\mathbf{x} = \mathbf{b}, \end{aligned}$$

where H is the covariance matrix for $r_1, \ldots, r_n$ and

$$A = \begin{pmatrix} 1 & 1 & \cdots & 1 \\ r_1 & r_2 & \cdots & r_n \end{pmatrix}, \qquad \mathbf{b} = \begin{pmatrix} C \\ R \end{pmatrix}.$$

If the joint probability distribution of $r_1, \ldots, r_n$ is nondegenerate, H is positive definite and so the total variance of the portfolio is the square of the associated H-norm. Consequently, if the expected returns and the variances and co-variances of these returns are specified, we can apply (4.4.3) to solve the corresponding portfolio problem.

Exercises

1. Find the least squares solution of the inconsistent linear system

$$\begin{aligned} x_1 + x_2 + x_3 &= 3, \\ x_3 &= 1, \\ x_1 + x_3 &= 2, \\ 2x_1 + 5x_3 &= 8, \\ -7x_1 + 8x_2 &= 0, \\ x_1 + 2x_2 - x_3 &= 1. \end{aligned}$$

2. Find the minimum norm solution of the underdetermined linear system

$$2x_1 + x_2 + x_3 + 5x_4 = 8,$$

$$-x_1 - x_2 + 3x_3 + 2x_4 = 0.$$

3. A surveyor, Cletus Hawkeye, of the Rezek Engineering Co. is assigned the task of determining the heights above sea level of three hills, H_1, H_2, H_3. He stands at sea level and measures the heights of H_1, H_2, H_3 as 1236 ft., 1941 ft., 2417 ft., respectively. Then to check his work the surveyor climbs Hill H_1 and measures the height of H_2 above H_1 as 711 ft. and the height of H_3 above H_1 as 1177 ft. After noting that the latter measurements are not consistent with those made at sea level, he utters a mild expletive and climbs Hill H_2 and measures the height of H_3 to be 475 ft. above H_2. Again, he notes the inconsistency of this measurement with those made earlier. Cletus knows that his boss, Joe Rezek, is a perfectionist who is certain to be displeased with a report containing inconsistent data so he worries about preparing his report as he drives back to the office. Suddenly he remembers the good old days in Math 384 and Mr. Rezek's fondness for things mathematical (Why, some of his best friends are mathematicians!), so he decides to compute the least squares estimates based on his data on the heights above sea level of H_1, H_2, H_3 and enter these in his report. Compute these estimates for him so that he can keep both hands on the steering wheel.

4. Find the vector $\mathbf{v}$ in R^3 of the form

$$\mathbf{v} = \alpha(1, 1, 2) + \beta(2, -1, 1)$$

that is closest to (1, 1, 1).

5. Find the minimum norm solution of the system

$$x_1 + x_2 + 5x_3 - 7x_4 = 1,$$

$$x_1 - 3x_2 - x_3 + x_4 = 2.$$

6. Find the point on the plane

$$x + 2y + 3z = 6$$

that is closest to the origin in R^3.

7. Compute the equation of the linear regression line corresponding to the data in the table below:

x	-2	-1	0	1	2	3
y	12	11	8	5	2	-3

8. (a) Suppose that the variable y is known to be a quadratic function of the variable x; that is,

$$y = ax^2 + bx + c,$$

but that the coefficients a, b, c are not known. Estimates of the coefficients a, b, c might be obtained by conducting an experiment in which values $y_1, \ldots, y_m$ of the variable y are measured for corresponding values $x_1, \ldots, x_m$ of the variable x. Find formulas for the coefficients a, b, c of the best least squares quadratic polynomial fit

$$y = ax^2 + bx + c$$

in terms of the data: $(x_1, y_1), (x_2, y_2), \ldots, (x_m, y_m)$.

(b) Apply these formulas to the data provided in the table below:

x	-2	-1	0	1	2	3	4
y	-5	-1	4	7	6	5	-1

9. Compare the solutions of

$$\begin{pmatrix} 1 & \frac{1}{2} & \frac{1}{3} \\ \frac{1}{2} & \frac{1}{3} & \frac{1}{4} \\ \frac{1}{3} & \frac{1}{4} & \frac{1}{5} \end{pmatrix} \begin{pmatrix} x \\ y \\ z \end{pmatrix} = \begin{pmatrix} 0 \\ 0 \\ 0 \end{pmatrix},$$

and of

$$\begin{pmatrix} 1 & \frac{1}{2} & \frac{1}{3} \\ \frac{1}{2} & \frac{1}{3} & \frac{1}{4} \\ \frac{1}{3} & \frac{1}{4} & \frac{1}{5} \end{pmatrix} \begin{pmatrix} x \\ y \\ z \end{pmatrix} = \begin{pmatrix} 0 \\ 0 \\ 0.01 \end{pmatrix}.$$

The matrix $\begin{pmatrix} 1 & \frac{1}{2} & \frac{1}{3} \\ \frac{1}{2} & \frac{1}{3} & \frac{1}{4} \\ \frac{1}{3} & \frac{1}{4} & \frac{1}{5} \end{pmatrix}$ is called the *Hilbert matrix* and is a prime example of an ill-conditioned matrix. The answers to the above computations highlight the extreme sensitivity of some matrices to round-off and other errors. For an in-depth look at ill-conditioned matrices, see Chapter 5 of *Introduction of Matrix Computations* by G. M. Stewart (Academic Press, New York, 1973).

10. (a) Find the linear regression line for the three data points

$$(-1, 2),\ (0, 1),\ (1, 0).$$

(b) Find the linear regression line for the four data points

$$(-1, 2),\ (0, 1),\ (1, 0),\ (7, 8).$$

Graph both lines on the same set of axes and comment.

11. (a) Compute the generalized inverse $A^\dagger$ of

$$A = \begin{pmatrix} 0 & 1 \\ 1 & 2 \\ 2 & 3 \end{pmatrix}.$$

(b) Compute $P_{\mathscr{R}(A)}$.

(c) Use the Gram–Schmidt Process to find $\mathscr{R}(A)^\perp$ and then compute $P_{\mathscr{R}(A)^\perp}$.

(d) Show that $P_{\mathscr{R}(A)} = I - P_{\mathscr{R}(A)^\perp}$.

12. Suppose that $\mathbf{x}^{(1)}, \ldots, \mathbf{x}^{(n)}$ are vectors in R^m and that G is the Gram matrix of the vectors $\{\mathbf{x}^{(1)}, \ldots, \mathbf{x}^{(n)}\}$, that is, $G = \{g_{ij}\}$ where

$$g_{ij} = \mathbf{x}^{(i)} \cdot \mathbf{x}^{(j)}$$

for $i, j = 1, \ldots, n$. Show that the set of vectors $\{\mathbf{x}^{(1)}, \ldots, \mathbf{x}^{(n)}\}$ is linearly independent if and only if G is invertible.

13. Let A be a matrix whose columns are linearly independent. Prove:
    (i) $AA^{\dagger}A = A$;
    (ii) $A^{\dagger}A = (A^{\dagger}A)^{\dagger}$;
    (iii) $P_{\mathscr{R}(A)}$ is symmetric;
    (iv) $P^2_{\mathscr{R}(A)} = P_{\mathscr{R}(A)}$.

14. Let A be a $m \times n$-matrix whose columns $\mathbf{a}^{(1)}, \mathbf{a}^{(2)}, \ldots, \mathbf{a}^{(n)}$ are orthonormal. Show that $A^{\dagger} = A^{\mathrm{T}}$ and $P_{\mathscr{R}(A)}\mathbf{x} = \sum_{j=1}^{n} (\mathbf{a}^{(j)} \cdot \mathbf{x})\mathbf{a}^{(j)}$ for all $\mathbf{x}$ in R^n.

15. Let
$$A_1 = \begin{pmatrix} 1.000 & 1.001 \\ 1.000 & 1.000 \\ 1.000 & 1.000 \end{pmatrix} \quad \text{and} \quad A_2 = \begin{pmatrix} 1.000 & 1.001 \\ 1.000 & 1.000 \\ 1.000 & 1.001 \end{pmatrix}.$$
Compute $P_{\mathscr{R}(A_1)}$ and $P_{\mathscr{R}(A_2)}$. It might be helpful to find $P_{\mathscr{R}(A_i)^{\perp}}$ and use the result of Exercise 16.

16. Let M be a subspace of R^n and let $M^{\perp}$ be its orthogonal complement.
    (a) Show that $P_{M^{\perp}} = I - P_M$ (here I is the identity matrix).
    (b) Show that
$$\mathbf{x} = P_M(\mathbf{x}) + P_{M^{\perp}}(\mathbf{x}),$$
and
$$\|\mathbf{x}\|^2 = \|P_M\mathbf{x}\|^2 + \|P_{M^{\perp}}\mathbf{x}\|^2$$
for all $\mathbf{x}$ in R^n.
    (c) Show that if $\mathbf{x} \in R^n$, then $\mathbf{x} \cdot P_M\mathbf{x} > 0$ unless $P_M(\mathbf{x}) = \mathbf{0}$.

17. (a) Let M be a subspace of R^n. Prove that the orthogonal complement $M^{\perp}$ of M is closed.
    (b) Use the fact that $M = (M^{\perp})^{\perp}$ to prove M is closed.

18. Suppose that H is an $n \times n$-matrix which is positive definite and that $\mathbf{x} \cdot_H \mathbf{y}$ and $\|\mathbf{x}\|_H$ are associated H-inner product and H-norm on R^n.
    (a) Verify that $|\mathbf{x} \cdot_H \mathbf{y}| \le \|\mathbf{x}\|_H \|\mathbf{y}\|_H$ for all $\mathbf{x}$, $\mathbf{y}$ in R^n (the Cauchy–Schwarz Inequality). (Hint: The quantity $(\mathbf{x} - \lambda\mathbf{y}) \cdot_H (\mathbf{x} - \lambda\mathbf{y})$ expands to a quadratic function of the real variable λ and this function is nonnegative for all real λ. What does this say about the discriminant of this quadratic function?)
    (b) Verify that $\|\mathbf{x} + \mathbf{y}\|_H \le \|\mathbf{x}\|_H + \|\mathbf{y}\|_H$ for all $\mathbf{x}$, $\mathbf{y}$ in R^n (the Triangle Inequality.) (Hint: Expand $\|\mathbf{x} + \mathbf{y}\|_H^2 = (\mathbf{x} + \mathbf{y}) \cdot_H (\mathbf{x} + \mathbf{y})$ and apply the Cauchy–Schwarz Inequality.)
    (c) Verify that equality holds in the Cauchy–Schwarz Inequality in (a) if and only if $\mathbf{x} = \lambda\mathbf{y}$ or $\mathbf{y} = \mathbf{0}$.

19. (a) Let $f(\mathbf{x})$ be a function on R^n with continuous first partial derivatives and let M be a subspace of R^n. Suppose $\mathbf{x}^* \in M$ minimizes $f(\mathbf{x})$ on M. Show $\nabla f(\mathbf{x}^*) \in M^{\perp}$. (Hint: Take any $\mathbf{x} \in M$ and consider $\phi(t) = f(\mathbf{x}^* + t\mathbf{x})$.)
    (b) If, in addition, $f(\mathbf{x})$ is convex, then show that any $\mathbf{x}^* \in M$ such that $\nabla f(\mathbf{x}^*) \in M^{\perp}$ is a global minimizer of $f(\mathbf{x})$ on M.

CHAPTER 5

Convex Programming and the Karush–Kuhn–Tucker Conditions

Many optimization problems of substantial practical interest involve the maximization or minimization of a function of several variables subject to one or more constraints. These constraints may be nonnegativity or interval restrictions on some of the variables, or they may be expressed as equations or inequalities involving functions of these variables. Such optimization problems will be referred to as *constrained optimization problems*.

Many constrained optimization problems can be expressed in the following form:

$$\text{Minimize} \quad f(\mathbf{x}) \quad \text{subject to}$$

$$g_1(\mathbf{x}) \le b_1, \quad g_2(\mathbf{x}) \le b_2, \ldots, \quad g_m(\mathbf{x}) \le b_m,$$

where $f(\mathbf{x})$, $g_1(\mathbf{x})$, $\ldots$, $g_m(\mathbf{x})$ are convex functions on R^n and $b_1, \ldots, b_m$ are fixed constants. Any such problem is called a *convex program*. All linear programming problems and the least squares problems considered in the last chapter either are or can be reformulated as convex programs.

The key to understanding convex programs is the Karush–Kuhn–Tucker Theorem. This result associates with a given convex program a system of algebraic equations and inequalities that often can be used to develop effective procedures for computing minimizers, and also can be used to obtain additional information about the sensitivity of the minimum value of the program to changes in the constraints.

There are at least three possible approaches to the development of the Karush–Kuhn–Tucker Theorem. One is based on separation and support theorems for convex sets, another on the use of penalty functions, and a third parallels the classical theory of Lagrange multipliers. Each of these approaches provides its own special insights concerning this important result so that it will be worth our while to consider all three as the story in this book unfolds.

In this text, we chose to begin with the development of the Karush–Kuhn–Tucker Theorem based on separation and support theorems for convex sets because it is the most geometric and intuitive of the three and therefore, the best place to start. In Chapter 6, we will reconsider the Karush–Kuhn–Tucker Theorem from the point of view of penalty functions and in Chapter 7 we will look at the result again within the context of the classical theory of Lagrange multipliers. Both of these alternative approaches will yield a rich harvest of additional insights and results.

The first section of this chapter provides the geometric tools for the development and understanding of the Karush–Kuhn–Tucker Theorem. This result is formulated and proved in Section 5.2. The remaining three sections of the chapter consider several important applications of this result including dual programs, quadratic programming, and constrained geometric programming.

5.1. Separation and Support Theorems for Convex Sets

The content of the two main results in this section on convex sets in R^n is described in the diagrams in R^3 that are given below:

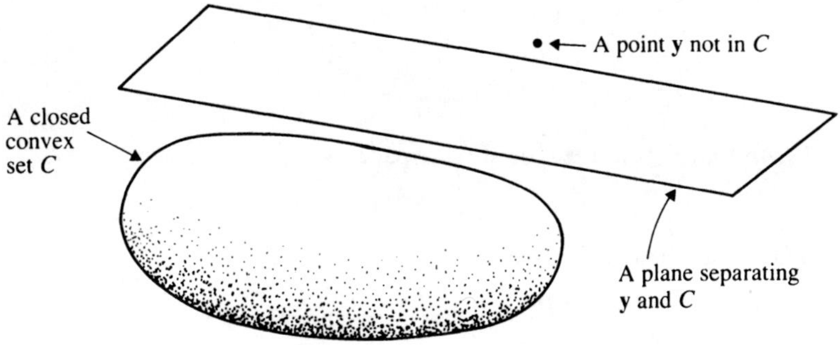

The Basic Separation Theorem

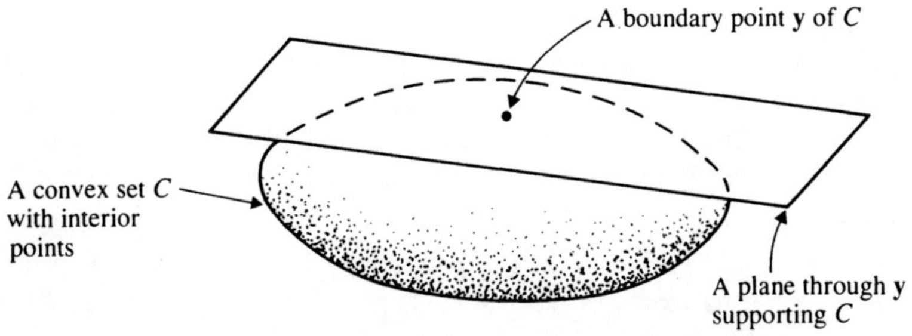

The Support Theorem

Any plane H in R^3 can be written as

$$H = \{\mathbf{x} \in R^3 \colon \mathbf{a} \cdot \mathbf{x} = \alpha\}$$

for a suitable choice of nonzero $\mathbf{a} \in R^3$ and $\alpha \in R$, since the vector equation $\mathbf{a} \cdot \mathbf{x} = \alpha$ simply reduces to the scalar equation

$$a_1 x_1 + a_2 x_2 + a_3 x_3 = \alpha$$

in that case. More generally, a *hyperplane*[1] H in R^n is any set of the form

$$H = \{\mathbf{x} \in R^n \colon \mathbf{a} \cdot \mathbf{x} = \alpha\}$$

for fixed $\mathbf{0} \neq \mathbf{a} \in R^n$, $\alpha \in R$. Note that a hyperplane in R^3 is a plane, in R^2 it is a line, and in R^1 it is a point.

The Basic Separation Theorem says that if C is a closed convex set in R^n and if $\mathbf{y} \in R^n$ does not belong to C, there is a hyperplane

$$H = \{\mathbf{x} \in R^n \colon \mathbf{a} \cdot \mathbf{x} = \alpha\}$$

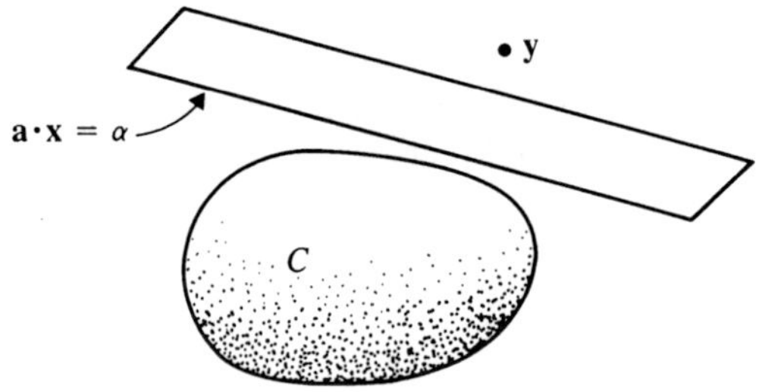

in R^n such that C is in the closed half-space

$$H^- = \{\mathbf{x} \in R^n \colon \mathbf{a} \cdot \mathbf{x} \leq \alpha\},$$

and $\mathbf{y}$ is in the open half-space

$$H^+ = \{\mathbf{x} \in R^n \colon \mathbf{a} \cdot \mathbf{x} > \alpha\}$$

determined by $\mathbf{a}$ and α (see (2.1.2)(d)); that is, there exist $\mathbf{a} \in R^n$, $\alpha \in R$ such that

$$\mathbf{a} \cdot \mathbf{x} \leq \alpha < \mathbf{a} \cdot \mathbf{y}$$

for all $\mathbf{x} \in C$.

A point $\mathbf{z}$ is a *boundary point* of a set C in R^n if for each $\varepsilon > 0$, the ball $B(\mathbf{z}, \varepsilon)$ centered at $\mathbf{z}$ of radius ε contains a point of C as well as a point that is not in C. The Support Theorem guarantees that if C is a convex set with interior points in R^n and if $\mathbf{z}$ is a boundary point of C, then there exists a hyperplane

$$H = \{\mathbf{x} \in R^n \colon \mathbf{a} \cdot \mathbf{x} = \alpha\}$$

[1] Although the term hyperplane sounds like a word borrowed from science fiction, precisely the opposite is true—science fiction writers borrowed it from mathematics!

that contains $\mathbf{z}$ such that C is contained in the closed half-space

$$F^- = \{\mathbf{x} \in R^n: \mathbf{a} \cdot \mathbf{x} \leq \alpha\}$$

determined by $\mathbf{a}$ and α, that is, there exist $\mathbf{a} \neq \mathbf{0}$ and $\alpha \in R$ such that

$$\mathbf{a} \cdot \mathbf{x} \leq \alpha = \mathbf{a} \cdot \mathbf{z}$$

for all $\mathbf{x} \in C$.

In R^n, the Support Theorem remains valid even if the hypothesis that the given convex set has interior points is dropped. However, the result as stated is adequate for the applications we have in mind so we will not pursue full generality here.

Both the Basic Separation Theorem and the Support Theorem are easy to believe because they square so well with our geometric intuition. Formal proofs of these results will require a substantial effort because the intuitive geometric pictures must be translated into statements of an algebraic or analytic nature. We shall now proceed to these derivations.

On the first time through this material, the reader may find it helpful to concentrate on the meaning and application of the main results and to postpone a detailed study of the proofs until a second reading of the chapter.

The first step is to extend to convex sets our earlier characterization (see (4.2.4)) of closest vectors to subspaces. The intuitive idea is very simple. Let C be a convex set in R^n and let $\mathbf{y} \in R^n$ be a vector that is not in C. Observe that $\mathbf{x}^* \in C$ should be the closest vector in C to $\mathbf{y}$ if and only if the angle $\theta_{\mathbf{x}}$ in the following figure

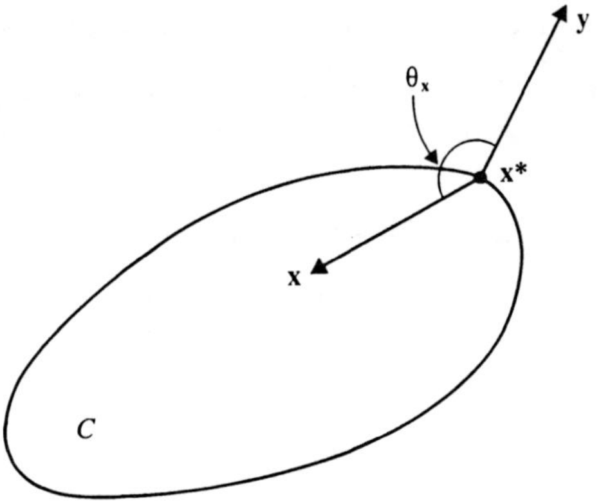

satisfies $\pi/2 \leq \theta_{\mathbf{x}} \leq \pi$ for all $\mathbf{x} \in C$. The requirement that $\pi/2 \leq \theta_{\mathbf{x}} \leq \pi$ is the same as $\cos(\theta_{\mathbf{x}}) \leq 0$ and this, in turn, is equivalent to

$$(\mathbf{y} - \mathbf{x}^*) \cdot (\mathbf{x} - \mathbf{x}^*) = \|\mathbf{y} - \mathbf{x}^*\| \, \|\mathbf{x} - \mathbf{x}^*\| \cos \theta_{\mathbf{x}} \leq 0.$$

Thus, our intuition tells us that $\mathbf{x}^*$ is the closest vector in C to $\mathbf{y}$ if and only if

$$(\mathbf{y} - \mathbf{x}^*) \cdot (\mathbf{x} - \mathbf{x}^*) \leq 0$$

for all $\mathbf{x} \in C$. This characterization of $\mathbf{x}^*$ does in fact hold for arbitrary convex sets in R^n. Of course, to verify this, it will be necessary for us to proceed from the intuitive discussion in R^2 that led us to a correct formulation of this characterization to a formal proof.

(5.1.1) Theorem. *Suppose that C is a convex set in R^n and that $\mathbf{y}$ is a vector in R^n that is not in C. Then $\mathbf{x}^* \in C$ is the closest vector in C to $\mathbf{y}$ (that is, $\|\mathbf{y} - \mathbf{x}^*\| \leq \|\mathbf{y} - \mathbf{x}\|$ for all $\mathbf{x} \in C$) if and only if*

$$(\mathbf{y} - \mathbf{x}^*)\cdot(\mathbf{x} - \mathbf{x}^*) \leq 0$$

for all $\mathbf{x} \in C$.

Proof. Suppose $\mathbf{x}^* \in C$ has the property that

$$(\mathbf{y} - \mathbf{x}^*)\cdot(\mathbf{x} - \mathbf{x}^*) \leq 0$$

for all $\mathbf{x} \in C$. To show that $\mathbf{x}^*$ is the unique closest vector in C to $\mathbf{y}$, it suffices to show that $\mathbf{x}^*$ is the strict global minimizer on C for the function

$$f(\mathbf{x}) = \|\mathbf{y} - \mathbf{x}\|^2 = (\mathbf{y} - \mathbf{x})\cdot(\mathbf{y} - \mathbf{x}).$$

To this end, we note that

$$\nabla f(\mathbf{x}) = -2(\mathbf{y} - \mathbf{x}),$$

$$Hf(\mathbf{x}) = 2I,$$

so $f(\mathbf{x})$ is strictly convex. Consequently, if $\mathbf{x} \in C$ and $\mathbf{x} \neq \mathbf{x}^*$, then

$$f(\mathbf{x}) > f(\mathbf{x}^*) + \nabla f(\mathbf{x}^*)\cdot(\mathbf{x} - \mathbf{x}^*) = f(\mathbf{x}^*) - 2(\mathbf{y} - \mathbf{x}^*)\cdot(\mathbf{x} - \mathbf{x}^*) \geq f(\mathbf{x}^*),$$

since $(\mathbf{y} - \mathbf{x}^*)\cdot(\mathbf{x} - \mathbf{x}^*) \leq 0$. Hence, $\mathbf{x}^*$ is the closest vector in C to $\mathbf{y}$.

Conversely, suppose that $\mathbf{x}^*$ is the closest vector in C to $\mathbf{y}$. For a given $\mathbf{x} \in C$, define a function $\varphi(t)$ for $0 \leq t \leq 1$ by

$$\varphi(t) = \|\mathbf{y} - (\mathbf{x}^* + t[\mathbf{x} - \mathbf{x}^*])\|^2.$$

The vector $\mathbf{x}^* + t[\mathbf{x} - \mathbf{x}^*]$ is on the line segment $[\mathbf{x}^*, \mathbf{x}]$ joining $\mathbf{x}^*$ to $\mathbf{x}$ whenever $0 \leq t \leq 1$, so any such vector is in C since $\mathbf{x}^* \in C$, $\mathbf{x} \in C$, and C is convex. Since $\mathbf{x}^*$ is the closest vector in C to $\mathbf{y}$, it follows that $\varphi(0)$ is the minimum value of $\varphi(t)$ on $0 \leq t \leq 1$. Therefore, $\varphi'(0) \geq 0$ (Why?). But

$$\varphi'(t) = -2(\mathbf{y} - (\mathbf{x}^* + t(\mathbf{x} - \mathbf{x}^*)))\cdot(\mathbf{x} - \mathbf{x}^*),$$

so that

$$0 \leq \varphi'(0) = -2(\mathbf{y} - \mathbf{x}^*)\cdot(\mathbf{x} - \mathbf{x}^*).$$

This proves that $(\mathbf{y} - \mathbf{x}^*)\cdot(\mathbf{x} - \mathbf{x}^*) \leq 0$, which completes the proof.

The following corollary shows that the preceding theorem is a genuine extension of (4.2.4).

(5.1.2) Corollary. *Suppose that M is a subspace of R^n and that $\mathbf{y} \in R^n$ is a vector not in M. Then $\mathbf{x}^* \in M$ is the closest vector in M to $\mathbf{y}$ if and only if $\mathbf{y} - \mathbf{x}^* \in M^{\perp}$.*

PROOF. If we apply (5.1.1) with $C = M$, we see that $\mathbf{x}^* \in M$ is the closest vector in M to $\mathbf{y}$ if and only if

$$(\mathbf{y} - \mathbf{x}^*)\cdot(\mathbf{x} - \mathbf{x}^*) \leq 0$$

for all $\mathbf{x} \in M$. However, since M is a subspace, both $\mathbf{x} + \mathbf{x}^*$ and $-\mathbf{x} + \mathbf{x}^*$ belong to M whenever $\mathbf{x} \in M$, so the preceding inequality reduces to

$$(\mathbf{y} - \mathbf{x}^*)\cdot\mathbf{x} \leq 0; \qquad (\mathbf{y} - \mathbf{x}^*)\cdot(-\mathbf{x}) \leq 0$$

for all $\mathbf{x} \in M$. These inequalities are in turn equivalent to

$$(\mathbf{y} - \mathbf{x}^*)\cdot\mathbf{x} = 0$$

for all $\mathbf{x} \in M$, which implies the desired result.

The preceding result begs the question of whether or not the closest vector in a given convex set to a given vector in R^n necessarily exists. The next theorem tells us that a closest vector always exists if C is closed. Closest vectors need not exist for arbitrary convex sets.

(5.1.3) Theorem. *If C is a closed (convex or not) subset of R^n and if $\mathbf{y} \in R^n$ does not belong to C, then there is a vector $\mathbf{x}^* \in C$ that is closest to $\mathbf{y}$, that is,*

$$\|\mathbf{y} - \mathbf{x}^*\| \leq \|\mathbf{y} - \mathbf{x}\|$$

for all $\mathbf{x} \in C$.

PROOF. Let α be the largest number such that $\alpha \leq \|\mathbf{y} - \mathbf{x}\|$ for all $\mathbf{x} \in C$. Then there is a sequence $\{\mathbf{x}^{(k)}\}$ of elements of C such that

$$\alpha = \lim_k \|\mathbf{y} - \mathbf{x}^{(k)}\|.$$

Because the sequence $\{\|\mathbf{y} - \mathbf{x}^{(k)}\|\}$ is a convergent sequence of real numbers, it must be bounded, that is, there is a positive number M such that

$$\|\mathbf{y} - \mathbf{x}^{(k)}\| \leq M$$

for all k. But then

$$\begin{aligned}\|\mathbf{x}^{(k)}\| = \|(\mathbf{x}^{(k)} - \mathbf{y}) + \mathbf{y}\| &\leq \|\mathbf{y} - \mathbf{x}^{(k)}\| + \|\mathbf{y}\| \\ &\leq M + \|\mathbf{y}\|\end{aligned}$$

for all k. Hence $\{\mathbf{x}^{(k)}\}$ is a bounded sequence in C. The Bolzano–Weierstrass Property yields a subsequence $\{\mathbf{x}^{(k_j)}\}$ that converges to a point $\mathbf{x}^*$ and $\mathbf{x}^* \in C$ because C is closed and all terms of the subsequence belong to C. Moreover,

$$\|\mathbf{y} - \mathbf{x}^*\| = \lim_j \|\mathbf{y} - \mathbf{x}^{(k_j)}\| = \alpha$$

so $\|\mathbf{y} - \mathbf{x}^*\| = \alpha \le \|\mathbf{y} - \mathbf{x}\|$ for all $\mathbf{x} \in C$. This proves that $\mathbf{x}^*$ is a closest vector in C to $\mathbf{y}$.

(5.1.4) Corollary. *Suppose that C is a closed convex subset of R^n and that $\mathbf{y}$ is a vector in R^n that is not in C. Then there is one and only one vector $\mathbf{x}^*$ that is closest to $\mathbf{y}$ in C.*

PROOF. We have just shown in (5.1.3) that there is at least one such vector $\mathbf{x}^*$. Suppose that $\mathbf{z}^* \in C$ is also closest to $\mathbf{y}$ in C. If we apply the criterion of (5.1.1) successively to $\mathbf{x}^*$ and $\mathbf{z}^*$, we obtain the inequalities

$$0 \ge (\mathbf{y} - \mathbf{x}^*)\cdot(\mathbf{z}^* - \mathbf{x}^*) = \mathbf{y}\cdot\mathbf{z}^* - \mathbf{x}^*\cdot\mathbf{z}^* - \mathbf{y}\cdot\mathbf{x}^* + \mathbf{x}^*\cdot\mathbf{x}^*,$$

$$0 \ge (\mathbf{y} - \mathbf{z}^*)\cdot(\mathbf{x}^* - \mathbf{z}^*) = \mathbf{y}\cdot\mathbf{x}^* - \mathbf{z}^*\cdot\mathbf{x}^* - \mathbf{y}\cdot\mathbf{z}^* + \mathbf{z}^*\cdot\mathbf{z}^*.$$

If we add these inequalities, we obtain

$$0 \ge \mathbf{x}^*\cdot\mathbf{x}^* - 2\mathbf{x}^*\cdot\mathbf{z}^* + \mathbf{z}^*\cdot\mathbf{z}^* = \|\mathbf{x}^* - \mathbf{z}^*\|^2,$$

which implies that $\mathbf{x}^* = \mathbf{z}^*$. Thus, there is precisely one vector in C that is closest to $\mathbf{y}$.

(5.1.5) The Basic Separation Theorem. *Suppose that C is a closed convex set in R^n and that $\mathbf{y}$ is a vector in R^n that is not in C. Then there are a nonzero vector $\mathbf{a} \in R^n$ and a real number α such that*

$$\mathbf{a}\cdot\mathbf{x} \le \alpha < \mathbf{a}\cdot\mathbf{y}$$

for all $\mathbf{x} \in C$.

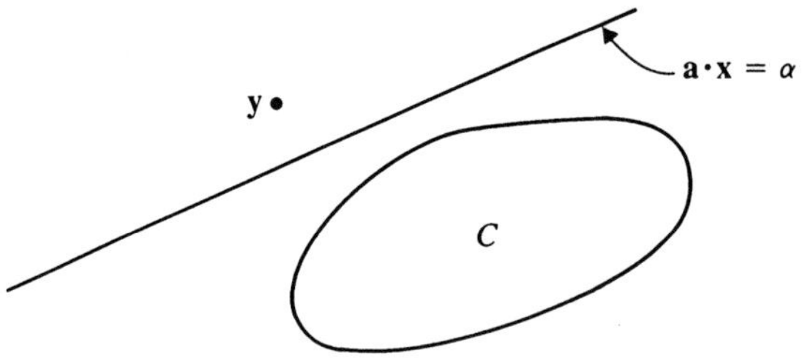

Two-Dimensional Diagram for the Basic Separation Theorem

PROOF. Let $\mathbf{x}^*$ be the closest vector in C to $\mathbf{y}$ (cf. (5.1.4)). Then, by (5.1.1),

$$(\mathbf{y} - \mathbf{x}^*)\cdot(\mathbf{x} - \mathbf{x}^*) \le 0$$

for all $\mathbf{x} \in C$. Therefore, if we set $\mathbf{a} = \mathbf{y} - \mathbf{x}^*$, then $\mathbf{a} \ne \mathbf{0}$ and

$$\mathbf{a}\cdot(\mathbf{x} - \mathbf{x}^*) \le 0$$

for all $\mathbf{x} \in C$, that is,

$$\mathbf{a}\cdot\mathbf{x} \le \mathbf{a}\cdot\mathbf{x}^*$$

for all $\mathbf{x} \in C$. Let $\alpha = \mathbf{a} \cdot \mathbf{x}^*$ and compute

$$\mathbf{a} \cdot \mathbf{y} - \alpha = \mathbf{a} \cdot (\mathbf{y} - \mathbf{x}^*) = \|\mathbf{a}\|^2 > 0.$$

Hence,

$$\mathbf{a} \cdot \mathbf{x} \leq \alpha < \mathbf{a} \cdot \mathbf{y}$$

for all $\mathbf{x} \in C$. This completes the proof.

Although the conclusion of the Basic Separation Theorem is stated in the useful form of an inequality, it is important to keep in mind the geometric meaning of this inequality: The point $\mathbf{y}$ and the closed convex set C are separated by the hyperplane

$$H = \{\mathbf{x} \in R^n : \mathbf{a} \cdot \mathbf{x} = \alpha\}$$

in the sense that C lies in the closed half-space

$$H^- = \{\mathbf{x} \in R^n : \mathbf{a} \cdot \mathbf{x} \leq \alpha\}$$

while $\mathbf{y}$ is in the opposite open half-space

$$H^+ = \{\mathbf{x} \in R^n : \mathbf{a} \cdot \mathbf{x} > \alpha\}$$

determined by this hyperplane. This geometric interpretation of the Basic Separation Theorem immediately yields the following corollary.

(5.1.6) Corollary. *A closed convex set in R^n is the intersection of all closed half-spaces containing it.*

For an arbitrary set A in R^n, the *closure* $\bar{A}$ of A is the set of all points $\mathbf{x}$ in R^n for which there is a sequence $\{\mathbf{x}^{(k)}\}$ of points of A with

$$\lim_k \|\mathbf{x}^{(k)} - \mathbf{x}\| = 0.$$

Thus, if F is a closed set in R^n, then $F = \bar{F}$, and if A is an arbitrary subset of R^n, then A is always a subset of its closure $\bar{A}$.

The next theorem shows that the closure operation on sets preserves convexity.

(5.1.7) Theorem. *If C is a convex set in R^n, then the closure $\bar{C}$ of C is also convex.*

PROOF. We must show that $\lambda\mathbf{x} + (1 - \lambda)\mathbf{y} \in \bar{C}$ for any $\mathbf{x}, \mathbf{y}$ in $\bar{C}$ and any choice of λ with $0 \leq \lambda \leq 1$. Choose $\{\mathbf{x}^{(k)}\}$, $\{\mathbf{y}^{(k)}\}$ in C so that

$$\lim_k \|\mathbf{x}^{(k)} - \mathbf{x}\| = 0, \qquad \lim_k \|\mathbf{y}^{(k)} - \mathbf{y}\| = 0.$$

Then, since

$$\begin{aligned}
&\|[\lambda \mathbf{x}^{(k)} + (1-\lambda)\mathbf{y}^{(k)}] - [\lambda \mathbf{x} + (1-\lambda)\mathbf{y}]\| \\
&= \|\lambda[\mathbf{x}^{(k)} - \mathbf{x}] + (1-\lambda)[\mathbf{y}^{(k)} - \mathbf{y}]\| \\
&\leq \lambda\|\mathbf{x}^{(k)} - \mathbf{x}\| + (1-\lambda)\|\mathbf{y}^{(k)} - \mathbf{y}\|,
\end{aligned}$$

it follows that

$$\lim_k \|[\lambda \mathbf{x}^{(k)} + (1-\lambda)\mathbf{y}^{(k)}] - [\lambda \mathbf{x} + (1-\lambda)\mathbf{y}]\| = 0,$$

so that $\lambda \mathbf{x} + (1-\lambda)\mathbf{y} \in \bar{C}$ as required.

Recall that a point $\mathbf{x}$ is an interior point of a subset A of R^n if there is an $r > 0$ such that the ball $B(\mathbf{x}, r)$ centered at $\mathbf{x}$ of radius r is contained in A. The *interior* A^0 *of* A is the set of all interior points of A. Of course, $A^0 \subset A$ but it may happen that the interior of A is empty even though A is not empty. For example, if M is a subspace of R^n and if $M \neq R^n$, then the interior M^0 of M is empty. (See Exercise 1.)

The following result, which is the technical basis for the proof of the Support Theorem, has a proof that is quite geometric and intuitive in character. As you read the computational details of the proof, you should study the accompanying diagrams to understand the motivation and intuition behind these computations.

(5.1.8) The Accessibility Lemma. *Suppose that C is a convex set with a nonempty interior in R^n. If $\mathbf{x} \in C^0$ and $\mathbf{y} \in \bar{C}$, then the "half-open" line segment from $\mathbf{x}$ to $\mathbf{y}$*

$$[\mathbf{x}, \mathbf{y}) = \{\lambda \mathbf{x} + (1-\lambda)\mathbf{y} : 0 < \lambda \leq 1\}$$

consists entirely of interior points of C.

PROOF. Suppose that $\mathbf{z}^{(0)} = \lambda_0 \mathbf{x} + (1-\lambda_0)\mathbf{y}$ is a fixed point of $[\mathbf{x}, \mathbf{y})$. We need to show that $\mathbf{z}^{(0)} \in C^0$. If $\lambda_0 = 1$, then $\mathbf{z}^{(0)} = \mathbf{x} \in C^0$ so we can restrict our attention to the case when $0 < \lambda_0 < 1$.

Since $\mathbf{x}$ is an interior point of C, there is an r_0 such that the ball $B(\mathbf{x}, r_0)$ centered at $\mathbf{x}$ of radius r_0 is contained in C. Now the function φ defined on R^n by

$$\varphi(\mathbf{w}) = \frac{1}{1-\lambda_0}[\mathbf{z}^{(0)} - \lambda_0 \mathbf{w}]$$

has the property that

$$\|\varphi(\mathbf{w}) - \varphi(\mathbf{x})\| = \frac{\lambda_0}{1-\lambda_0}\|\mathbf{w} - \mathbf{x}\|$$

and that $\varphi(\mathbf{x}) = \mathbf{y}$. Therefore, φ establishes a one-to-one correspondence between the points of $B(\mathbf{x}, r_0)$ and those of $B(\mathbf{y}, [\lambda_0/(1-\lambda_0)]r_0)$. Since $\mathbf{y}$ is in the closure of C, the ball $B(\mathbf{y}, [\lambda_0/(1-\lambda_0)]r_0)$ must intersect C; in particular, there must be a $\mathbf{w} \in B(\mathbf{x}, r_0)$ such that $\varphi(\mathbf{w}) \in C \cap B(\mathbf{y}, [\lambda_0/(1-\lambda_0)]r_0)$. Note that

$$z^{(0)} = \lambda_0 \mathbf{w} + (1-\lambda_0)\varphi(\mathbf{w}).$$

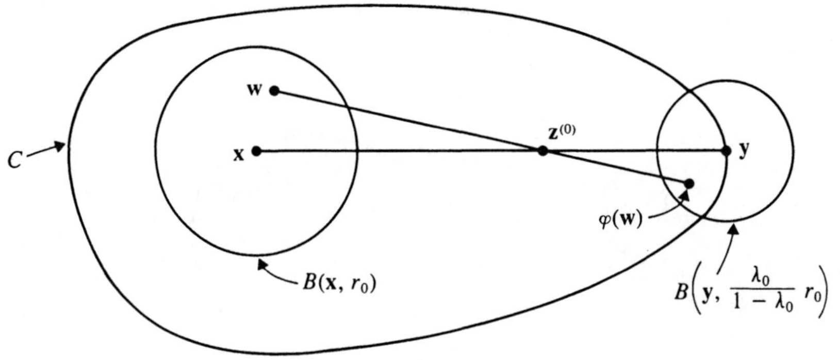

The function ψ defined on R^n by

$$\psi(\mathbf{u}) = \lambda_0 \mathbf{u} + (1-\lambda_0)\varphi(\mathbf{w})$$

has the property that

$$(*) \qquad \|\psi(\mathbf{u}) - \psi(\mathbf{w})\| = \lambda_0 \|\mathbf{u} - \mathbf{w}\|$$

and that $\psi(\mathbf{w}) = \mathbf{z}^{(0)}$. Since $\mathbf{w} \in B(\mathbf{x}, r_0)$, there is an $r_1 > 0$ such that

$$B(\mathbf{w}, r_1) \subset B(\mathbf{x}, r_0) \subset C.$$

Equation $(*)$ implies that ψ establishes a one-to-one correspondence between the points of $B(\mathbf{w}, r_1)$ and those of $B(\mathbf{z}^{(0)}, \lambda_0 r_1)$. But if $\mathbf{u} \in B(\mathbf{w}, r_1)$ then $\psi(\mathbf{u}) \in C$ since $\mathbf{u} \in C$, $\varphi(\mathbf{w}) \in C$, and $0 < \lambda_0 < 1$. Therefore, $B(\mathbf{z}^{(0)}, \lambda_0 r_1) \subset C$ so that $\mathbf{z}^{(0)} \in C^0$.

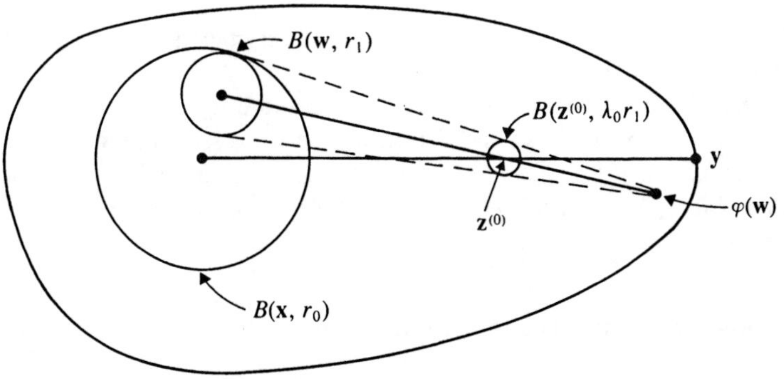

This completes the proof.

We now have the tools available to establish the Support Theorem.

(5.1.9) Support Theorem. *Suppose that C is a convex set with interior points in R^n and that $\mathbf{z}$ is a boundary point of C. Then there is a nonzero vector $\mathbf{a} \in R^n$ such that*

$$\mathbf{a} \cdot \mathbf{x} \leq \mathbf{a} \cdot \mathbf{z}$$

for all $\mathbf{x} \in C$.

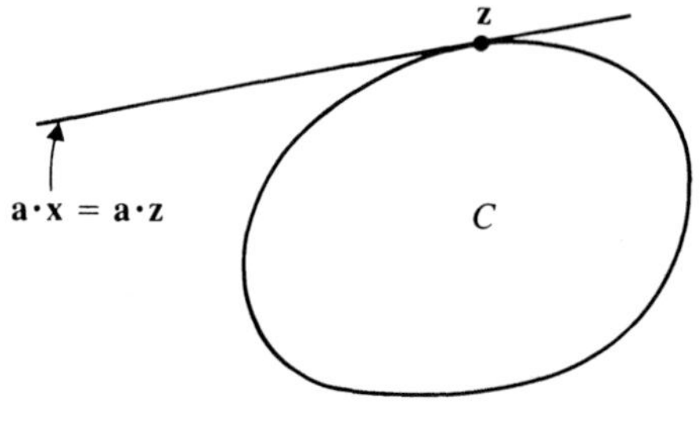

Two-Dimensional Diagram of the Support Theorem

PROOF. Let $\mathbf{y}$ be an interior point of C and, for each $s > 1$, let

$$\mathbf{z}_s = \mathbf{y} + s(\mathbf{z} - \mathbf{y}).$$

First note that $\mathbf{z}_s \notin \bar{C}$ for all $s > 1$. For if $\mathbf{z}_s \in \bar{C}$ then the Accessibility Lemma implies that the line segment

$$[\mathbf{y}, \mathbf{z}_s) = \{\mathbf{y} + t(\mathbf{z} - \mathbf{y}) : 0 \leq t < s\}$$

consists entirely of interior points of C; in particular, $\mathbf{z} = \mathbf{y} + (\mathbf{z} - \mathbf{y}) \in C^0$, contrary to the fact that $\mathbf{z}$ is a boundary point of C.

Since $\mathbf{z}_s \notin \bar{C}$ for each $s > 1$, the Basic Separation Theorem implies that there exist $\mathbf{b}_s \neq \mathbf{0}$ such that

$$\mathbf{b}_s \cdot \mathbf{x} < \mathbf{b}_s \cdot \mathbf{z}_s$$

for all $\mathbf{x} \in C$ and for all $s > 1$. Since this inequality persists if we replace $\mathbf{b}_s$ by $\mathbf{b}_s/\|\mathbf{b}_s\|$, we can require that $\|\mathbf{b}_s\| = 1$.

Let $\{s_k\}$ be any sequence for which $s_k > 1$ and $\lim_k s_k = 1$ and let $\mathbf{a}_k = \mathbf{b}_{s_k}$. Then $\|\mathbf{a}_k\| = 1$ for all k, so the Bolzano–Weierstrass Theorem implies that some subsequence $\{\mathbf{a}_{k_p}\}$ of $\{\mathbf{a}_k\}$ converges to some $\mathbf{a} \in R^n$; moreover, since $1 = \lim_p \|\mathbf{a}_{k_p}\| = \|\mathbf{a}\|$, it follows that $\mathbf{a} \neq \mathbf{0}$. Also, for all $\mathbf{x} \in C$,

$$\mathbf{a} \cdot \mathbf{x} = \lim_p (\mathbf{a}_{k_p} \cdot \mathbf{x}) \leq \lim_p (\mathbf{a}_{k_p} \cdot \mathbf{z}_{s_{k_p}}) = \mathbf{a} \cdot \mathbf{z}.$$

This shows that $\mathbf{a}$ has the required properties.

The next result paves the way for the Karush–Kuhn–Tucker Theorem. It is a close relative of an old friend from Chapter 2—the "tangent plane below" characterization of convex functions. More precisely, we proved in (2.3.5) that a function with continuous first partial derivatives on a convex set C in R^n is convex if and only if

$$(*) \qquad f(\mathbf{x}) \geq f(\mathbf{y}) + \nabla f(\mathbf{y}) \cdot (\mathbf{x} - \mathbf{y})$$

for all $\mathbf{x}$, $\mathbf{y}$ in C. If the differentiability hypothesis for $f(\mathbf{x})$ is dropped, the inequality $(*)$ may not make sense since the gradient $\nabla f(\mathbf{x})$ may not be defined. However, we will now show that all is not lost when this differentiability hypothesis on $f(\mathbf{x})$ is dropped provided that C has a nonempty interior.

(5.1.10) Theorem. *Suppose that $f(\mathbf{x})$ is a convex function defined on a convex set C with nonempty interior in R^n. If $\mathbf{x}^{(0)}$ is an interior point of C, there is a vector $\mathbf{d}$ in R^n such that*

$$f(\mathbf{x}) \geq f(\mathbf{x}^{(0)}) + \mathbf{d} \cdot (\mathbf{x} - \mathbf{x}^{(0)})$$

for all $\mathbf{x} \in C$.

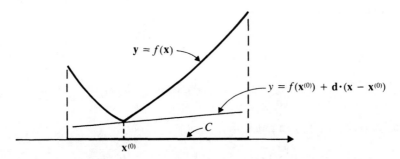

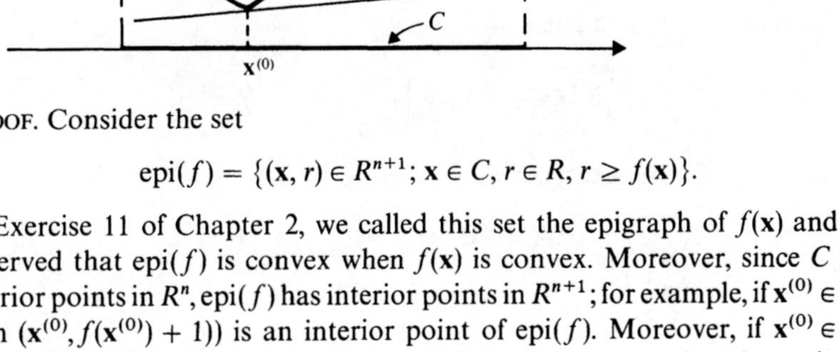

PROOF. Consider the set

$$\text{epi}(f) = \{(\mathbf{x}, r) \in R^{n+1}; \mathbf{x} \in C, r \in R, r \geq f(\mathbf{x})\}.$$

In Exercise 11 of Chapter 2, we called this set the epigraph of $f(\mathbf{x})$ and we observed that epi(f) is convex when $f(\mathbf{x})$ is convex. Moreover, since C has interior points in R^n, epi(f) has interior points in R^{n+1}; for example, if $\mathbf{x}^{(0)} \in C^0$, then $(\mathbf{x}^{(0)}, f(\mathbf{x}^{(0)}) + 1))$ is an interior point of epi(f). Moreover, if $\mathbf{x}^{(0)} \in C^0$, then $(\mathbf{x}^{(0)}, f(\mathbf{x}^{(0)}))$ clearly belongs to epi(f) but it is not an interior point of epi(f).

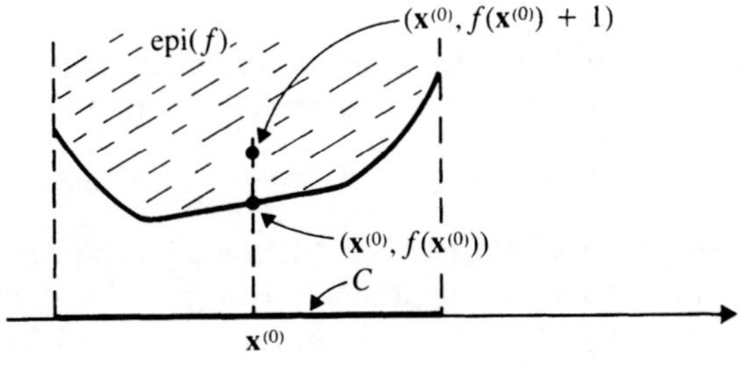

Thus epi(f) is a convex set with nonempty interior and $(\mathbf{x}^{(0)}, f(\mathbf{x}^{(0)}))$ is a boundary point of epi(f) whenever $\mathbf{x}^{(0)}$ is an interior point of C. Consequently, for a given $\mathbf{x}^{(0)} \in C^0$, we can apply the Support Theorem to obtain a vector in R^{n+1}, say $\mathbf{a} = (\mathbf{b}, c)$ with $\mathbf{b} \in R^n$, $c \in R$, such that

$$\mathbf{b}\cdot\mathbf{x} + cr = \mathbf{a}\cdot(\mathbf{x}, r) \leq \mathbf{a}\cdot(\mathbf{x}^{(0)}, f(\mathbf{x}^{(0)})) = \mathbf{b}\cdot\mathbf{x}^{(0)} + cf(\mathbf{x}^{(0)})$$

for all $(\mathbf{x}, r)$ in epi(f).

We shall now show that the component c of the vector $\mathbf{a}$ must be negative. For if $c > 0$, then $\mathbf{b}\cdot\mathbf{x} + cr$ cannot have $\mathbf{b}\cdot\mathbf{x}^{(0)} + cf(\mathbf{x}^{(0)})$ as an upper bound for all $(\mathbf{x}, r)$ in epi(f) since $(\mathbf{x}, s) \in$ epi(f) whenever $(\mathbf{x}, r) \in$ epi(f) and $s \geq r$. Also, if $c = 0$, then $\mathbf{b}\cdot\mathbf{x} \leq \mathbf{b}\cdot\mathbf{x}^{(0)}$ for all $\mathbf{x} \in C$. Moreover, $\mathbf{b} \neq \mathbf{0}$ since $\mathbf{a} = (\mathbf{b}, c) \neq \mathbf{0}$. Since $\mathbf{x}^{(0)} \in C^0$, there is a $t > 0$ such that $\mathbf{x} = \mathbf{x}^{(0)} + t\mathbf{b} \in C$. But then

$$\mathbf{b}\cdot\mathbf{x} = \mathbf{b}\cdot\mathbf{x}^{(0)} + t\|\mathbf{b}\|^2 \leq \mathbf{b}\cdot\mathbf{x}^{(0)},$$

which is impossible. This proves that $c \neq 0$ and hence $c < 0$.

Since $c < 0$ and

$$\mathbf{b}\cdot\mathbf{x} + cr \leq \mathbf{b}\cdot\mathbf{x}^{(0)} + cf(\mathbf{x}^{(0)})$$

for all $(\mathbf{x}, r) \in$ epi(f), it follows that

$$\left(\frac{1}{c}\mathbf{b}\right)\cdot\mathbf{x} + r \geq \left(\frac{1}{c}\mathbf{b}\right)\cdot\mathbf{x}^{(0)} + f(\mathbf{x}^{(0)})$$

for all $(\mathbf{x}, r) \in$ epi(f); in particular,

$$\left(\frac{1}{c}\mathbf{b}\right)\cdot\mathbf{x} + f(\mathbf{x}) \geq \left(\frac{1}{c}\mathbf{b}\right)\cdot\mathbf{x}^{(0)} + f(\mathbf{x}^{(0)})$$

for all $\mathbf{x} \in C$. Consequently, if we set $\mathbf{d} = -(1/c)\mathbf{b}$, then we obtain

$$f(\mathbf{x}) \geq f(\mathbf{x}^{(0)}) + \mathbf{d}\cdot(\mathbf{x} - \mathbf{x}^{(0)})$$

for all $\mathbf{x} \in C$, which is the desired conclusion.

The geometric content of (5.1.10) can be simply stated as follows: Even if a convex function $f(\mathbf{x})$ is not differentiable, there are planes that play the role of tangent planes to the graph of $f(\mathbf{x})$ (more precisely, tangent hyperplanes supporting epi $f(\mathbf{x})$).

A vector $\mathbf{d} \in R^n$ with the stated property of (5.1.10) is called a *subgradient of* $f(\mathbf{x})$ *at* $\mathbf{x}^{(0)}$ and the set of all subgradients of $f(\mathbf{x})$ at $\mathbf{x}^{(0)}$, that is,

$$\{\mathbf{d} \in R^n: f(\mathbf{x}) \geq f(\mathbf{x}^{(0)}) + \mathbf{d}\cdot(\mathbf{x} - \mathbf{x}^{(0)}) \text{ for all } \mathbf{x} \in C\}$$

is called the *subdifferential of* $f(\mathbf{x})$ *at* $\mathbf{x}^{(0)}$. It can be shown that the subdifferential of $f(\mathbf{x})$ at $\mathbf{x}^{(0)}$ reduces to a single vector $\mathbf{d}$ if and only if the first partial derivatives of $f(\mathbf{x})$ exist at $\mathbf{x}^{(0)}$; in this case, $\mathbf{d} = \nabla f(\mathbf{x}^{(0)})$. (See Exercise 12).

5.2. Convex Programming; The Karush–Kuhn–Tucker Theorem

One of the most highly developed areas of study in nonlinear optimization is convex programming which is concerned with the minimization of convex functions subject to inequality constraints on other convex functions. The whole area of linear programming falls under this heading, as does all of the theory of least squares optimization. The basic tool for the development of convex programming is the subgradient support theorem (5.1.10) for the epigraph of a convex function, a result which rests in turn on the Support Theorem.

Let us begin by defining the context of convex programming. Suppose that $f(\mathbf{x}), g_1(\mathbf{x}), \ldots, g_m(\mathbf{x})$ are real-valued functions defined on a subset C of R^n. We are interested in the following *program*:

$$(P) \quad \begin{cases} \text{Minimize} \quad f(\mathbf{x}) \quad \text{subject to} \\ \quad g_1(\mathbf{x}) \le 0, \quad g_2(\mathbf{x}) \le 0, \ldots, \quad g_m(\mathbf{x}) \le 0, \\ \text{where} \quad \mathbf{x} \in C \subset R^n. \end{cases}$$

The function $f(\mathbf{x})$ is called the *objective function* of (P) and the function inequalities $g_1(\mathbf{x}) \le 0, \ldots, g_m(\mathbf{x}) \le 0$ are called the *(inequality) constraints* for (P). A point $\mathbf{x} \in C$ that satisfies all of the constraints of the program (P) is called a *feasible point for* (P), and the set F of all feasible points for (P) is the *feasibility region for* (P). If the feasibility region for (P) is not empty, we say that (P) is *consistent*. If there is a feasible point $\mathbf{x}$ for (P) such that $g_i(\mathbf{x}) < 0$ for $i = 1, \ldots, m$, then (P) is *superconsistent* and the point $\mathbf{x}$ is called a *Slater point* for (P). If (P) is a consistent program and if $\mathbf{x}^*$ is a feasible point for (P) such that $f(\mathbf{x}^*) \le f(\mathbf{x})$ for all feasible points $\mathbf{x}$ for (P), then $\mathbf{x}^*$ is a *solution* for (P).

We call (P) a *convex program* if the objective function $f(\mathbf{x})$, the constraint functions $g_1(\mathbf{x}), \ldots, g_m(\mathbf{x})$, and the underlying set C are all convex. In this case, the feasibility region F for (P) is a convex set since

$$G_i = \{\mathbf{x} \in C \colon g_i(\mathbf{x}) \le 0\}$$

is easily seen to be convex for $i = 1, \ldots, m$ and

$$F = \bigcap_{i=1}^{m} G_i.$$

The following example should help to clarify these concepts.

(5.2.1) Example. Consider the program

$$(P) \quad \begin{cases} \text{Minimize} \quad f(x, y) = x^4 + y^4 \\ \text{subject to the constraints} \\ \quad x^2 - 1 \le 0, \quad y^2 - 1 \le 0, \quad e^{x+y} - 1 \le 0, \\ \text{where} \quad (x, y) \in R^2. \end{cases}$$

The feasibility region F for (P) is given by

$$F = \{(x, y) \in R^2 : x + y \leq 0, |x| \leq 1, |y| \leq 1\}$$

so (P) is consistent. Moreover, there are many Slater points for (P); in fact, any point interior to the triangular region bounded by the lines $y = -x$, $y = -1$, $x = -1$ is a Slater point. In particular, (P) is superconsistent. Since the objective and constraint functions are easily seen to be convex, (P) is a convex program. It is easy to see that $(0, 0)$ is the unique solution of (P).

To proceed with our study of convex programs, we need to explore two concepts from real analysis—the supremum and infimum of a real-valued function defined on a subset of R^n.

(5.2.2) Definitions. Suppose that $f(\mathbf{x})$ is a real-valued function defined on a subset C of R^n. If there is a smallest real number β such that $f(\mathbf{x}) \leq \beta$ for all $\mathbf{x} \in C$, then β is called *the supremum of* $f(\mathbf{x})$ *on* C and we write

$$\sup_{\mathbf{x} \in C} f(\mathbf{x}) = \beta.$$

If there is a largest real number α such that $f(\mathbf{x}) \geq \alpha$ for all $\mathbf{x} \in C$, then α is called *the infimum of* $f(\mathbf{x})$ *on* C and we write

$$\inf_{\mathbf{x} \in C} f(\mathbf{x}) = \alpha.$$

(5.2.3) Examples and Remarks

(a) Note that if $\mathbf{x}^*$ is a global maximizer of $f(\mathbf{x})$ on C, then

$$\sup_{\mathbf{x} \in C} f(\mathbf{x}) = f(\mathbf{x}^*).$$

Similarly, if $\mathbf{x}^*$ is a global minimizer of $f(\mathbf{x})$ on C, then

$$\inf_{\mathbf{x} \in C} f(\mathbf{x}) = f(\mathbf{x}^*).$$

(b) For a real number β to be the supremum of $f(\mathbf{x})$ on C, it must be true that β is an upper bound for $f(\mathbf{x})$ on C, that is,

$$f(\mathbf{x}) \leq \beta$$

for all $\mathbf{x} \in C$. Consequently, if there are no upper bounds for $f(\mathbf{x})$ on C then the supremum of $f(\mathbf{x})$ on C cannot exist. For example, if

$$C = \{(x, y) \in R^2 : 0 < x < 1, 0 < y < 1\},$$

and if

$$f(x, y) = \frac{1}{x + y},$$

then the value of $f(x, y)$ can be made as large as desired by choosing x and y

to be sufficiently small positive numbers. Thus, there are no upper bounds for $f(x, y)$ on C.

Similarly, if there are no lower bounds for $f(\mathbf{x})$ on C, then the infimum of $f(\mathbf{x})$ on C does not exist.

(c) If $f(\mathbf{x})$ is a real-valued function defined on a subset C of R^n and if there are upper bounds for $f(\mathbf{x})$ on C, then the supremum of $f(\mathbf{x})$ on C exists even though there may not be a global maximizer for $f(\mathbf{x})$ on C. For example, the function $f(x)$ defined on R^1 by

$$f(x) = \arctan x$$

has all of its values in the open interval $-\pi/2 < y < \pi/2$.

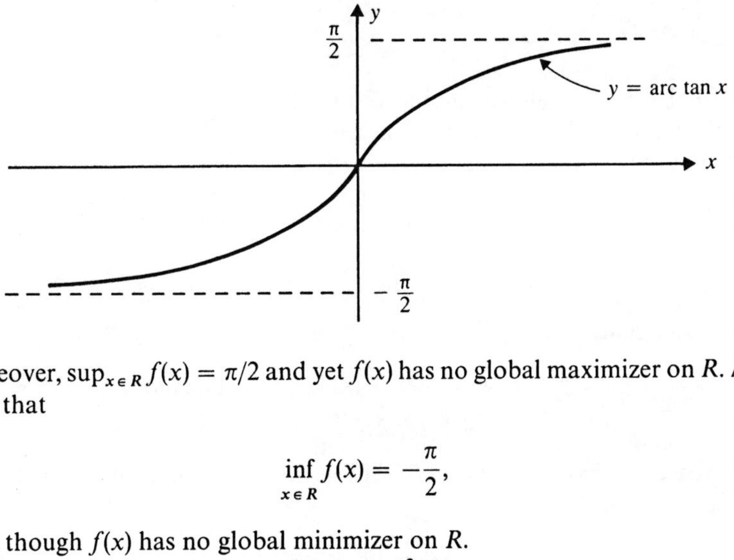

Moreover, $\sup_{x \in R} f(x) = \pi/2$ and yet $f(x)$ has no global maximizer on R. Also note that

$$\inf_{x \in R} f(x) = -\frac{\pi}{2},$$

even though $f(x)$ has no global minimizer on R.

(d) Consider the function defined on R^2 by

$$f(x, y) = e^{-(x+y)^2}.$$

The graph of $f(x, y)$ is a bell-shaped wave cresting at a height of 1 above the line $y = -x$ with the wave tails approaching the xy-plane from above. In this case,

$$\inf_{(x,y) \in R^2} f(x, y) = 0, \qquad \sup_{(x,y) \in R^2} f(x, y) = 1.$$

Any point on the line $y = -x$ is a global maximizer but there are no global minimizers for $f(x, y)$.

The preceding examples should serve to clarify the meaning of supremum and infimum of a function on a set. We now return to a discussion of convex programming.

Given a program (P), denote the infimum, if it exists, of the objective function $f(\mathbf{x})$ on the feasibility region F for (P) by MP, that is,

$$MP = \inf_{\mathbf{x} \in F} \{f(\mathbf{x})\}.$$

This infimum exists whenever $f(\mathbf{x})$ has a lower bound on the feasibility region for (P). If we use the symbol $\mathbf{g}(\mathbf{x})$ for the vector function $(g_1(\mathbf{x}), g_2(\mathbf{x}), \ldots, g_m(\mathbf{x}))$ and if we agree that the notation

$$\mathbf{g}(\mathbf{x}) \leq \mathbf{0}$$

means $g_1(\mathbf{x}) \leq 0, g_2(\mathbf{x}) \leq 0, \ldots, g_m(\mathbf{x}) \leq 0$, then the following more suggestive description of MP results:

$$MP = \inf_{\substack{\mathbf{x} \in C \\ \mathbf{g}(\mathbf{x}) \leq \mathbf{0}}} \{f(\mathbf{x})\}.$$

Note that if $\mathbf{x}^*$ is a global minimizer for the objective function $f(\mathbf{x})$ on the feasibility region F for (P), then $\mathbf{x}^*$ is a solution of (P) and $MP = f(\mathbf{x}^*)$.

It may happen, even for convex programs, that the objective function $f(\mathbf{x})$ has a lower bound on the feasibility region for (P) (so that MP exists) and yet (P) has no solution. This situation is illustrated by the following example.

(5.2.4) Example. Consider the convex program

$$(P) \quad \begin{cases} \text{Minimize} \quad f(x) = e^x \quad \text{subject to} \\ \quad g_1(x) = x \leq 0 \quad \text{for } x \in R^1. \end{cases}$$

In this case, the feasibility region F for (P) is the set of nonpositive real numbers and so

$$MP = \inf_{x \leq 0} \{e^x\} = 0$$

but there is no feasible point $\mathbf{x}^*$ for which $f(\mathbf{x}^*) = 0$, because $e^x > 0$ for all real numbers x.

Linear programming is one of the most important areas in the field of constrained optimization. Its importance stems in part from the wide range of applied problems in which linear programs arise, and in part from the effectiveness of the mathematical methods that have been developed to solve linear programs. The following example will show how linear programming fits into the context of convex programming. We shall return to this example later to apply results on convex programming such as the Karush–Kuhn–Tucker Theorem to derive conclusions concerning the solution of linear programming problems.

(5.2.5) Example (Linear Programming). The familiar minimum standard form for a linear programming problem is formulated as follows: Given a $m \times n$-matrix $A = (a_{ij})$ and constants $b_1, \ldots, b_n; c_1, \ldots, c_m$, we seek to

$$(LP)\quad \begin{cases} \text{Minimize} \quad b_1x_1 + b_2x_2 + \cdots + b_nx_n \\ \text{subject to the constraints} \\ \quad a_{11}x_1 + a_{12}x_2 + \cdots + a_{1n}x_n \geq c_1, \\ \quad a_{21}x_1 + a_{22}x_2 + \cdots + a_{2n}x_n \geq c_2, \\ \quad \vdots \qquad\qquad\qquad\qquad \vdots \quad \vdots \\ \quad a_{m1}x_1 + a_{m2}x_2 + \cdots + a_{mn}x_n \geq c_m, \\ \text{where} \quad x_1 \geq 0, \quad x_2 \geq 0, \ldots, \quad x_n \geq 0. \end{cases}$$

An illustration of a concrete problem that can readily be described in terms of a linear program in minimum standard form is the classical Diet Problem. Suppose that we seek to plan a diet using n foods $F_1, \ldots, F_n$ that will provide the minimum daily requirements of m nutrients $N_1, \ldots, N_m$ at minimum cost. If we let

b_i = cost (in cents per ounce) of food F_i for $i = 1, \ldots, n$,

c_j = minimum daily requirement (in milligrams) of nutrient N_j for $j = 1, \ldots, m$,

a_{ij} = number of milligrams of nutrient N_j in one ounce of food F_i for $i = 1, \ldots, n$ and $j = 1, \ldots, m$,

x_i = number of ounces of food F_i in a given diet for $i = 1, \ldots, n$,

then (LP) is precisely the mathematical formulation of the Diet Problem.

If we agree that $\mathbf{u} \geq \mathbf{v}$ means that $u_i \geq v_i$ for all i, then we can use vector notation to write the minimum standard form of a linear program in the following compact form: Given an $m \times n$-matrix A and vectors $\mathbf{b} \in R^n$, $\mathbf{c} \in R^m$:

$$(LP)\quad \begin{cases} \text{Minimize} \quad \mathbf{b} \cdot \mathbf{x} \quad \text{subject to the} \\ \text{constraints} \quad A\mathbf{x} \geq \mathbf{c}, \quad \mathbf{x} \geq \mathbf{0}. \end{cases}$$

If the ith row vector of A is denoted by $\mathbf{a}^{(i)}$, then the constraints $A\mathbf{x} \geq \mathbf{c}$ can be written as

$$\mathbf{a}^{(i)} \cdot \mathbf{x} \geq c_i, \qquad i = 1, 2, \ldots, m.$$

The functions $f(\mathbf{x}) = \mathbf{b} \cdot \mathbf{x}$ and $g_i(\mathbf{x}) = c_i - \mathbf{a}^{(i)} \cdot \mathbf{x}$, $i = 1, 2, \ldots, m$, are linear and therefore convex. Also, the set $C = \{\mathbf{x} \in R^n : \mathbf{x} \geq \mathbf{0}\}$ is convex, so (LP) can be reformulated as a convex program as follows:

$$\begin{cases} \text{Minimize} \quad f(\mathbf{x}) = \mathbf{b}\cdot\mathbf{x} \\ \text{subject to the constraints} \\ \quad g_1(\mathbf{x}) = c_1 - \mathbf{a}^{(1)}\cdot\mathbf{x} \leq 0, \\ \quad \vdots \qquad \vdots \quad \vdots \qquad \vdots \\ \quad g_m(\mathbf{x}) = c_m - \mathbf{a}^{(m)}\cdot\mathbf{x} \leq 0, \\ \text{where} \quad \mathbf{x} \in C = \{\mathbf{y} \in R: \mathbf{y} \geq \mathbf{0}\}. \end{cases}$$

Thus, every linear program is also a convex program.

Our first objective in our study of the general convex program,

$$(P) \quad \begin{cases} \text{Minimize} \quad f(\mathbf{x}) \quad \text{subject to} \\ \quad g_1(\mathbf{x}) \leq 0, \ldots, \quad g_m(\mathbf{x}) \leq 0, \\ \text{where} \quad \mathbf{x} \in C \subset R^n \text{ and } f(\mathbf{x}), g_1(\mathbf{x}), \ldots, g_m(\mathbf{x}) \text{ are convex functions} \\ \text{defined on a convex set } C, \end{cases}$$

is to investigate the sensitivity of the value of

$$MP = \inf_{\mathbf{x} \in F} \{f(\mathbf{x})\}$$

to slight changes in the constraints. For this purpose, we define for each $\mathbf{z} \in R^m$ a program $(P(\mathbf{z}))$ as follows:

$$(P(\mathbf{z})) \quad \begin{cases} \text{Minimize} \quad f(\mathbf{x}) \quad \text{subject to} \\ \quad g_1(\mathbf{x}) \leq z_1, \ldots, \quad g_m(\mathbf{x}) \leq z_m, \\ \text{where} \quad \mathbf{x} \in C \subset R^n \text{ and } f(\mathbf{x}), g_1(\mathbf{x}), \ldots, g_m(\mathbf{x}) \text{ are convex functions} \\ \text{defined on a convex set } C. \end{cases}$$

Since the function $\hat{g}_i(\mathbf{x}) = g_i(\mathbf{x}) - z_i$ is convex (Why?) for $i = 1, \ldots, m$, it is an easy matter to rewrite $(P(\mathbf{z}))$ in the standard form (P) for a convex program. Also, note that (P) is identical to $(P(\mathbf{0}))$ and that $(P(\mathbf{z}))$ can be thought of as a "*perturbation*" of (P).

Suppose that we denote the feasibility region of $(P(\mathbf{z}))$ by $F(\mathbf{z})$ and set

$$MP(\mathbf{z}) = \inf_{\mathbf{x} \in F(\mathbf{z})} \{f(\mathbf{x})\}.$$

We are interested in investigating questions of the following sort: If $\mathbf{z}$ is a vector in R^m that is close to $\mathbf{0}$, how close is $MP(\mathbf{z})$ to $MP(\mathbf{0}) = MP$? Which of the constraints $g_i(\mathbf{x}) \leq z_i$ have the greatest effect on the value of $MP(\mathbf{z})$ as $\mathbf{z}$ varies? which have the least effect? Are there any constraints $g_i(\mathbf{x}) \leq z_i$ that have no effect at all on $MP(\mathbf{z})$ as $\mathbf{z}$ varies near $\mathbf{0}$? As we will see, the answers to these and other related questions will flow from the convexity of the function

$$\mathbf{z} \to MP(\mathbf{z})$$

on its domain in R^m.

It is convenient to regard the domain of the function $MP(\mathbf{z})$ to be the set of all $\mathbf{z} \in R^m$ for which the feasibility region for $(P(\mathbf{z}))$ is not empty. This is a somewhat unconventional choice because the objective function $f(\mathbf{x})$ may have no lower bound on the feasibility region for $(P(\mathbf{z}))$; in this case, we assign the value $-\infty$ to $MP(\mathbf{z})$. Of course, if $f(\mathbf{x})$ does have a lower bound on the feasibility region for $(P(\mathbf{z}))$ and this region is nonempty, then $MP(\mathbf{z})$ is a finite real number. Thus, $MP(\mathbf{z})$ is either equal to $-\infty$ or to a finite real number at any point $\mathbf{z}$ in its domain.

(5.2.6) Theorem. *If (P) is a convex program and if $(P(\mathbf{z}))$ is the perturbation of (P) by $\mathbf{z} \in R^m$, then the function $MP(\mathbf{z})$ is convex and its domain is a convex subset of R^m. If (P) is superconsistent, then $\mathbf{0}$ is an interior point of the domain of $MP(\mathbf{z})$.*

PROOF. Suppose that $\mathbf{z}^{(1)}$, $\mathbf{z}^{(2)}$ belong to the domain of $MP(\mathbf{z})$ and that $0 \leq \lambda \leq 1$. Then, in terms of the vector function notation $\mathbf{g}(\mathbf{w}) \leq \mathbf{u}$ for the constraints $g_1(\mathbf{w}) \leq u_1, \ldots, g_m(\mathbf{w}) \leq u_m$, we see that $\lambda\mathbf{x}^{(1)} + [1 - \lambda]\mathbf{x}^{(2)}$ is feasible for $(P(\lambda\mathbf{z}^{(1)} + [1 - \lambda]\mathbf{z}^{(2)}))$ whenever $\mathbf{x}^{(1)}, \mathbf{x}^{(2)}$ are feasible for $(P(\mathbf{z}^{(1)}))$, $(P(\mathbf{z}^{(2)}))$, respectively, because

$$\begin{aligned}\mathbf{g}(\lambda\mathbf{x}^{(1)} + [1 - \lambda]\mathbf{x}^{(2)}) &\leq \lambda\mathbf{g}(\mathbf{x}^{(1)}) + [1 - \lambda]\mathbf{g}(\mathbf{x}^{(2)}) \\ &\leq \lambda\mathbf{z}^{(1)} + [1 - \lambda]\mathbf{z}^{(2)}.\end{aligned}$$

Thus, the domain of $MP(\mathbf{z})$ is a convex subset of R^m. Also, if $MP(\lambda\mathbf{z}^{(1)} + [1 - \lambda]\mathbf{z}^{(2)})$ is a finite real number, then

$$\begin{aligned}MP(\lambda\mathbf{z}^{(1)} + [1 - \lambda]\mathbf{z}^{(2)}) &= \inf\{f(\mathbf{x}) \colon \mathbf{x} \in C, \mathbf{g}(\mathbf{x}) \leq \lambda\mathbf{z}^{(1)} + [1 - \lambda]\mathbf{z}^{(2)}\} \\ &= \inf\{f(\mathbf{x}) \colon \mathbf{x} = \lambda\mathbf{x}^{(1)} + [1 - \lambda]\mathbf{x}^{(2)} \text{ where} \\ &\qquad \mathbf{x}^{(1)}, \mathbf{x}^{(2)} \in C \text{ and } \mathbf{g}(\mathbf{x}) \leq \lambda\mathbf{z}^{(1)} + [1 - \lambda]\mathbf{z}^{(2)}\} \\ &\leq \inf\{f(\mathbf{x}) \colon \mathbf{x} = \lambda\mathbf{x}^{(1)} + [1 - \lambda]\mathbf{x}^{(2)} \text{ where} \\ &\qquad \mathbf{x}^{(1)}, \mathbf{x}^{(2)} \in C \text{ and } \lambda\mathbf{g}(\mathbf{x}^{(1)}) + [1 - \lambda]\mathbf{g}(\mathbf{x}^{(2)}) \\ &\qquad \leq \lambda\mathbf{z}^{(1)} + [1 - \lambda]\mathbf{z}^{(2)}\} \\ &\leq \inf\{f(\lambda\mathbf{x}^{(1)} + [1 - \lambda]\mathbf{x}^{(2)}) \colon \mathbf{x}^{(1)}, \mathbf{x}^{(2)} \in C \text{ and} \\ &\qquad \mathbf{g}(\mathbf{x}^{(1)}) \leq \mathbf{z}^{(1)}, \mathbf{g}(\mathbf{x}^{(2)}) \leq \mathbf{z}^{(2)}\} \\ &\leq \inf\{\lambda f(\mathbf{x}^{(1)}) + [1 - \lambda]f(\mathbf{x}^{(2)}) \colon \mathbf{x}^{(1)}, \mathbf{x}^{(2)} \in C \text{ and} \\ &\qquad \mathbf{g}(\mathbf{x}^{(1)}) \leq \mathbf{z}^{(1)}, \mathbf{g}(\mathbf{x}^{(2)}) \leq \mathbf{z}^{(2)}\} \\ &= \lambda MP(\mathbf{z}^{(1)}) + [1 - \lambda]MP(\mathbf{z}^{(2)}).\end{aligned}$$

On the other hand, if $MP(\lambda \mathbf{z}^{(1)} + [1 - \lambda]\mathbf{z}^{(2)}) = -\infty$, then surely $MP(\lambda \mathbf{z}^{(1)} + [1 - \lambda]\mathbf{z}^{(2)}) \leq \lambda MP(\mathbf{z}^{(1)}) + [1 - \lambda]MP(\mathbf{z}^{(2)})$. Consequently, $MP(\mathbf{z})$ is convex on its domain.

Now suppose that (P) is superconsistent, that is, suppose that there is a $\mathbf{w} \in R^n$ *such that* $g_i(\mathbf{w}) < 0$ for $i = 1, \ldots, m$. Let

$$r = \min\{-g_i(\mathbf{w}) \colon 1 \leq i \leq m\},$$

then $r > 0$, and for any $\mathbf{z}$ in the ball $B(\mathbf{0}, r)$ centered at $\mathbf{0}$ of radius r, $-r < z_i < r$ for all i so that

$$g_i(\mathbf{w}) \leq -r < z_i, \qquad i = 1, 2, \ldots, m.$$

Therefore, $(P(\mathbf{z}))$ is a consistent program for all $\mathbf{z} \in B(\mathbf{0}, r)$. This shows that $\mathbf{0}$ is an interior point of the domain of $MP(\mathbf{z})$.

(5.2.7) Remarks. The following comments are in order with regard to the conclusion of (5.2.6).

(1) If $\mathbf{z}^{(1)}$ and $\mathbf{z}^{(2)}$ are points in the domain of $MP(\mathbf{z})$ and if $MP(\mathbf{z}^{(1)}) = -\infty$, then $MP(\mathbf{z})$ is equal to $-\infty$ on the entire half-open line segment

$$[\mathbf{z}^{(1)}, \mathbf{z}^{(2)}) = \{\lambda \mathbf{z}^{(1)} + [1 - \lambda]\mathbf{z}^{(2)} \colon 0 < \lambda \leq 1\}$$

joining $\mathbf{z}^{(1)}$ to $\mathbf{z}^{(2)}$. This follows at once from the inequality

$$MP(\lambda \mathbf{z}^{(1)} + [1 - \lambda]\mathbf{z}^{(2)}) \leq \lambda MP(\mathbf{z}^{(1)}) + [1 - \lambda]MP(\mathbf{z}^{(2)}).$$

Thus, if $MP(\mathbf{z})$ assumes the value $-\infty$ at any point of its domain, it is identically equal to $-\infty$ except possibly at boundary points of its domain.

(2) If $MP(\mathbf{z}^{(0)})$ is finite at an interior point of the domain of $MP(\mathbf{z})$, then $MP(\mathbf{z})$ is finite on its entire domain. For if $MP(\mathbf{z}^{(1)}) = -\infty$ at some point $\mathbf{z}^{(1)}$ in the domain of $MP(\mathbf{z})$, then there exist $\mathbf{z}^{(2)}$ in the domain of $MP(\mathbf{z})$ and λ_0 such that $0 < \lambda_0 < 1$ and

$$\mathbf{z}^{(0)} = \lambda_0 \mathbf{z}^{(1)} + [1 - \lambda_0]\mathbf{z}^{(2)}.$$

Thus, $\mathbf{z}^{(0)}$ is on the half-open line segment $[\mathbf{z}^{(1)}, \mathbf{z}^{(2)})$ joining $\mathbf{z}^{(1)}$ to $\mathbf{z}^{(2)}$, so $MP(\mathbf{z}^{(0)}) = -\infty$ by (1) above. This contradicts the assumption that $MP(\mathbf{z}^{(0)})$ is finite, so $MP(\mathbf{z})$ must be finite on its entire domain.

In combination with (5.2.6) and (5.1.10), Remark (5.2.7)(2) yields the following result.

(5.2.8) Theorem. *If (P) is a superconsistent convex program such that $MP = MP(\mathbf{0})$ is finite, then $MP(\mathbf{z})$ is finite on its entire domain and there exists a vector $\boldsymbol{\lambda} \in R^m$ such that $\boldsymbol{\lambda} \geq \mathbf{0}$ and*

$$MP(\mathbf{z}) \geq MP(\mathbf{0}) - \boldsymbol{\lambda} \cdot \mathbf{z}$$

for all $\mathbf{z}$ in the domain of $MP(\mathbf{z})$.

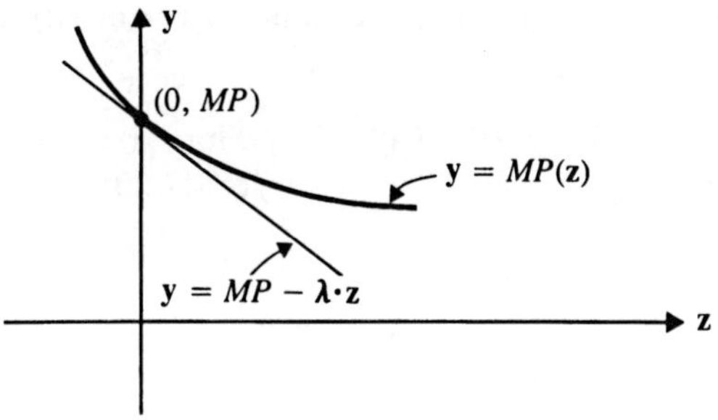

PROOF. The first assertion is an immediate consequence of (5.2.6) and (5.2.7)(2) in the paragraph preceding the statement of this theorem. Moreover, these results show that $\mathbf{0}$ is an interior point of the domain of $MP(\mathbf{z})$ and that $MP(\mathbf{z})$ is a convex real-valued function on this domain so Theorem (5.1.10) implies that there is a vector $-\boldsymbol{\lambda} \in R^m$ such that

$$MP(\mathbf{z}) \geq MP(\mathbf{0}) - \boldsymbol{\lambda} \cdot \mathbf{z}$$

for all $\mathbf{z}$ in the domain of $MP(\mathbf{z})$.

All we have left to do is to prove that $\boldsymbol{\lambda} \geq \mathbf{0}$. To this end, suppose that some component λ_i of $\boldsymbol{\lambda}$ is negative. Since $\mathbf{0}$ is an interior point of the domain of $MP(\mathbf{z})$, there is a positive number r such that the ball $B(\mathbf{0}, r)$ centered at $\mathbf{0}$ radius r is contained in the domain of $MP(\mathbf{z})$. In particular, if $\mathbf{z}^{(i)}$ is the vector with $r/2$ at the ith component and 0's elsewhere, then $\mathbf{z}^{(n)}$ is in the domain of $MP(\mathbf{z})$ and

$$MP(\mathbf{z}^{(i)}) \geq MP(\mathbf{0}) - \boldsymbol{\lambda} \cdot \mathbf{z}^{(i)} = MP - \frac{r}{2}\lambda_i.$$

But $\mathbf{z}^{(i)} \geq \mathbf{0}$, so $MP(\mathbf{z}^{(i)}) \leq MP(\mathbf{0}) = MP$, which contradicts the preceding inequality since $\lambda_i < 0$. Therefore, $\boldsymbol{\lambda} \geq \mathbf{0}$ and the proof is complete.

If (P) is a convex program for which MP is finite and for which there is a $\boldsymbol{\lambda} \in R^m$ such that $\boldsymbol{\lambda} \geq \mathbf{0}$ and

$$MP(\mathbf{z}) \geq MP(0) - \boldsymbol{\lambda} \cdot \mathbf{z}$$

for all $\mathbf{z}$ in the domain of $MP(\mathbf{z})$, then $\boldsymbol{\lambda}$ is called a *sensitivity vector* for (P). Theorem (5.2.8) simply guarantees that superconsistent convex programs always have sensitivity vectors.

Keep in mind that if $MP(\mathbf{z})$ is differentiable at $\mathbf{z} = \mathbf{0}$, then the vector $\boldsymbol{\lambda}$ in (5.2.8) can be taken to be $\nabla MP(0)$ since $MP(\mathbf{z})$ is a convex function. However, the following examples show that $MP(\mathbf{z})$ may fail to be differentiable at $\mathbf{z} = \mathbf{0}$.

(5.2.9) Examples

(a) Consider the following convex program:

$$(P) \quad \text{Minimize} \quad \sqrt{x^2 + y^2} \quad \text{subject to} \quad x + y \leq 0.$$

In this case, it is evident that the corresponding perturbation

$$(P(z))\quad \text{Minimize} \quad \sqrt{x^2+y^2} \quad \text{subject to} \quad x+y\le z$$

is minimized at (0, 0) for $z \ge 0$ and at $(z/2, z/2)$ for $z < 0$. Therefore $MP(z) = 0$ for $z \ge 0$ and $MP(z) = -z/\sqrt{2}$ for $z < 0$, so $MP(z)$ is not differentiable at $z = 0$.

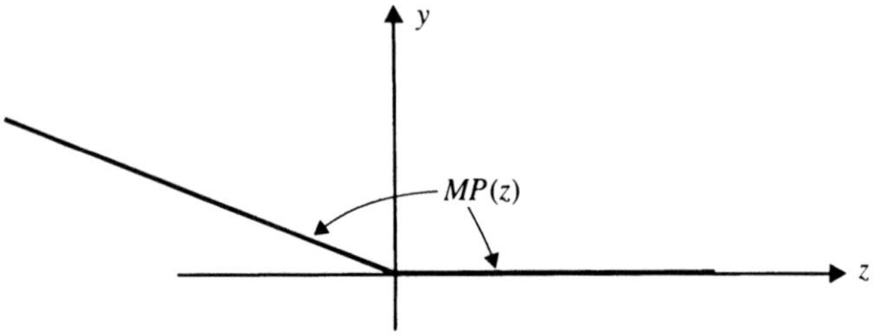

(b) Consider the linear program

$$(P)\quad \begin{cases} \text{Minimize} \quad -2x - y \quad \text{subject to} \\ \quad x+y\le 1, \quad 0\le x\le 1, \quad 0\le y. \end{cases}$$

We will investigate the variation in $MP(\mathbf{z})$ for $\mathbf{z}$ that vary the constraint $x + y - 1 \le 0$ but hold the constraints $0 \le x \le 1, 0 \le y$ fixed. The feasibility region F and the level lines for the objective function are displayed in the following diagram:

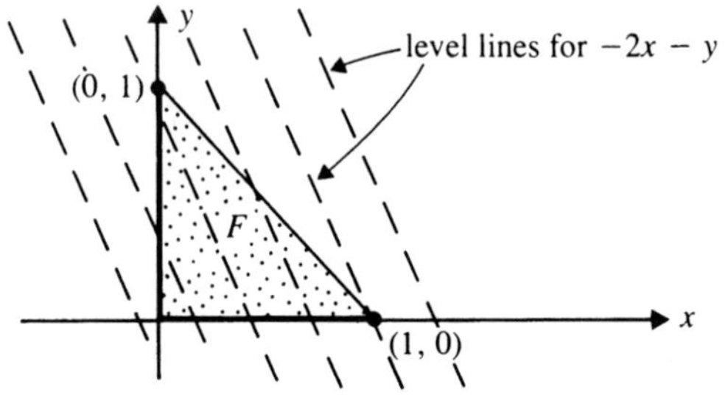

If we replace the constraint $x + y - 1 \le 0$ by $x + y - 1 \le z$ and hold the other constraints fixed (which amounts to perturbing (P) by the vector $\mathbf{z} = (z, 0, 0, 0)$ (Why?)), we obtain the feasibility regions displayed below:

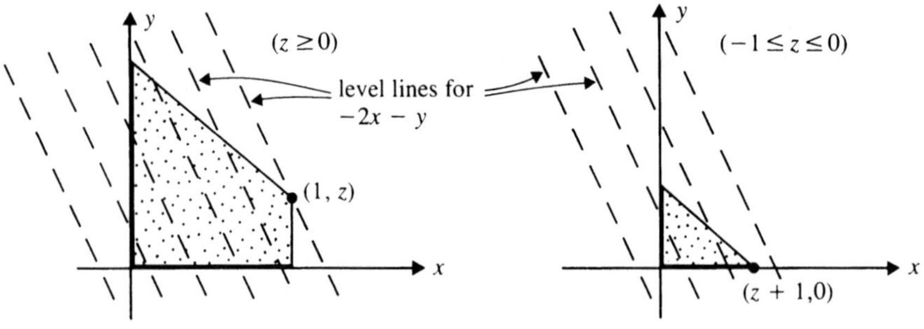

The objective function is minimized at $(1, z)$ for $z \geq 0$ and at $(z + 1, 0)$ for $-1 \leq z \leq 0$. Therefore, $MP(\mathbf{z}) = -2(1) - z = -2 - z$ for $z \geq 0$ and $MP(\mathbf{z}) = -2(z + 1) = -2z - 2$ for $-1 \leq z \leq 0$. It follows that the "cross section" of the graph of $MP(\mathbf{z})$ in the direction of these $\mathbf{z}$ at $\mathbf{0}$ looks like

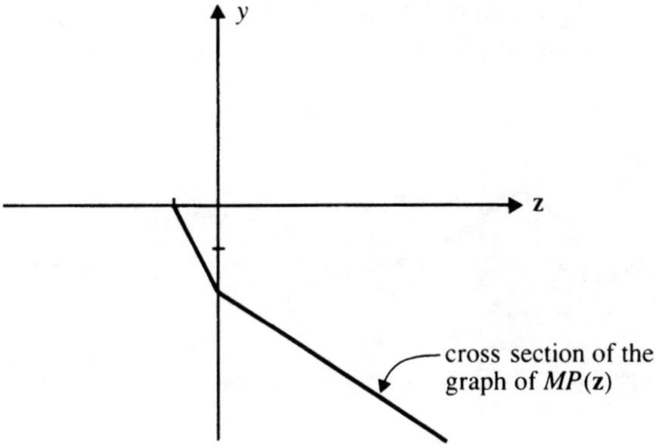

Evidently, $MP(\mathbf{z})$ is not a differentiable function at $\mathbf{z} = \mathbf{0}$.

The next example gives a convex program (P) such that $MP(\mathbf{z})$ is not even continuous at 0.

(c) (Duffin) Consider the convex program

$$(P) \quad \begin{cases} \text{Minimize} \quad e^{-y} \quad \text{subject to} \\ \quad \sqrt{x^2 + y^2} - x \leq 0. \end{cases}$$

Note that the given constraint implies that $y = 0$ and $x \geq 0$ because $\sqrt{x^2 + y^2} \geq x$ for all x and y. Consequently, $MP(0) = e^{-0} = 1$. On the other hand, if $z > 0$, then

$$\sqrt{x^2 + y^2} - x \leq z$$

implies that $y^2 \leq 2zx + z^2$ so the feasibility region for $(P(z))$ has the following form:

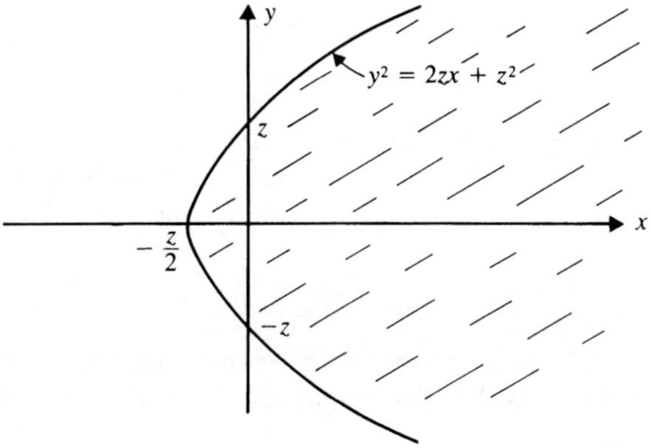

Therefore, $MP(z) = \inf_{y \in R} e^{-y} = 0$, so $MP(z)$ is the discontinuous convex function defined for $z \geq 0$ by

$$MP(z) = \begin{cases} 1 & \text{if } z = 0, \\ 0 & \text{if } z > 0. \end{cases}$$

In particular, $MP(z)$ is not differentiable at $z = 0$.

The following diagram for the single constraint case suggests the general situation:

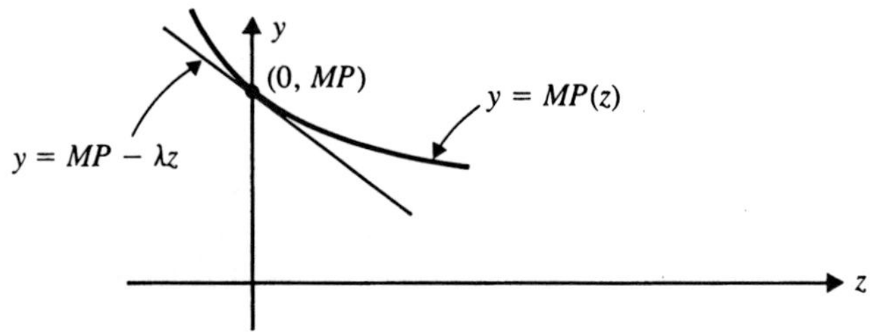

Note that $MP(z)$ is a decreasing function of z. In fact, it is true in general that if $\mathbf{z}^{(1)}, \mathbf{z}^{(2)}$ are in R^m and if $\mathbf{z}^{(1)} \leq \mathbf{z}^{(2)}$, then $MP(\mathbf{z}^{(2)}) \leq MP(\mathbf{z}^{(1)})$.

The following example suggests the rationale for the term "sensitivity vector" for $\boldsymbol{\lambda}$.

(5.2.10) Example. Suppose that for a given convex program with three constraints

$$(P) \quad \begin{cases} \text{Minimize} \quad f(\mathbf{x}) \quad \text{subject to} \\ \quad g_1(\mathbf{x}) \leq 0, \quad g_2(\mathbf{x}) \leq 0, \quad g_3(\mathbf{x}) \leq 0, \end{cases}$$

we find a sensitivity vector to be $\boldsymbol{\lambda} = (100, 1, 0)$. Then if $\mathbf{z} = (z_1, z_2, z_3)$ is in the domain of $MP(\mathbf{z})$ we have

$$(*) \qquad MP(\mathbf{z}) \geq MP(\mathbf{0}) - \boldsymbol{\lambda} \cdot \mathbf{z} = MP - 100z_1 - z_2.$$

The inequality $(*)$ shows that if the first constraint $g_1(\mathbf{x}) \leq 0$ is relaxed to $g_1(\mathbf{x}) \leq 1$ while the other two constraints are held at 0, then

$$MP(1, 0, 0) \geq MP - 100.$$

Thus, there is a potential improvement of 100 in the infimum of the objective function if we relax the first constraint. On the other hand, if we relax the second constraint $g_2(\mathbf{x}) \leq 0$ to $g_2(\mathbf{x}) \leq 1$ and hold the first and third constraints at zero, then

$$MP(0, 1, 0) \geq MP - 1.$$

Therefore, the relaxation of the second constraint by 1 can result in an improvement of at most 1 in the infimum of the objective function. Finally, if

$g_3(\mathbf{x}) \leq 0$ is relaxed to $g_3(\mathbf{x}) \leq 1$ while the first two constraints are held fixed, then

$$MP(0, 0, 1) \geq MP,$$

that is, there is no potential improvement in the infimum of the objective function.

In general, the sizes of the components of a sensitivity vector $\boldsymbol{\lambda}$ for (P) provide a measure of the sensitivity of the infimum of the objective function to small relaxations in the corresponding constraints. Large components correspond to the constraints that are the most important for the value of this infimum and the smallest components to the constraints that are the least important. Relaxations of constraints corresponding to zero components have no effect at all on this infimum.

With the help of the next theorem, we will see that if (P) is a superconsistent convex program, then (P) can be replaced with an equivalent *unconstrained* minimization problem.

(5.2.11) Theorem. *Suppose that*

$$(P) \quad \begin{cases} \textit{Minimize} \quad f(\mathbf{x}) \quad \textit{subject to} \\ \quad g_1(\mathbf{x}) \leq 0, \ldots, \quad g_m(\mathbf{x}) \leq 0, \\ \textit{where} \quad \mathbf{x} \in C \end{cases}$$

is a convex program for which there is a sensitivity vector $\boldsymbol{\lambda}$. *Then*

$$MP = \inf_{\mathbf{x} \in C} \left\{ f(\mathbf{x}) + \sum_{i=1}^{m} \lambda_i g_i(\mathbf{x}) \right\}.$$

PROOF. For each $\mathbf{z}$ in the domain of $MP(\mathbf{z})$, we know that

$$MP(\mathbf{z}) \geq MP - \boldsymbol{\lambda} \cdot \mathbf{z}.$$

In particular, if $\mathbf{x} \in C$, then $\mathbf{z} = (g_1(\mathbf{x}), \ldots, g_m(\mathbf{x})) = \mathbf{g}(\mathbf{x})$ is in the domain of $MP(\mathbf{z})$, so

$$MP(\mathbf{g}(\mathbf{x})) + \sum_{i=1}^{m} \lambda_i g_i(\mathbf{x}) \geq MP.$$

It follows immediately from the definition of $MP(\mathbf{g}(\mathbf{x}))$ that $f(\mathbf{x}) \geq MP(\mathbf{g}(\mathbf{x}))$. Consequently,

$$f(\mathbf{x}) + \sum_{i=1}^{m} \lambda_i g_i(\mathbf{x}) \geq MP$$

for all $\mathbf{x} \in C$ and so

$$\inf_{\mathbf{x} \in C} \left\{ f(\mathbf{x}) + \sum_{i=1}^{m} \lambda_i g_i(\mathbf{x}) \right\} \geq MP.$$

On the other hand, because $\lambda_i \geq 0$ for $i = 1, \ldots, m$,

$$\inf\left\{f(\mathbf{x}) + \sum_{i=1}^{m} \lambda_i g_i(\mathbf{x}) \colon \mathbf{x} \in C\right\} \leq \inf\left\{f(\mathbf{x}) + \sum_{i=1}^{m} \lambda_i g_i(\mathbf{x}) \colon \mathbf{x} \in C, \mathbf{g}(\mathbf{x}) \leq \mathbf{0}\right\}$$
$$\leq \inf\{f(\mathbf{x})\colon \mathbf{x} \in C, \mathbf{g}(\mathbf{x}) \leq \mathbf{0}\} = MP.$$

We conclude that

$$MP = \inf_{\mathbf{x} \in C} \left\{f(\mathbf{x}) + \sum_{i=1}^{m} \lambda_i g_i(\mathbf{x})\right\},$$

which completes the proof.

Of course, if (P) is a superconsistent convex program such that MP is finite, then (5.2.8) implies that there is a sensitivity vector $\boldsymbol{\lambda}$ for (P) so the conclusion of (5.2.11) holds in this case.

To pursue the connection established in (5.2.11) between constrained and unconstrained problems a bit further, it is useful to introduce the Lagrangian function.

(5.2.12) Definition. The *Lagrangian* $L(\mathbf{x}, \boldsymbol{\lambda})$ of a convex program

$$(P) \quad \begin{cases} \text{Minimize} \quad f(\mathbf{x}) \quad \text{subject to} \\ \quad g_1(\mathbf{x}) \leq 0, \ldots, \quad g_m(\mathbf{x}) \leq 0, \\ \text{where} \quad \mathbf{x} \in C \end{cases}$$

is the function defined by

$$L(\mathbf{x}, \boldsymbol{\lambda}) = f(\mathbf{x}) + \sum_{i=1}^{m} \lambda_i g_i(\mathbf{x})$$

for $\mathbf{x} \in C$ and $\boldsymbol{\lambda} \geq \mathbf{0}$.

Notice that the Lagrangian $L(\mathbf{x}, \boldsymbol{\lambda})$ is a function of $m + n$ variables where m is the number of inequality constraints and n is the number of variables involved in the objective and constraint functions. Also note that if (P) is a superconsistent convex program such that MP is finite, then a sensitivity vector $\boldsymbol{\lambda}$ exists and

$$MP = \inf_{\mathbf{x} \in C} \{L(\mathbf{x}, \boldsymbol{\lambda})\}$$

by (5.2.11).

The following theorem is the central result in the theory of convex programming.

(5.2.13) The Karush–Kuhn–Tucker Theorem (Saddle Point Form). *Suppose that (P) is a superconsistent convex program. Then $\mathbf{x}^* \in C$ is a solution of (P) if and only if there is a $\boldsymbol{\lambda}^* \in R^m$ such that:*

(1) $\boldsymbol{\lambda}^* \geq \mathbf{0}$;
(2) $L(\mathbf{x}^*, \boldsymbol{\lambda}) \leq L(\mathbf{x}^*, \boldsymbol{\lambda}^*) \leq L(\mathbf{x}, \boldsymbol{\lambda}^*)$ *for all* $\mathbf{x} \in C$ *and all* $\boldsymbol{\lambda} \geq \mathbf{0}$;
(3) $\lambda_i^* g_i(\mathbf{x}^*) = 0$ *for* $i = 1, 2, \ldots, m$.

PROOF. If $\mathbf{x}^*$ is a solution of (P), then $\mathbf{x}^* \in C$, $\mathbf{g}(\mathbf{x}^*) \leq \mathbf{0}$ and $f(\mathbf{x}^*) = MP$. According to (5.2.8), there is a sensitivity vector $\boldsymbol{\lambda}^* \geq \mathbf{0}$ in R^m, consequently, (5.2.11) asserts that

$$f(\mathbf{x}^*) = \inf_{\mathbf{x} \in C} L(\mathbf{x}, \boldsymbol{\lambda}^*).$$

Also, since $\boldsymbol{\lambda}^* \geq \mathbf{0}$ and $g_i(\mathbf{x}^*) \leq 0$ for $i = 1, \ldots, m$, it follows that

$$f(\mathbf{x}^*) = f(\mathbf{x}^*) + \sum_{i=1}^{m} \lambda_i^* g_i(\mathbf{x}^*) = L(\mathbf{x}^*, \lambda^*).$$

Consequently,

$$L(\mathbf{x}^*, \boldsymbol{\lambda}^*) \leq L(\mathbf{x}, \boldsymbol{\lambda}^*)$$

for all $\mathbf{x} \in C$ and $\lambda_i^* g_i(\mathbf{x}^*) = 0$ for $i = 1, 2, \ldots, m$.

To prove the left-hand inequality in (2) we note that for any $\boldsymbol{\lambda} \geq \mathbf{0}$ in R^m,

$$\begin{aligned} L(\mathbf{x}^*, \boldsymbol{\lambda}^*) - L(\mathbf{x}^*, \boldsymbol{\lambda}) &= \sum_{i=1}^{m} (\lambda_i^* - \lambda_i) g_i(\mathbf{x}^*) \\ &= -\sum_{i=1}^{m} \lambda_i g_i(\mathbf{x}^*) \geq 0 \end{aligned}$$

because $\lambda_i^* g_i(\mathbf{x}^*) = 0$, $\lambda_i \geq 0$ and $g_i(\mathbf{x}^*) \leq 0$ for $i = 1, 2, \ldots, m$. Therefore

$$L(\mathbf{x}^*, \boldsymbol{\lambda}^*) - L(\mathbf{x}^*, \boldsymbol{\lambda}) \geq 0$$

for all $\boldsymbol{\lambda} \geq \mathbf{0}$ in R^m. This completes the proof of (1), (2), and (3) when $\mathbf{x}^*$ is given to be a solution of (P).

On the other hand, suppose $\mathbf{x}^* \in C$ and $\boldsymbol{\lambda}^* \geq \mathbf{0}$ in R^m satisfy (2) for all $\mathbf{x} \in C$ and all $\boldsymbol{\lambda} \geq \mathbf{0}$ in R^m. For a given i such that $1 \leq i \leq m$, let

$$\lambda_j^{(i)} = \begin{cases} \lambda_j^* & \text{if } j \neq i, \\ \lambda_j^* + 1 & \text{if } j = i. \end{cases}$$

Then $\boldsymbol{\lambda}^{(i)} = (\lambda_j^{(i)}) \geq \mathbf{0}$ and (2) implies that

$$0 \geq L(\mathbf{x}^*, \boldsymbol{\lambda}^{(i)}) - L(\mathbf{x}^*, \boldsymbol{\lambda}^*) = g_i(\mathbf{x}^*),$$

so that $\mathbf{x}^*$ is feasible for (P). Moreover, (2) also implies that

$$f(\mathbf{x}^*) = L(\mathbf{x}^*, \mathbf{0}) \leq L(\mathbf{x}^*, \boldsymbol{\lambda}^*) = f(\mathbf{x}^*) + \sum_{i=1}^{m} \lambda_i^* g_i(\mathbf{x}^*).$$

But $\boldsymbol{\lambda}_i^* \geq 0$, $g_i(\mathbf{x}^*) \leq 0$ for $i = 1, 2, \ldots, m$, so that

$$f(\mathbf{x}^*) + \sum_{i=1}^{m} \lambda_i^* g_i(\mathbf{x}^*) \leq f(\mathbf{x}^*).$$

We conclude that

$$f(\mathbf{x}^*) = f(\mathbf{x}^*) + \sum_{i=1}^{m} \lambda_i^* g_i(\mathbf{x}^*) = L(\mathbf{x}^*, \boldsymbol{\lambda}^*)$$

and also that $\lambda_i^* g_i(\mathbf{x}^*) = 0$ for $i = 1, 2, \ldots, m$. If we apply (2) and use the following simple facts about the infima of functions on sets:

(a) if $A \subset B$, then $\inf_{\mathbf{x} \in B}\{h(\mathbf{x})\} \le \inf_{\mathbf{x} \in A}\{h(\mathbf{x})\}$;
(b) if $h(\mathbf{x}) \le k(\mathbf{x})$ for all $\mathbf{x} \in A$, then

$$\inf_{\mathbf{y} \in A} \{h(\mathbf{x})\} \le \inf_{\mathbf{x} \in A} \{k(\mathbf{x})\},$$

we see that

$$\begin{aligned} f(\mathbf{x}^*) = L(\mathbf{x}^*, \boldsymbol{\lambda}^*) &= \inf\{L(\mathbf{x}, \boldsymbol{\lambda}^*) \colon \mathbf{x} \in C\} \\ &\le \inf\{L(\mathbf{x}, \boldsymbol{\lambda}^*) \colon \mathbf{x} \in C, g_1(\mathbf{x}) \le 0, \ldots, g_m(\mathbf{x}) \le 0\} \\ &= \inf\left\{f(\mathbf{x}) + \sum_{i=1}^{m} \lambda_i^* g_i(\mathbf{x}) \colon \mathbf{x} \in C, g_1(\mathbf{x}) \le 0, \ldots, g_m(\mathbf{x}) \le 0\right\} \\ &\le \inf\{f(\mathbf{x}) \colon \mathbf{x} \in C, g_1(\mathbf{x}) \le 0, \ldots, g_m(\mathbf{x}) \le 0\} = MP. \end{aligned}$$

We conclude that $\mathbf{x}^*$ is a feasible point for (P) such that $f(\mathbf{x}^*) = MP$, that is, $\mathbf{x}^*$ is a solution of (P).

A point $(\mathbf{x}^*, \boldsymbol{\lambda}^*)$ such that $\mathbf{x}^* \in C$, $\boldsymbol{\lambda}^* \ge 0$ and that satisfies the inequality

(1) $L(\mathbf{x}^*, \boldsymbol{\lambda}) \le L(\mathbf{x}^*, \boldsymbol{\lambda}^*) \le L(\mathbf{x}, \boldsymbol{\lambda}^*)$ for all $\mathbf{x} \in C$, $\boldsymbol{\lambda} \ge \mathbf{0}$

is called a *saddle point for the Lagrangian of* (P). Condition (3), in (5.2.13):

(3) $\lambda_i^* g_i(\mathbf{x}^*) = 0$, $i = 1, 2, \ldots, m$

is referred to as the *complementary slackness condition.* The Karush–Kuhn–Tucker Theorem (5.2.13) asserts that *if* (P) *is a superconsistent convex program, then* $\mathbf{x}^* \in C$ *is a solution for* (P) *if and only if there is a* $\boldsymbol{\lambda} \ge \mathbf{0}$ *in* R^m *such that* $(\mathbf{x}^*, \boldsymbol{\lambda}^*)$ *is a saddle point of the Lagrangian of* (P) *and such that the complementary slackness condition is satisfied.* However, the proof of (5.2.13) yields even more information. It shows that if $(\mathbf{x}^*, \boldsymbol{\lambda}^*)$ is a saddle point of the Lagrangian of any convex program (P), then:

(1) MP is finite and $\mathbf{x}^*$ is a solution of (P).
(2) The complementary slackness condition

$$\lambda_i^* g_i(\mathbf{x}^*) = 0, \qquad i = 1, 2, \ldots, m$$

is satisfied.

If we impose the restriction that the objective and constraint functions have continuous first partial derivatives in (P), we obtain the following version of the Karush–Kuhn–Tucker Theorem.

(5.2.14) The Karush–Kuhn–Tucker Theorem (Gradient Form). *Suppose that* (P) *is a superconsistent convex program such that the objective function* $f(\mathbf{x})$ *and the constraint functions* $g_1(\mathbf{x}), \ldots, g_m(\mathbf{x})$ *have continuous first partial derivatives on the underlying set* C *for* (P). *If* $\mathbf{x}^*$ *is feasible for* (P) *and an* interior point *of* C, *then* $\mathbf{x}^*$ *is a solution of* (P) *if and only if there is a* $\boldsymbol{\lambda}^* \in R^m$ *such that*:

(1) $\lambda_i^* \geq 0$ *for* $i = 1, 2, \ldots, m$;
(2) $\lambda_i^* g_i(\mathbf{x}^*) = 0$ *for* $i = 1, 2, \ldots, m$;
(3) $\nabla f(\mathbf{x}^*) + \sum_{i=1}^m \lambda_i^* \nabla g_i(\mathbf{x}^*) = \mathbf{0}$.

PROOF. If $\mathbf{x}^*$ is a solution of (P), then (5.2.13) asserts that there is a $\boldsymbol{\lambda}^*$ in R^m satisfying (1) and (2) for which $(\mathbf{x}^*, \boldsymbol{\lambda}^*)$ is a saddle point of the Lagrangian of (P). But then

$$L(\mathbf{x}^*, \boldsymbol{\lambda}^*) \leq L(\mathbf{x}, \boldsymbol{\lambda}^*)$$

for all $\mathbf{x} \in C$, so $\mathbf{x}^*$ is a global minimizer for $h(\mathbf{x}) = L(\mathbf{x}, \boldsymbol{\lambda}^*)$ on C. Since $\mathbf{x}^*$ is an interior point of C and since $h(\mathbf{x})$ has continuous first partial derivatives on C, it follows that $\nabla h(\mathbf{x}^*) = 0$, that is, the gradient condition (3) holds.

Conversely, suppose that $\mathbf{x}^* \in C$ and $\boldsymbol{\lambda}^* \in R^m$ satisfy conditions (1), (2), and (3). If $\mathbf{x}$ is any feasible point for (P), then

$$\begin{aligned} f(\mathbf{x}) &\geq f(\mathbf{x}) + \sum_{i=1}^m \lambda_i^* g_i(\mathbf{x}) \\ &\geq [f(\mathbf{x}^*) + \nabla f(\mathbf{x}^*)\cdot(\mathbf{x} - \mathbf{x}^*)] + \sum_{i=1}^m \lambda_i^* [g_i(\mathbf{x}^*) + \nabla g_i(\mathbf{x}^*)\cdot(\mathbf{x} - \mathbf{x}^*)] \end{aligned}$$

because $\boldsymbol{\lambda}_i^* \geq 0$, $g_i(\mathbf{x}) \leq 0$ for $i = 1, \ldots, m$ and because $f(\mathbf{x}), g_1(\mathbf{x}), \ldots, g_m(\mathbf{x})$ are convex functions (cf. (2.3.5)). It follows that

$$\begin{aligned} f(\mathbf{x}) &\geq \left[f(\mathbf{x}^*) + \sum_{i=1}^m \lambda_i^* g_i(\mathbf{x}^*)\right] + \left[\nabla f(\mathbf{x}^*) + \sum_{i=1}^m \lambda_i^* \nabla g_i(\mathbf{x}^*)\right]\cdot(\mathbf{x} - \mathbf{x}^*) \\ &= f(\mathbf{x}^*) \end{aligned}$$

because of (2) and (3), so $\mathbf{x}^*$ is a global minimizer for $f(\mathbf{x})$ on the feasibility region for (P), that is, $\mathbf{x}^*$ is a solution for (P).

The following example illustrates how the gradient form of the Karush–Kuhn–Tucker Theorem can be applied to help locate solutions of convex programming problems.

(5.2.15) Example. Consider the program

$$(P) \quad \begin{cases} \text{Minimize} \quad f(x_1, x_2) = x_1^2 - 2x_1 + x_2^2 + 1 \\ \text{subject to the constraints} \\ \quad g_1(x_1, x_2) = x_1 + x_2 \leq 0, \\ \quad g_2(x_1, x_2) = x_1^2 - 4 \leq 0. \end{cases}$$

It is evident that (P) is a superconsistent convex program satisfying the differentiability conditions of (5.2.14). To find a solution $\mathbf{x}^* = (x_1^*, x_2^*)$ for (P), we note that there must be a $\boldsymbol{\lambda}^* = (\lambda_1^*, \lambda_2^*)$ such that

(1) $\lambda_1^* \geq 0, \lambda_2^* \geq 0$;

(2) $\lambda_1^*(x_1^* + x_2^*) = 0,$

$\lambda_2^*[(x_1^*)^2 - 4] = 0;$

(3) $2x_1^* - 2 + \lambda_1^* + 2\lambda_2^* x_1 = 0,$

$2x_2^* \qquad + \lambda_1^* \qquad = 0.$

It is not difficult to check that the system (1), (2), and (3) has only two solutions

$$x_1^* = 1, \qquad x_2^* = 0, \qquad \lambda_1^* = 0, \qquad \lambda_2^* = 0,$$

$$x_1^* = \tfrac{1}{2}, \qquad x_2^* = -\tfrac{1}{2}, \qquad \lambda_1^* = 1, \qquad \lambda_2^* = 0.$$

The first of these must be discarded as a possible solution because $\mathbf{x}^* = (1, 0)$ is not feasible for (P). On the other hand, $\mathbf{x}^* = (\frac{1}{2}, -\frac{1}{2})$ is feasible so (5.2.14) implies that this is the one and only solution of (P).

The following result shows that the vectors $\boldsymbol{\lambda}^*$ produced in the Karush–Kuhn–Tucker Theorem are precisely the same as the sensitivity vectors introduced in the remarks following (5.2.8).

(5.2.16) Theorem. *Suppose that (P) is a convex program and that $\mathbf{x}^*$ is a solution of (P). If $\boldsymbol{\lambda}^*$ is a vector in R^m such that $(\mathbf{x}^*, \boldsymbol{\lambda}^*)$ satisfy the Karush–Kuhn–Tucker Theorem conditions:*

(1) $\boldsymbol{\lambda}^* \geq \mathbf{0}$;

(2) $L(\mathbf{x}^*, \boldsymbol{\lambda}) \leq L(\mathbf{x}^*, \boldsymbol{\lambda}^*) \leq L(\mathbf{x}, \boldsymbol{\lambda}^*)$ *for all* $\mathbf{x} \in C$ *and all* $\boldsymbol{\lambda} \geq \mathbf{0}$;

(3) $\lambda_i^* g_i(\mathbf{x}^*) = 0$ *for* $i = 1, 2, \ldots, m$;

then $\boldsymbol{\lambda}^*$ is a sensitivity vector for (P); that is,

$$MP(\mathbf{z}) \geq MP - \boldsymbol{\lambda}^* \cdot \mathbf{z}$$

for all $\mathbf{z}$ in the domain of $MP(\mathbf{z})$.

Proof. First, we note that MP is finite because $\mathbf{x}^*$ is a solution to (P). If $\mathbf{z}$ is in the domain of $MP(\mathbf{z})$, then there is an $\mathbf{x} \in C$ such that $g_i(\mathbf{x}) \leq z_i$ for $i = 1, 2, \ldots, m$. By virtue of (3), we see that

$$MP = f(\mathbf{x}^*) = f(\mathbf{x}^*) + \sum_{i=1}^{m} \lambda_i^* g_i(\mathbf{x}^*)$$

$$= L(\mathbf{x}^*, \boldsymbol{\lambda}^*),$$

and (2) yields

$$L(\mathbf{x}^*, \boldsymbol{\lambda}^*) \leq L(\mathbf{x}, \boldsymbol{\lambda}^*) = f(\mathbf{x}) + \sum_{i=1}^{m} \lambda_i^* g_i(\mathbf{x}).$$

Therefore, because $\sum_{i=1}^{m} \lambda_i^* g_i(\mathbf{x}) \leq \boldsymbol{\lambda}^* \cdot \mathbf{z}$, it follows that

$$MP \leq f(\mathbf{x}) + \boldsymbol{\lambda}^* \cdot \mathbf{z}$$

for any $\mathbf{x}$ that is feasible for $P(\mathbf{z})$. But

$$MP(\mathbf{z}) = \inf\{f(\mathbf{x}): \mathbf{x} \in C, g_1(\mathbf{x}) \leq z_1, \ldots, g_m(\mathbf{x}) \leq z_m)$$

so we conclude that

$$MP \leq MP(\mathbf{z}) + \boldsymbol{\lambda}^* \cdot \mathbf{z}$$

for all $\mathbf{z}$ in the domain of $MP(\mathbf{z})$ so $\boldsymbol{\lambda}^*$ is a sensitivity vector for (P).

The components $\lambda_1^*, \ldots, \lambda_m^*$ of a vector $\boldsymbol{\lambda}^* \geq \mathbf{0}$ satisfying (1), (2), and (3) of (5.2.13) are often referred to as *Karush–Kuhn–Tucker multipliers* for the corresponding convex program (P). The connection established in (5.2.16) between vectors of the Karush–Kuhn–Tucker multipliers and sensitivity vectors provides us with a two-edged sword. On the one hand, the theoretical existence of the Karush–Kuhn–Tucker multipliers sets up the transfer from a given constrained problem to the corresponding unconstrained problem

$$\underset{\mathbf{x} \in C}{\text{Minimize}} \quad \left[f(\mathbf{x}) + \sum_{i=1}^{m} \lambda_i^* g_i(\mathbf{x}) \right]$$

via (5.2.11). On the other hand, if we use the Karush–Kuhn–Tucker Theorem to find $\boldsymbol{\lambda}^*$, then $\boldsymbol{\lambda}^*$ provides information about the sensitivity of (P) to its constraints (cf. (5.2.10)). For instance, the convex program

$$(P) \quad \begin{cases} \text{Minimize} \quad f(x_1, x_2) = x_1^2 - 2x_1 + x_2^2 + 1 \\ \text{subject to the constraints} \\ \quad g_1(x_1, x_2) = x_1 + x_2 \leq 0, \\ \quad g_2(x_1, x_2) = x_1^2 - 4 \leq 0, \end{cases}$$

solved in (5.2.15) produced the Karush–Kuhn–Tucker multipliers $\lambda_1^* = 1$, $\lambda_2^* = 0$. Therefore, according to (5.2.11), the minimum MP of (P) is also the minimum of the unconstrained problem

$$\text{Minimize} \quad h(x_1, x_2) = x_1^2 - x_1 + x_2^2 + x_2 + 1.$$

Also, by virtue of our interpretation of sensitivity vectors in (5.2.10) we see that relaxations of the constraint $x_1^2 - 4 \leq 0$ have no effect on this minimum. Note that these two observations are just two ways of describing the same feature of this problem.

5.3. The Karush–Kuhn–Tucker Theorem and Constrained Geometric Programming

In Section 2.5, we saw how the Arithmetic–Geometric Mean Inequality can be used to solve unconstrained geometric programs; that is, programs in which we seek to minimize a posynomial

$$g(\mathbf{t}) = \sum_{i=1}^{n} c_i \prod_{j=1}^{m} t_j^{\alpha_{ij}}$$

over all $\mathbf{t} = (t_1, \ldots, t_m)$ such that $t_j > 0$ for $j = 1, \ldots, m$, where $c_i > 0$ for $i = 1, \ldots, n$ and $\alpha_{ij} \in R$ for all i, j. This technique extends to constrained problems in which we seek to minimize a posynomial objective function subject to posynomial constraints. Again, the Arithmetic–Geometric Mean Inequality is the star of the show, but we will see that the Karush–Kuhn–Tucker Theorem plays a strong supporting role by providing the theoretical justification that the technique works in general.

Our development of constrained geometric programming begins with the following useful variant of the Arithmetic–Geometric Mean Inequality (2.4.1).

(5.3.1) Theorem (Extended Arithmetic–Geometric Mean Inequality). *Suppose that $x_1, \ldots, x_n$ are positive numbers. If $\delta_1, \ldots, \delta_n$ are numbers that are all positive or all zero and if $\lambda = \delta_1 + \cdots + \delta_n$, then*

$$\left(\sum_{i=1}^{n} x_i\right)^{\lambda} \geq \lambda^{\lambda}\left(\prod_{i=1}^{n}\left(\frac{x_i}{\delta_i}\right)^{\delta_i}\right)$$

under the conventions $0^0 = 1$ and $(x_i/0)^0 = 1$.

Equality holds in this inequality if and only if $\delta_1 = \delta_2 = \cdots = \delta_n = 0$ or

$$x_i = \frac{\delta_i}{\lambda}\left(\sum_{j=1}^{n} x_j\right)$$

for $i = 1, \ldots, n$.

PROOF. Suppose first that all of the δ_i's are positive numbers. Note that δ_i/λ is positive for $i = 1, 2, \ldots, n$ and that

$$\frac{\delta_1}{\lambda} + \frac{\delta_2}{\lambda} + \cdots + \frac{\delta_n}{\lambda} = 1.$$

Consequently, if we apply the Arithmetic–Geometric Mean Inequality (2.4.1) to $(\lambda x_i)/\delta_i$ and δ_i/λ for $i = 1, 2, \ldots, n$, we obtain

$$\sum_{i=1}^{n} x_i = \sum_{i=1}^{n}\left(\frac{\delta_i}{\lambda}\right)\left(\frac{\lambda x_i}{\delta_i}\right) \geq \prod_{i=1}^{n}\left(\frac{\lambda x_i}{\delta_i}\right)^{\delta_i/\lambda},$$

with equality holding if and only if

$$\frac{\lambda x_1}{\delta_1} = \frac{\lambda x_2}{\delta_2} = \cdots = \frac{\lambda x_n}{\delta_n}.$$

But then

$$\left(\sum_{i=1}^{n} x_i\right)^{\lambda} \geq \prod_{i=1}^{n} \left(\frac{\lambda x_i}{\delta_i}\right)^{\delta_i} = \lambda^{\lambda} \prod_{i=1}^{n} \left(\frac{x_i}{\delta_i}\right)^{\delta_i}.$$

Moreover, if $\lambda x_i/\delta_i = M$ for $i = 1, 2, \ldots, n$, then

$$\sum_{i=1}^{n} x_i = \sum_{i=1}^{n} \frac{M\delta_i}{\lambda} = \frac{M}{\lambda} \sum_{i=1}^{n} \delta_i = M,$$

that is,

$$x_i = \frac{M\delta_i}{\lambda} = \frac{\delta_i}{\lambda}\left(\sum_{i=1}^{n} x_i\right),$$

so the equality condition holds.

Finally, note that if all of the δ_i's are equal to zero, then both sides of the prescribed inequality are equal to 1. This completes the proof.

The next definition describes the context of constrained geometric programming.

(5.3.2) Definition. Suppose that $g_0(\mathbf{t})$, $g_1(\mathbf{t})$, $\ldots$, $g_k(\mathbf{t})$ are posynomials in m positive real variables $\mathbf{t} = (t_1, t_2, \ldots, t_m)$. Then the program

$$(GP) \quad \begin{cases} \text{Minimize} \quad g_0(\mathbf{t}) \quad \text{subject to the constraints} \\ \quad g_1(\mathbf{t}) \leq 1, \quad g_2(\mathbf{t}) \leq 1, \ldots, \quad g_k(\mathbf{t}) \leq 1, \\ \text{where} \quad t_1 > 0, \quad t_2 > 0, \ldots, \quad t_m > 0 \end{cases}$$

is called a *constrained geometric program.*

We will now show how the Arithmetic–Geometric Mean Inequalities (2.4.1) and (5.3.1) can be used to solve concrete geometric programs.

(5.3.3) Example. The following simple problem is discussed in *Geometric Programming: Theory and Applications* by R. J. Duffin, E. L. Peterson, and C. Zener (Wiley, New York, 1967), a book by the founders of geometric programming which provided the first systematic treatment of the subject.

"Suppose that 400 cubic yards of gravel must be ferried across a river. Suppose the gravel is to be shipped in an open box of length t_1, width t_2, and height t_3. The sides and bottom of the box cost \$10 per square yard and the ends of the box cost \$20 per square yard. The box will have no salvage value and each round trip of the box on the ferry will cost 10 cents. What is the minimum total cost of transporting 400 cubic yards of gravel?

As stated, this problem can be formulated as the unconstrained geometric program

$$\begin{cases} \text{Minimize} \quad \dfrac{40}{t_1 t_2 t_3} + 40t_2 t_3 + 20t_1 t_3 + 10t_1 t_2 \\ \text{where} \quad t_1 > 0, \quad t_2 > 0, \quad t_3 > 0. \end{cases}$$

We urge you to solve this unconstrained problem by the methods developed in Section 2.5 just to refresh your memory of the unconstrained case. You should come up with a minimum total cost of \$100 and optimal dimensions of the box as $t_1^* = 2$ yards, $t_2^* = 1$ yard, $t_3^* = \frac{1}{2}$ yard.

Suppose that we now consider the following variant of the above problem (which was also discussed by Duffin, Peterson, and Zener in their book): It is required that the sides and bottom of the box should be made from scrap material but only 4 square yards of this scrap material are available.

This variation of the problem leads us to the following constrained geometric program:

$$\begin{cases} \text{Minimize} \quad \dfrac{40}{t_1 t_2 t_3} + 40t_2 t_3 \\ \text{subject to} \quad 2t_1 t_3 + t_1 t_2 \leq 4, \\ \text{where} \quad t_1 > 0, \quad t_2 > 0, \quad t_3 > 0. \end{cases}$$

If we divide the constraint inequality by 4, we obtain a geometric program in the standard form of (5.3.2) with

$$g_0(t_1, t_2, t_3) = \frac{40}{t_2 t_2 t_3} + 40t_2 t_3,$$

$$g_1(t_1, t_2, t_3) = \frac{t_1 t_3}{2} + \frac{t_1 t_2}{4}.$$

Now proceed as follows: For any $\lambda > 0$, $[g_1(\mathbf{t})]^\lambda \leq 1$ so for any $\delta_1 > 0, \delta_2 > 0$, we have

$$g_0(\mathbf{t}) \geq g_0(\mathbf{t})[g_1(\mathbf{t})]^\lambda = \left(\frac{40}{t_1 t_2 t_3} + 40t_2 t_3\right)(g_1(\mathbf{t}))^\lambda$$

$$= \left(\delta_1\left(\frac{40}{\delta_1 t_1 t_2 t_3}\right) + \delta_2\left(\frac{40t_2 t_3}{\delta_2}\right)\right)(g_1(\mathbf{t}))^\lambda.$$

If we now impose the restriction that $\delta_1 + \delta_2 = 1$ and apply the Arithmetic–Geometric Mean Inequality (2.5.1), we obtain

$$g_0(\mathbf{t}) \geq \left(\left(\frac{40}{\delta_1 t_1 t_2 t_3}\right)^{\delta_1}\left(\frac{40t_2 t_3}{\delta_2}\right)^{\delta_2}\right)\left(g_1(\mathbf{t})\right)^\lambda$$

$$= \left(\left(\frac{40}{\delta_1}\right)^{\delta_1}\left(\frac{40}{\delta_2}\right)^{\delta_2}(t_1)^{-\delta_1}(t_2)^{(-\delta_1+\delta_2)}(t_3)^{(-\delta_1+\delta_2)}\right)(g_1(\mathbf{t}))^\lambda.$$

Next, we focus on the factor $[g_1(\mathbf{t})]^\lambda$ in the last expression and apply the

Extended Arithmetic–Geometric Mean Inequality (5.3.1) to it to obtain

$$\begin{aligned} g_1(\mathbf{t})^\lambda &= \left(\frac{t_1 t_3}{2} + \frac{t_1 t_2}{4}\right)^\lambda \\ &\geq \lambda^\lambda \left(\left(\frac{t_1 t_3}{2\delta_3}\right)^{\delta_3} \left(\frac{t_1 t_2}{4\delta_4}\right)^{\delta_4}\right) \\ &= \lambda^\lambda \left(\left(\frac{1}{2\delta_3}\right)^{\delta_3} \left(\frac{1}{4\delta_4}\right)^{\delta_4}\right) t_1^{(\delta_3+\delta_4)} t_2^{\delta_4} t_3^{\delta_3}, \end{aligned}$$

provided that $\lambda = \delta_3 + \delta_4$. If we combine the conclusions of the preceding calculations, we see that

$$\begin{aligned} g_0(\mathbf{t}) &\geq \left(\left(\frac{40}{\delta_1}\right)^{\delta_1} \left(\frac{40}{\delta_2}\right)^{\delta_2} \left(\frac{1}{2\delta_3}\right)^{\delta_3} \left(\frac{1}{4\delta_4}\right)^{\delta_4}\right) (\delta_3 + \delta_4)^{\delta_3+\delta_4} \\ &= v(\delta_1, \delta_2, \delta_3, \delta_4) \end{aligned}$$

provided that

$$\begin{aligned} \delta_1 + \delta_2 \qquad\qquad &= 1, \\ \delta_3 + \delta_4 &= \lambda, \\ -\delta_1 \qquad + \delta_3 + \delta_4 &= 0, \\ -\delta_1 + \delta_2 \qquad + \delta_4 &= 0, \\ -\delta_1 + \delta_2 + \delta_3 \qquad &= 0, \\ \delta_1 > 0, \delta_2 > 0, \delta_3 \geq 0, \delta_4 &\geq 0. \end{aligned}$$

(The last three equations result from equating the exponents of t_1, t_2, t_3 to zero.) This system has the unique solution

$$\delta_1^* = \tfrac{2}{3}, \qquad \delta_2^* = \tfrac{1}{3}, \qquad \delta_3^* = \tfrac{1}{3}, \qquad \delta_4^* = \tfrac{1}{3},$$

and the corresponding value of $v(\delta_1, \delta_2, \delta_3, \delta_4)$ is

$$v(\tfrac{2}{3}, \tfrac{1}{3}, \tfrac{1}{3}, \tfrac{1}{3}) = 60.$$

Thus, 60 is a lower bound for the value of $g_0(\mathbf{t})$ subject to the constraint $g_1(\mathbf{t}) \leq 1$. To find the minimum value of $g_0(\mathbf{t})$ and a minimizer $\mathbf{t}^* = (t_1^*, t_2^*, t_3^*)$ yielding this minimum value, we seek those values t_1^*, t_2^*, t_3^* that force equality in the Arithmetic–Geometric Mean Inequalities (2.4.1) and (5.3.1) and simultaneously in the constraint $g_1(\mathbf{t}) \leq 1$. If we can find such t_1^*, t_2^*, t_3^*, then we will know that the minimum of $g_0(\mathbf{t})$ subject to this constraint is actually 60 and that $\mathbf{t}^* = (t_1^*, t_2^*, t_3^*)$ is the minimizing point.

The equality conditions for the Arithmetic–Geometric Mean Inequalities (2.4.1) and (5.3.1) imply that

$$\frac{40}{\frac{2}{3} t_1^* t_2^* t_3^*} = \frac{40 t_2^* t_3^*}{\frac{1}{3}} = 60,$$

$$\frac{t_1^* t_3^*}{\frac{2}{3}} = \frac{t_1^* t_2^*}{\frac{4}{3}} = K.$$

(Why are the two terms in the first equation equal to 60?) Since δ_3^*, δ_4^* are positive, equality must hold in the constraint $g_1(\mathbf{t}) \leq 1$ so

$$1 = \delta_3 K + \delta_4 K = \tfrac{2}{3}K,$$

and hence $K = \frac{3}{2}$. These considerations lead us to the following system of equations:

$$\begin{aligned} t_1^* t_2^* t_3^* &= 1, \\ 2t_2^* t_3^* &= 1, \\ t_1^* t_3^* &= 1, \\ t_1^* t_2^* &= 2. \end{aligned}$$

We can convert this system to a system of linear equations in $\log t_1^*$, $\log t_2^*$, $\log t_3^*$ by taking logarithms

$$\begin{aligned} \log t_1^* + \log t_2^* + \log t_3^* &= 0, \\ \log t_2^* + \log t_3^* &= -\log 2, \\ \log t_1^* \qquad\qquad + \log t_3^* &= 0, \\ \log t_1^* + \log t_2^* \qquad\qquad &= \log 2. \end{aligned}$$

It is easy to check that $t_1^* = 2$, $t_2^* = 1$, $t_3^* = \frac{1}{2}$ is the unique solution of this system. Consequently, the minimum cost for ferrying the gravel is \$60 and the optimal dimensions of the box are

$$t_1^* = 2, \qquad t_2^* = 1, \qquad t_3^* = \tfrac{1}{2}.$$

We solved the geometric program in the preceding example by applying the Arithmetic–Geometric Mean Inequality (2.4.1) to the objective function $g_0(\mathbf{t})$ and the Extended Arithmetic–Geometric Mean Inequality (5.3.1) to the constraint $g_1(\mathbf{t}) \leq 1$. What if we had applied (2.4.1) to both functions? This would have resulted in the following system of equations for $\delta_1, \delta_2, \delta_3, \delta_4$:

$$\begin{aligned} \delta_1 + \delta_2 \qquad\qquad &= 1, \\ \delta_3 + \delta_4 &= 1, \\ -\delta_1 \qquad + \delta_3 + \delta_4 &= 0, \\ -\delta_4 + \delta_2 \qquad + \delta_4 &= 0, \\ -\delta_1 + \delta_2 + \delta_3 \qquad &= 0. \end{aligned}$$

This is an inconsistent system of equations. In other words, if we use (2.4.1) on both the objective and the constraint functions, and if we then add the equations resulting from setting the exponents of t_1, t_2, t_3 equal to zero, we have imposed too many restrictions on $\delta_1, \delta_2, \delta_3, \delta_4$.

It would seem from this that our solution of the constrained geometric program in (5.3.3) was a stroke of good fortune, a lucky break in a special situation. However, with the help of the Karush–Kuhn–Tucker Theorem, we can show that the technique applied there actually works in general. We will now develop this technique in the context of the general constrained geometric program (*GP*) defined in (5.3.2).

Because each of the $k + 1$ posynomials in the standard constrained geometric program (*GP*) (see (5.3.2)) may consist of several terms, we need to arrange all of the terms so that the resulting notation is simple and descriptive. One reasonable way to do this is to begin by counting the terms of the objective function $g_0(\mathbf{t})$ from left to right, listing them from 1 to say n_0, and then continuing by counting the terms of the first constraint posynomial $g_1(\mathbf{t})$ from left to right, listing them from $n_0 + 1$ to say n_1, and so on until we count the terms of the last constraint posynomial $g_k(\mathbf{t})$ from left to right, listing them from $n_{k-1} + 1$ to $n_k = p$. The jth term in this counting scheme will be denoted as follows:

$$u_j(\mathbf{t}) = c_j t_1^{\alpha_{j1}} t_2^{\alpha_{j2}} \dots t_m^{\alpha_{jm}}.$$

With this notational scheme, we can rewrite the standard constrained geometric program (*GP*) as

$$(GP)\quad \begin{cases} \text{Minimize} \quad g_0(\mathbf{t}) = u_1(\mathbf{t}) + \cdots + u_{n_0}(\mathbf{t}) \\ \text{subject to the constraints} \\ \quad g_1(\mathbf{t}) = u_{n_0+1}(\mathbf{t}) + \cdots + u_{n_1}(\mathbf{t}) \le 1, \\ \quad \cdots\cdots\cdots\cdots\cdots\cdots\cdots\cdots \\ \quad \cdots\cdots\cdots\cdots\cdots\cdots\cdots\cdots \\ \quad g_k(\mathbf{t}) = u_{n_{k-1}+1}(\mathbf{t}) + \cdots + u_{n_k}(\mathbf{t}) \le 1, \\ \text{where} \quad t_1 > 0, \quad t_2 > 0, \dots, \quad t_m > 0, \quad \text{and} \quad n_k = p. \end{cases}$$

Application of the Arithmetic–Geometric Mean Inequalities (2.4.1) and (5.3.1) to the functions $g_0(t), g_1(t), \dots, g_k(t)$ just as in Example (5.3.3) leads us to consider the following program:

$$(DGP)\quad \begin{cases} \text{Maximize} \quad v(\boldsymbol{\delta}) = \left(\prod_{j=1}^{p} \left(\dfrac{c_j}{\delta_j}\right)^{\delta_j}\right) \prod_{i=1}^{k} \lambda_i(\boldsymbol{\delta})^{\lambda_i(\boldsymbol{\delta})} \\ \text{subject to the constraints} \\ \qquad \delta_1 + \cdots + \quad \delta_{n_0} = 1, \\ \quad \alpha_{11}\delta_1 + \cdots + \alpha_{p1}\delta_p \ = 0, \\ \quad \vdots \\ \quad \alpha_{1m}\delta_1 + \cdots + \alpha_{pm}\delta_p \ = 0, \\ \text{where} \quad \delta_i > 0 \text{ for } i = 1, \dots, n_0, \text{ and for each } k \ge 1, \text{ either} \\ \delta_i > 0 \text{ for all } i \text{ with } n_{k-1} + 1 \le i \le n_k \text{ or } \delta_i = 0 \text{ for all } i \\ \text{with } n_{k-1} + 1 \le i \le n_k. \text{ Here, } \lambda_i(\boldsymbol{\delta}) = \delta_{n_{i-1}+1} + \cdots + \delta_{n_i}. \end{cases}$$

The program (DGP) is called the *dual program* of (GP), the function $v(\boldsymbol{\delta})$ is the *dual objective function* and the constraints in (DGP) are the *dual constraints.* If $\delta_1, \delta_2, \ldots, \delta_p$ are numbers that satisfy the dual constraints in (DGP), then $\boldsymbol{\delta} = (\delta_1, \ldots, \delta_p)$ is a *feasible vector* for (DGP) and (DGP) is said to be *consistent*. A vector $\boldsymbol{\delta}^* = (\delta_1^*, \ldots, \delta_p^*)$ that maximizes $v(\boldsymbol{\delta})$ on the set of feasible vectors for (DGP) is a *solution* to (DGP).

For the constrained geometric program considered in Example (5.3.3), there are four dual variables $\delta_1, \delta_2, \delta_3, \delta_4$ because the objective function and the single constraint function each contain two terms. There is one and only one vector $\boldsymbol{\delta} = (\frac{2}{3}, \frac{1}{3}, \frac{1}{3}, \frac{1}{3})$ that is feasible for (DGP) so it is automatically a solution for (DGP).

In general, the number of dual variables in (DGP) is equal to the total number $p\ (= n_k)$ of terms in the objective and constraint functions for (GP). Note that the constraint equations in (DGP) are *linear*, so the problem of identifying feasible vectors for (DGP) reduces to that of finding the solutions of a linear system of equations that have nonnegative components. Also note that $\lambda_i(\boldsymbol{\delta})$ is simply the sum of the components of $\boldsymbol{\delta}$ that correspond to terms of the ith constraint function $g_i(\mathbf{t})$ for $i = 1, 2, \ldots, k$.

Before we proceed to the development of the relationship between the solutions of the program (GP) and the corresponding dual program (DGP), we will work out the dual of another concrete geometric program.

(5.3.4) Example. Consider the geometric program

$$\begin{cases} \text{Minimize} \quad t_1 t_2^{-1} t_3^2 \\ \text{subject to the constraints} \\ \qquad\qquad \frac{1}{2} t_1^3 t_2 t_3^{-1} \le 1, \\ \frac{1}{4} t_1^{-1/2} + \frac{1}{4} t_2^{-1/2} + \frac{1}{4} t_3 \le 1, \\ \text{where} \quad t_1 > 0, \quad t_2 > 0, \quad t_3 > 0. \end{cases}$$

The objective function and the two constraints contain a total of five terms so there are five dual variables. The dual constraints are

$$\begin{aligned} \delta_1 \qquad\qquad\qquad\qquad\qquad &= 1, \\ \delta_1 + 3\delta_2 - \tfrac{1}{2}\delta_3 \qquad\qquad\quad &= 0, \\ -\delta_1 + \delta_2 \qquad\quad - \tfrac{1}{2}\delta_4 \qquad &= 0, \\ 2\delta_1 - \delta_2 \qquad\qquad\quad + \delta_5 &= 0, \end{aligned}$$

where

$$\delta_1 > 0, \qquad \delta_2 \ge 0, \qquad \delta_3 \ge 0, \qquad \delta_4 \ge 0, \qquad \delta_5 \ge 0,$$

and the dual objective function is

$$v(\boldsymbol{\delta}) = \left(\frac{1}{\delta_1}\right)^{\delta_1} \left(\frac{1}{2\delta_2}\right)^{\delta_2} \left(\frac{1}{4\delta_3}\right)^{\delta_3} \left(\frac{1}{4\delta_4}\right)^{\delta_4} \left(\frac{1}{4\delta_5}\right)^{\delta_5} (\delta_2)^{\delta_2} (\delta_3 + \delta_4 + \delta_5)^{\delta_3 + \delta_4 + \delta_5}$$

The given program is consistent; in fact, it is even superconsistent because, for example, the values $t_1 = 1, t_2 = 1, t_3 = 1$ satisfy both constraints with strict inequalities.

The dual program is consistent; in fact, it is a routine matter to verify that the general solution of the four dual constraint equations is

$$\delta_1 = 1, \quad \delta_2 = -\tfrac{1}{3} + \tfrac{1}{6}r, \quad \delta_3 = r, \quad \delta_4 = -\tfrac{8}{3} + \tfrac{1}{3}r, \quad \delta_5 = -\tfrac{7}{3} + \tfrac{1}{6}r,$$

where r is an arbitrary real number. The requirement that $\delta_i \geq 0$ for $i = 1, 2, 3, 4, 5$ adds the further restriction that $r \geq 14$. Thus, the set of points feasible for the dual program is

$$F = \{(1, [-\tfrac{1}{3} + \tfrac{1}{6}r], r, [-\tfrac{8}{3} + \tfrac{1}{3}r], [-\tfrac{7}{3} + \tfrac{1}{6}r]) \colon r \geq 14\}.$$

The Karush–Kuhn–Tucker Theorem would not seem to be applicable to geometric programs because posynomials need not be convex functions. For example, the posynomial in one variable

$$g(t) = t^{1/2}, \qquad t > 0,$$

is not convex. However, any posynomial $g(\mathbf{t})$ can be transformed into a convex function $h(\mathbf{x})$ by the change of variables

$$(*) \qquad t_j = e^{x_j}, \qquad j = 1, 2, \ldots, m.$$

More precisely, if the change of variables $(*)$ is applied to the posynomial

$$g(\mathbf{t}) = \sum_{i=1}^{n} c_i t_1^{\alpha_{i1}} t_2^{\alpha_{i2}} \ldots t_m^{\alpha_{im}}, \qquad c_i > 0,$$

the corresponding function of $\mathbf{x} = (x_1, x_2, \ldots, x_m)$ is

$$h(\mathbf{x}) = \sum_{i=1}^{n} c_i e^{\sum_{j=1}^{m} \alpha_{ij} x_j},$$

and $h(\mathbf{x})$ is convex on R^m by virtue of (2.3.10).

This observation enables us to transform the standard constrained geometric program

$$(GP) \quad \begin{cases} \text{Minimize} \quad g_0(\mathbf{t}) \quad \text{subject to the constraints} \\ \quad g_1(\mathbf{t}) \leq 1, \quad g_2(\mathbf{t}) \leq 1, \ldots, g_k(\mathbf{t}) \leq 1, \\ \text{where} \quad t_1 > 0, \quad t_2 > 0, \ldots, \quad t_m > 0 \end{cases}$$

into an *associated convex program*

$$(GP)^* \quad \begin{cases} \text{Minimize} \quad h_0(\mathbf{x}) \quad \text{subject to the constraints} \\ \quad h_1(\mathbf{x}) - 1 \leq 0, \quad h_2(\mathbf{x}) - 1 \leq 0, \ldots, h_k(\mathbf{x}) - 1 \leq 0, \\ \text{where} \quad \mathbf{x} \in R^m, \end{cases}$$

via the change of variables $t_i = e^{x_i}$ for $i = 1, 2, \ldots, m$. Moreover, because $t = e^x$ is a strictly increasing function, the programs (GP) and $(GP)^*$ are equivalent in the sense that $\mathbf{t}^* = (t_1^*, t_2^*, \ldots, t_m^*)$ is a solution to (GP) if and only if $\mathbf{x}^* = (x_1^*, \ldots, x_m^*)$ is a solution to $(GP)^*$ where $t_i^* = e^{x_i^*}$ for $i = 1, 2, \ldots, m$.

We are now prepared to state and prove the central result in the theory of constrained geometric programming.

(5.3.5) Theorem.

(1) *If* $\mathbf{t}$ *is a feasible vector for the constrained geometric program* (GP) *and if* $\boldsymbol{\delta}$ *is a feasible vector for the corresponding dual program* (DGP), *then*

$$g_0(\mathbf{t}) \geq v(\boldsymbol{\delta}) \qquad \text{(the Primal–Dual Inequality).}$$

(2) *Suppose that the constrained geometric program* (GP) *is superconsistent and that* $\mathbf{t}^*$ *is a solution for* (GP). *Then the corresponding dual program* (DGP) *is consistent and has a solution* $\boldsymbol{\delta}^*$ *which satisfies*

$$g_0(\mathbf{t}^*) = v(\boldsymbol{\delta}^*),$$

and

$$\delta_i^* = \begin{cases} \dfrac{u_i(\mathbf{t}^*)}{g_0(\mathbf{t}^*)}, & i = 1, \ldots, n_0, \\ \lambda_j(\boldsymbol{\delta}^*) u_i(\mathbf{t}^*), & i = n_{j-1} + 1, \ldots, n_j; \quad j = 1, \ldots, k. \end{cases}$$

PROOF. (1) The Primal–Dual Inequality follows immediately from the definition of the dual program (DGP) and the Arithmetic–Geometric Mean Inequalities (2.4.1) and (5.3.1).

(2) Since (GP) is superconsistent, so is the associated convex program $(GP)^*$. Also since (GP) has a solution $\mathbf{t}^* = (t_1^*, t_2^*, \ldots, t_m^*)$, the associated convex program $(GP)^*$ has a solution $\mathbf{x}^* = (x_1^*, x_2^*, \ldots, x_m^*)$ given by

$$x_i^* = \ln t_i^*, \qquad i = 1, 2, \ldots, m.$$

According to the Karush–Kuhn–Tucker Theorem (5.2.14), there is a vector $\boldsymbol{\lambda}^* = (\lambda_1^*, \ldots, \lambda_k^*)$ such that:

(a) $\lambda_i^* \geq 0$ for $i = 1, 2, \ldots, k$;
(b) $\lambda_i^*(h_i(\mathbf{x}^*) - 1) = 0$ for $i = 1, 2, \ldots, k$;
(c) $\dfrac{\partial h_0}{\partial x_j}(\mathbf{x}^*) + \sum_{i=1}^{k} \lambda_i^* \dfrac{\partial h_i}{\partial x_j}(\mathbf{x}^*) = 0$ for $j = 1, \ldots, m$.

Because $t_i = e^{x_i}$ for $i = 1, \ldots, m$, it follows that for $i = 0, 1, \ldots, k$

$$\frac{\partial h_i}{\partial x_j} = \frac{\partial h_i}{\partial t_j} \frac{dt_j}{dx_j} = \frac{\partial g_i}{\partial t_j} e^{x_j},$$

so condition (c) is equivalent to

$$\text{(c}'\text{)} \qquad \frac{\partial g_0}{\partial t_j}(\mathbf{t}^*) + \sum_{i=1}^{k} \lambda_i^* \frac{\partial g_i}{\partial t_j}(\mathbf{t}^*) = 0, \qquad j = 1, 2, \ldots, m$$

since $e^{x_j} > 0$ for $j = 1, \ldots, m$. But $t_j^* > 0$ for $j = 1, 2, \ldots, m$, so (c′) is equivalent to

$$\text{(c}''\text{)} \qquad t_j^* \frac{\partial g_0}{\partial t_j}(\mathbf{t}^*) + \sum_{i=1}^{k} \lambda_i^* t_j^* \frac{\partial g_i}{\partial t_j}(\mathbf{t}^*) = 0, \qquad j = 1, \ldots, m.$$

Because the terms of $g_i(\mathbf{t})$ are of the form

$$u_q(\mathbf{t}) = c_q t_1^{\alpha_{q1}} t_2^{\alpha_{q2}} \ldots t_m^{\alpha_{qm}},$$

it is clear that

$$t_j^* \frac{\partial g_i}{\partial t_j}(\mathbf{t}^*) = \sum_{q=n_{i-1}+1}^{n_i} \alpha_{qj} u_q(\mathbf{t}^*), \qquad j = 1, \ldots, m,$$

so (c)″ implies that

$$0 = \sum_{q=1}^{n_0} \alpha_{qj} u_q(\mathbf{t}^*) + \sum_{r=1}^{k} \sum_{q=n_{r-1}+1}^{n_r} \lambda_r^* \alpha_{qj} u_q(\mathbf{t}^*).$$

If we divide the last equation by

$$g_0(\mathbf{t}^*) = \sum_{q=1}^{n_0} u_q(\mathbf{t}^*),$$

we obtain

$$0 = \sum_{q=1}^{n_0} \alpha_{qj} \left(\frac{u_q(\mathbf{t}^*)}{g_0(\mathbf{t}^*)} \right) + \sum_{r=1}^{k} \sum_{q=n_{r-1}+1}^{n_r} \alpha_{qj} \left(\frac{\lambda_r^* u_q(\mathbf{t}^*)}{g_0(\mathbf{t}^*)} \right).$$

Define the vector $\boldsymbol{\delta}^*$ by

$$\delta_q^* = \begin{cases} \dfrac{u_q(\mathbf{t}^*)}{g_0(\mathbf{t}^*)}, & q = 1, 2, \ldots, n_0, \\[2ex] \dfrac{\lambda_r^* u_q(\mathbf{t}^*)}{g_0(\mathbf{t}^*)}, & q = n_{r-1} + 1, \ldots, n_r, \quad r = 1, \ldots, k. \end{cases}$$

Note that $\delta_q^* > 0$ for $q = 1, 2, \ldots, n_0$ and that, for each $r \geq 1$, either $\delta_i^* > 0$ for all i with $n_{r-1} + 1 \leq i \leq n_r$ or $\delta_i^* = 0$ for all i with $n_{r-1} + 1 \leq i \leq n_r$ according as the corresponding Karush–Kuhn–Tucker multiplier λ_r^* is positive or zero. Also observe that the vector $\boldsymbol{\delta}^*$ satisfies all of the m exponents constraint equations in (DGP) as well as the constraint

$$\sum_{q=1}^{n_0} \delta_q^* = \sum_{q=1}^{n_0} \frac{u_q(\mathbf{t}^*)}{g_0(\mathbf{t}^*)} = 1.$$

Therefore, $\boldsymbol{\delta}^* = (\delta_1^*, \ldots, \delta_p^*)$ is a feasible vector for (DGP).

The Karush–Kuhn–Tucker multipliers λ_r^* are related to the corresponding $\lambda_r(\boldsymbol{\delta}^*)$ in (DGP) as follows:

$$\lambda_r(\boldsymbol{\delta}^*) = \sum_{q=n_{r-1}+1}^{n_r} \delta_q^* \sum_{q=n_{r-1}+1}^{n_r} \lambda_r^* \frac{u_q(\mathbf{t}^*)}{g_0(\mathbf{t}^*)} = \lambda_r^* \frac{g_r(\mathbf{t}^*)}{g_0(\mathbf{t}^*)}$$

for $r = 1, \ldots, k$. The Karush–Kuhn–Tucker condition (b) yields

$$\text{(b}'\text{)} \qquad \lambda_r^*(g_r(t^*) - 1) = 0, \qquad r = 1, \ldots, k,$$

so $\lambda_r^* g_r(\mathbf{t}^*) = \lambda_r^*$ for $r = 1, \ldots, k$. Therefore, for $r = 1, \ldots, k$ and $q = n_{r-1} + 1, \ldots, n_r$, we see that

$$(*) \qquad \delta_q^* = \frac{\lambda_r^* u_q(\mathbf{t}^*)}{g_0(\mathbf{t}^*)} = \frac{\lambda_r^* g_r(\mathbf{t}^*) u_q(\mathbf{t}^*)}{g_0(\mathbf{t}^*)} = \lambda_r(\boldsymbol{\delta}^*) u_q(\mathbf{t}^*).$$

The fact that $\boldsymbol{\delta}^*$ is feasible for (DGP) and that $\mathbf{t}^*$ is feasible for (GP) implies that

$$g_0(\mathbf{t}^*) \geq v(\boldsymbol{\delta}^*)$$

because of the Primal–Dual Inequality (1). Moreover, the values of $\boldsymbol{\delta}_q^*$ for $q = 1, 2, \ldots, p$ are precisely those that force equality in the Arithmetic–Geometric Mean Inequalities (2.4.1) and (5.3.1) that were used to obtain the Duality Inequality. Finally, equation (b′) shows that either $g_r(t^*) = 1$ or $\lambda_r^* = 0$ for $r = 1, 2, \ldots, k$ and equation $(*)$ shows that $\lambda_r^* = 0$ if and only if $\lambda_r(\boldsymbol{\delta}^*) = 0$ for $r = 1, \ldots, k$. This means that the values of δ_q^* actually force equality in the Primal–Dual Inequality. This completes the proof.

The second statement of the preceding theorem implies that if (GP) is a superconsistent geometric program for which the dual (DGP) is not consistent, then (GP) has no solution. It can also be shown that if (GP) and (DGP) are both consistent, then (GP) (and hence (DGP)) must have a solution. (For a proof of this latter result, see *Geometric Programming: Theory and Applications* by R. J. Duffin, E. L. Peterson, and C. Zener. (Wiley, New York, 1967))

Some philosophical comments concerning the last theorem and its proof are in order. First of all, the theorem itself is a very practical result since it justifies the computational procedure used to solve constrained geometric programs. More precisely, the theoretical existence of the Karush–Kuhn–Tucker multipliers is all that is needed to prove that the very practical calculation of the solution $\boldsymbol{\delta}^*$ of (DGP) and hence the solution $\mathbf{t}^*$ of (GP) can be made. This illustrates how in mathematics theoretical facts can have very practical implications. Second, the proof of Theorem (5.3.5) shows that the Karush–Kuhn–Tucker multipliers λ_r^* are related to the numbers $\lambda_r(\boldsymbol{\delta}^*)$ prescribed in (DGP) by the formula

$$\lambda_r(\boldsymbol{\delta}^*) = \frac{\lambda_r^*}{g_0(\mathbf{t}^*)}. \tag{1}$$

Since the Karush–Kuhn–Tucker multiplier λ_r^* measures the sensitivity of the solution of the associated convex program to changes in the rth constraint, the formula (1) shows that the size of $\lambda_r(\boldsymbol{\delta}^*)$ gives a qualitative measure of the sensitivity of the solution of the constrained geometric program to its rth constraint. In short, the larger the value of $\lambda_r(\boldsymbol{\delta}^*)$, the greater the decrease of the minimum value of the geometric program that results from relaxation of the kth constraint.

The constrained geometric program considered in Example (5.3.3) was particularly simple to solve because the dual had a unique feasible vector and so that vector is necessarily the solution of the dual program. On the other hand, the set of feasible vectors for the dual of the geometric program considered in Example (5.3.4) is

$$\{(1, [-\tfrac{1}{3} + \tfrac{1}{6}r], r, [-\tfrac{8}{3} + \tfrac{1}{3}r], [-\tfrac{7}{3} + \tfrac{1}{6}r]): r \geq 14\}.$$

Since this program is superconsistent, we know from the remark after (5.3.5) that both the program and its dual have solutions $\mathbf{t}^*$ and $\boldsymbol{\delta}^*$, but calculation of these solutions is not quite as easy as the program in Example (5.3.3). If we replace the variables in the dual objective function $v(\boldsymbol{\delta})$ by

$$\delta_1 = 1, \quad \delta_2 = -\tfrac{1}{3} + \tfrac{1}{6}r, \quad \delta_3 = r, \quad \delta_4 = -\tfrac{8}{3} + \tfrac{1}{3}r, \quad \delta_5 = -\tfrac{7}{3} + \tfrac{1}{6}r,$$

then we reduce the problem of finding a solution δ^* of (DGP) to that of maximizing the function of one variable $v(r)$ on the interval $r \geq 14$.

The Primal–Dual Inequality provides a means for estimating the minimum value of the program (GP) (or the maximum value of the dual program (DGP)). In fact, if $\mathbf{t}$ is a feasible vector for (GP) and $\boldsymbol{\delta}$ is feasible for (DGP), then

$$g_0(\mathbf{t}) \geq \min(GP) \geq \max(DGP) \geq v(\boldsymbol{\delta}).$$

5.4. Dual Convex Programs

In this section, we will associate with a given convex program (P) a new program (DP) called the *dual* of (P). The dual program (DP) is an unconstrained program which is sometimes easier to solve than the given (primal) program (P) and yet the solutions of (DP) can be used to generate solutions of (P). Thus, the consideration of the dual program (DP) provides one approach to the solution of the primal program (P). Moreover, if the given program (P) arises from some economic or physical context, then it is often possible to interpret the dual program (DP) in economic or physical terms and this interpretation may provide new insights into the context of (P).

To set the stage, recall our formulation of the general convex program

$$(P) \quad \begin{cases} \text{Minimize} \quad f(\mathbf{x}) \quad \text{subject to} \\ g_1(\mathbf{x}) \leq 0, \ldots, \quad g_m(\mathbf{x}) \leq 0; \qquad \mathbf{x} \in C, \\ \text{where} \quad f(\mathbf{x}), g_1(\mathbf{x}), \ldots, g_m(\mathbf{x}) \text{ are convex functions} \\ \text{defined on a convex set } C. \end{cases}$$

The quantity MP is defined for (P) as follows:

$$MP = \inf\{f(\mathbf{x}): \mathbf{x} \in C, g_1(\mathbf{x}) \leq 0, \ldots, g_m(\mathbf{x}) \leq 0\},$$

that is, MP is the infimum of the objective function $f(\mathbf{x})$ on the set F of feasible points for (P). If $\mathbf{x}^* \in F$ and $f(\mathbf{x}^*) = MP$, then $\mathbf{x}^*$ is a solution of (P). If $f(\mathbf{x})$ is not bounded below on F, then $MP = -\infty$ and (P) has no solutions.

Suppose that $\mathbf{x}^*$ is a solution to (P) and that there is a vector $\boldsymbol{\lambda}^* = (\lambda_1^*, \ldots, \lambda_m^*)$ of Karush–Kuhn–Tucker multipliers associated with $\mathbf{x}^*$ as in the statement of the Saddle Point Form of the Karush–Kuhn–Tucker Theorem (5.2.13), that is,

(1) $\boldsymbol{\lambda}^* \geq 0$;
(2) $L(\mathbf{x}^*, \boldsymbol{\lambda}) \leq L(\mathbf{x}^*, \boldsymbol{\lambda}^*) \leq L(\mathbf{x}, \boldsymbol{\lambda}^*)$ for all $\mathbf{x} \in C$ and all $\boldsymbol{\lambda} \geq \mathbf{0}$;
(3) $\lambda_i^* g_i(\mathbf{x}^*) = 0$ for $i = 1, 2, \ldots, m$.

Let us concentrate, for the moment, on the implications of the saddle point condition (2). For any given $\boldsymbol{\lambda} \geq \mathbf{0}$, it is clear from (2) that

$$\inf_{\mathbf{x} \in C} L(\mathbf{x}, \boldsymbol{\lambda}) \leq L(\mathbf{x}^*, \boldsymbol{\lambda}) \leq L(\mathbf{x}^*, \boldsymbol{\lambda}^*).$$

Consequently,

$$\sup_{\boldsymbol{\lambda} \geq 0} \left\{ \inf_{\mathbf{x} \in C} L(\mathbf{x}, \boldsymbol{\lambda}) \right\} \leq L(\mathbf{x}^*, \boldsymbol{\lambda}^*).$$

On the other hand, since $L(\mathbf{x}^*, \boldsymbol{\lambda}^*) \leq L(\mathbf{x}, \boldsymbol{\lambda}^*)$ for all $\mathbf{x} \in C$ by (2), it follows that

$$L(\mathbf{x}^*, \boldsymbol{\lambda}^*) \leq \inf_{\mathbf{x} \in C} L(\mathbf{x}, \boldsymbol{\lambda}^*) \leq \sup_{\boldsymbol{\lambda} \geq \mathbf{0}} \left\{ \inf_{\mathbf{x} \in C} L(\mathbf{x}, \boldsymbol{\lambda}) \right\}.$$

We conclude that

$$L(\mathbf{x}^*, \boldsymbol{\lambda}^*) = \sup_{\boldsymbol{\lambda} \geq 0} \left\{ \inf_{\mathbf{x} \in C} L(\mathbf{x}, \boldsymbol{\lambda}) \right\}.$$

But now observe that (3) yields

$$MP = f(\mathbf{x}^*) = f(\mathbf{x}^*) + \sum_{i=1}^{m} \lambda_i^* g_i(\mathbf{x}^*) = L(\mathbf{x}^*, \boldsymbol{\lambda}^*),$$

so

$$MP = \sup_{\boldsymbol{\lambda} \geq \mathbf{0}} \left\{ \inf_{\mathbf{x} \in C} L(\mathbf{x}, \boldsymbol{\lambda}) \right\}.$$

These considerations lead us to the following definitions.

(5.4.1) Definitions. Given the convex program

$$(P) \quad \begin{cases} \text{Minimize} \quad f(\mathbf{x}) \quad \text{subject to} \\ \quad g_1(\mathbf{x}) \leq 0, \ldots, \quad g_m(\mathbf{x}) \leq 0; \\ \text{where} \quad f(\mathbf{x}), g_1(\mathbf{x}), \ldots, g_m(\mathbf{x}) \text{ are convex functions with} \\ \text{continuous first partial derivatives on a convex set } C \end{cases}$$

define the *dual program*

$$(DP)\quad \begin{cases} \text{Maximize} \quad h(\boldsymbol{\lambda}) = \inf_{\mathbf{x}\in C} L(\mathbf{x}, \boldsymbol{\lambda}), \\ \text{where} \quad \boldsymbol{\lambda} \geq \mathbf{0} \text{ and } L(\mathbf{x}, \boldsymbol{\lambda}) \text{ is the Lagrangian of } (P). \end{cases}$$

The value of

$$\sup_{\boldsymbol{\lambda}\geq\mathbf{0}} \left\{ \inf_{\mathbf{x}\in C} L(\mathbf{x}, \boldsymbol{\lambda}) \right\}$$

will be denoted by MD. A vector $\boldsymbol{\lambda} \geq \mathbf{0}$ in R^m is *feasible* for (DP) if $h(\lambda) = \inf_{\mathbf{x}\in C} L(\mathbf{x}, \boldsymbol{\lambda}) > -\infty$. The dual program (DP) is *consistent* if there is at least one feasible vector. If $\boldsymbol{\lambda}^*$ is a feasible vector for (DP) such that $h(\boldsymbol{\lambda}^*) = \sup_{\boldsymbol{\lambda}\geq\mathbf{0}} h(\boldsymbol{\lambda})$, then $\boldsymbol{\lambda}^*$ is a *solution* of (DP).

The computations preceding the definition of the dual program (DP) for (P) show that if $\mathbf{x}^*$ is a solution to (P) and if $\boldsymbol{\lambda}^*$ is a corresponding vector of the Karush–Kuhn–Tucker multipliers, then $\boldsymbol{\lambda}^*$ is a solution of (DP) and $MD = MP$. In particular, if a superconsistent convex program (P) has a solution $\mathbf{x}^*$, then there is a vector $\boldsymbol{\lambda}^*$ of the Karush–Kuhn–Tucker multipliers that is a solution of the dual program (DP). Thus, the Karush–Kuhn–Tucker vectors for the primal program (P) are solutions of the dual program (DP). Moreover, if $\boldsymbol{\lambda}^*$ is a Karush–Kuhn–Tucker vector for (P) and if (P) is known to have a solution $\mathbf{x}^*$, then since

$$f(\mathbf{x}^*) = MP = L(\mathbf{x}^*, \boldsymbol{\lambda}^*) = \inf_{\mathbf{x}\in C} L(\mathbf{x}, \boldsymbol{\lambda}^*)$$

it follows that $\mathbf{x}^*$ can be found by minimizing $L(\mathbf{x}, \boldsymbol{\lambda}^*)$ over C. Thus, at least in theory, the duality approach to the solution of a given convex program is quite simple:

Step 1. Given a convex program (P), construct its dual (DP).
Step 2. Find the solutions $\boldsymbol{\lambda}^*$ of (DP).
Step 3. Find the corresponding solutions to (P) by minimizing $L(\mathbf{x}, \boldsymbol{\lambda}^*)$ on C.

Of course, this approach is only useful if the dual program (DP) is easier to solve than (P). Although this is not always the case, it happens often enough to make the approach worthwhile.

We will now illustrate the duality approach to the solution of convex programs with some examples, beginning with the important special case of linear programs.

(5.4.2) Example (Linear Programming Revisited). In Example (5.2.5), we demonstrated that if A is an $m \times n$-matrix, if $\mathbf{b} \in R^n$, and if $\mathbf{c} \in R^m$, then the general linear program

$$(LP)\quad \begin{cases} \text{Minimize} \quad \mathbf{b}\cdot\mathbf{x} \quad \text{subject to} \\ A\mathbf{x} \geq \mathbf{c} \quad \text{where} \quad \mathbf{x} \geq \mathbf{0} \end{cases}$$

can be reformulated as a convex program (P) by setting

$$f(\mathbf{x}) = \mathbf{b}\cdot\mathbf{x}, \qquad C = \{\mathbf{x} \in R^n : \mathbf{x} \geq \mathbf{0}\}$$

$$g_i(\mathbf{x}) = c_i - \mathbf{a}^{(i)}\cdot\mathbf{x} \quad \text{where} \quad \mathbf{a}^{(i)} = i\text{th row of } A \text{ for } i = 1, 2, \ldots, m.$$

In this case, the Lagrangian of (P) is given by

$$\begin{aligned} L(\mathbf{x}, \boldsymbol{\lambda}) &= \mathbf{b}\cdot\mathbf{x} + \sum_{i=1}^{m} \lambda_i(c_i - \mathbf{a}^{(i)}\cdot\mathbf{x}) \\ &= \left(\mathbf{b} - \sum_{i=1}^{m} \lambda_i \mathbf{a}^{(i)}\right)\cdot\mathbf{x} + \sum_{i=1}^{m} \lambda_i c_i \\ &= (\mathbf{b} - A^{\mathrm{T}}\boldsymbol{\lambda})\cdot\mathbf{x} + \boldsymbol{\lambda}\cdot\mathbf{c}. \end{aligned}$$

To construct the dual of (LP) we begin by identifying the set of feasible points for the dual, that is, the set of all $\boldsymbol{\lambda} \geq 0$ for which

$$\inf_{\mathbf{x}\in C} L(\mathbf{x}, \boldsymbol{\lambda}) > -\infty.$$

To this end, we note that if $b_{i_0} - (A^{\mathrm{T}}\boldsymbol{\lambda})_{i_0} < 0$ for an i_0 and if $\mathbf{x}^{(t)}$ is defined for $t < 0$ by

$$x_j^{(t)} = \begin{cases} t(b_{i_0} - (A^{\mathrm{T}}\boldsymbol{\lambda})_{i_0}) & \text{if } j = i_0, \\ 0 & \text{if } j \neq i_0, \end{cases}$$

then $\mathbf{x}^{(t)} \geq \mathbf{0}$ for $t < 0$ and

$$L(\mathbf{x}^{(t)}, \boldsymbol{\lambda}) = t(b_{i_0} - (A^{\mathrm{T}}\boldsymbol{\lambda})_{i_0})^2 + \boldsymbol{\lambda}\cdot\mathbf{c}.$$

Thus $L(\mathbf{x}^{(t)}, \boldsymbol{\lambda}) \to -\infty$ as $t \to -\infty$ and so

$$\inf_{\mathbf{x}\geq 0} L(\mathbf{x}, \boldsymbol{\lambda}) = -\infty.$$

Hence, $\boldsymbol{\lambda}$ is not feasible for the dual. On the other hand, if $\mathbf{b} - A^{\mathrm{T}}\boldsymbol{\lambda} \geq \mathbf{0}$, then

$$L(\mathbf{x}, \boldsymbol{\lambda}) = (\mathbf{b} - A^{\mathrm{T}}\boldsymbol{\lambda})\cdot\mathbf{x} + \boldsymbol{\lambda}\cdot\mathbf{c} \geq \boldsymbol{\lambda}\cdot\mathbf{c}$$

for all $\mathbf{x} \geq 0$, so the set of feasible vectors $\boldsymbol{\lambda}$ for the dual of L is precisely

$$F = \{\boldsymbol{\lambda} \in R^m : \boldsymbol{\lambda} \geq 0 \text{ and } A^{\mathrm{T}}\boldsymbol{\lambda} \leq \mathbf{b}\}.$$

Next, we note that if $\boldsymbol{\lambda} \in F$, then

$$h(\boldsymbol{\lambda}) = \inf_{\mathbf{x}\geq 0} L(\mathbf{x}, \boldsymbol{\lambda}) = \boldsymbol{\lambda}\cdot\mathbf{c}$$

because $L(\mathbf{0}, \boldsymbol{\lambda}) = \boldsymbol{\lambda}\cdot\mathbf{c}$ and $L(\mathbf{x}, \boldsymbol{\lambda}) \geq \boldsymbol{\lambda}\cdot\mathbf{c}$ for all $\mathbf{x} \geq \mathbf{0}$. Consequently, the dual program for (LP) can be formulated as

$$(DLP) \quad \begin{cases} \text{Maximize} \quad h(\boldsymbol{\lambda}) = \boldsymbol{\lambda}\cdot\mathbf{c} \\ \text{subject to } A^{\mathrm{T}}\boldsymbol{\lambda} \leq \mathbf{b}, \quad \boldsymbol{\lambda} \geq 0, \end{cases}$$

which is a linear program in standard maximum form.

This discussion sets up

(5.4.3) Theorem (Linear Programming Duality). *If (LP) is superconsistent and if* $\mathbf{x}^*$ *is a solution for (LP), then the dual program (DLP) has a solution* $\boldsymbol{\lambda}^*$ *and*

$$\mathbf{b}\cdot\mathbf{x}^* = f(\mathbf{x}^*) = h(\boldsymbol{\lambda}^*) = \mathbf{c}\cdot\boldsymbol{\lambda}^*.$$

Moreover, if $\mathbf{x}$ *is any feasible vector for (LP) and* $\boldsymbol{\lambda}$ *is any feasible vector for (DLP), then*

$$\mathbf{b}\cdot\mathbf{x} \geq \boldsymbol{\lambda}\cdot\mathbf{c}.$$

PROOF. The first statement is already established in the discussion on the previous two pages. The second statement follows from the inequalities

$$\mathbf{b}\cdot\mathbf{x} = f(\mathbf{x}) \geq f(\mathbf{x}^*) = h(\boldsymbol{\lambda}^*) \geq h(\boldsymbol{\lambda}) = \mathbf{c}\cdot\boldsymbol{\lambda}.$$

This completes the proof.

The theorem above is not the end of the story of duality in linear programming. In fact, the following stronger result is known:

The Duality Theorem of Linear Programming. *If either (LP) or (DLP) has a solution, then the other has a solution and the corresponding values of the objective functions are equal.*

The preceding theorem is true whether (*LP*) is superconsistent or not. Thus if a linear program has a solution then it also has a sensitivity vector which is nothing but a solution of the dual linear program. One popular proof of the Duality Theorem of Linear Programming is based on Farkas's Lemma. The latter result is discussed in Exercise 2.

The theorem above represents the best we can do by specializing the nonlinear methods of this chapter to the linear case. The full truth can be best uncovered by the use of purely linear methods. For more on this, see *Introduction to Linear and Nonlinear Programming* by D. G. Luenberger (Addison-Wesley, 1973). For another view of how nonlinear theory can establish another special case of this theorem, see the version of the Karush–Kuhn–Tucker Theorem studied in Remarks (7.2.6) of Chapter 7.

In Example (5.2.5), we pointed out that the classical Diet Problem provided an illustration of a linear program (*LP*) in minimum standard form. Recall that this problem asks that we plan a diet using n foods $F_1, \ldots, F_n$ that will provide at least the minimum daily requirements of m nutrients $N_1, \ldots, N_m$ at minimum cost. In the notation of (*LP*), we let

b_i = cost per unit of food F_i for $i = 1, \ldots, n$,

c_j = minimum number of units of nutrient N_j needed per day for $j = 1, \ldots, m$,

a_{ij} = number of units of nutrient N_j in one unit of food F_i for $i = 1, \ldots, n$ and $j = 1, \ldots, m$,

x_i = number of units of food F_i in a given diet for $i = 1, \ldots, n$.

What is the economic interpretation of the dual program of the Diet Problem? One way to think of it goes like this: Suppose that a manufacturer of nutrient pills proposes that we use his products to meet our nutritional needs instead of a diet consisting of the foods $F_1, \ldots, F_n$. Even though we find the idea of using pills instead of food to be revolting, we ask the manufacturer to send us a price list for his pills so that we can compare the cost of the cheapest adequate pill diet to that of the cheapest adequate food diet. The manufacturer agrees to "sharpen his pencil and come up with some really attractive prices." Of course, his notion of "attractive prices" is different from ours. He intends to set his price per unit of a given nutrient as high as he can so that he can maximize his revenue from us. However, he realizes that if he sets these prices so high that there is a cheaper adequate food diet, he will lose our business. How should he set the price λ_j of a unit of nutrient N_j so that he will maximize his revenue for an adequate pill diet and yet be sure that no adequate food diet is cheaper? In terms of the notation we have introduced, his problem can be formulated as follows:

$$\begin{cases} \text{Maximize } \lambda_1 c_1 + \lambda_2 c_2 + \cdots + \lambda_m c_m \\ \text{subject to} \\ \quad a_{11}\lambda_1 + a_{21}\lambda_2 + \cdots + a_{m1}\lambda_m \le b_1, \\ \quad \vdots \qquad\qquad\qquad\qquad\qquad\qquad \vdots \\ \quad a_{1n}\lambda_1 + a_{2n}\lambda_2 + \cdots + a_{mn}\lambda_m \le b_n, \\ \text{where } \quad \lambda_1 \ge 0, \ldots, \lambda_m \ge 0. \end{cases}$$

The typical constraint

$$a_{1i}\lambda_1 + a_{2i}\lambda_2 + \cdots + a_{mi}\lambda_m \le b_i$$

simply requires that the "value" of a unit of the food F_i, in terms of the prices $\lambda_1, \ldots, \lambda_m$ that he assigns to the nutrients contained in it, does not exceed the actual cost per unit of that food.

Observe that this maximization problem for the pill manufacturer is precisely the dual of the Diet Problem. Using this fact, the Duality Theorem for Linear Programming, and the Karush–Kuhn–Tucker Theorem, we can enjoy the cheapest possible food diet without doing the work necessary to figure out the diet ourselves. We simply let the pill manufacturer do most of the work for us! When he submits the price list $\lambda_1^*, \lambda_2^*, \ldots, \lambda_m^*$ per unit for the nutrients $N_1, N_2, \ldots, N_m$, we will know from the Duality Theorem that the cheapest adequate food diet $x_1^*, x_2^*, \ldots, x_n^*$ made up of the foods $F_1, F_2, \ldots, F_n$ satisfies

$$b_1 x_1^* + \cdots + b_n x_n^* = \lambda_1^* c_1 + \cdots + \lambda_m^* c_m,$$

and for any price λ_i^* which is positive, we know that

$$a_{i1} x_1^* + a_{i2} x_2^* + \cdots + a_{in} x_n^* = c_i$$

by the Karush–Kuhn–Tucker Theorem. We are able to use these relationships to solve for $x_1^*, \ldots, x_n^*$ without going to all of the trouble of solving the primal Diet Problem by some technique such as the Simplex Method.

Before we leave the special case of linear programs, let us work out a concrete example.

(5.4.4) Example. Consider the linear program

$$(LP)\quad \begin{cases} \text{Minimize} \quad 4x_1 + 15x_2 + 12x_3 + 2x_4, \\ \text{subject to the constraints} \\ \qquad 2x_2 + 3x_3 + x_4 \geq 1, \\ \quad x_1 + 3x_2 + \ x_3 - x_4 \geq 1, \\ \text{where} \quad x_1 \geq 0, \quad x_2 \geq 0, \quad x_3 \geq 0, \quad x_4 \geq 0. \end{cases}$$

In terms of the notation of the preceding example, we see that

$$\mathbf{b} = (4, 15, 12, 2), \qquad A = \begin{pmatrix} 0 & 2 & 3 & 1 \\ 1 & 3 & 1 & -1 \end{pmatrix},$$
$$\mathbf{c} = (1, 1),$$

and the dual program is

$$(DLP)\quad \begin{cases} \text{Maximize} \quad \lambda_1 + \lambda_2 \\ \text{subject to} \\ \qquad \lambda_2 \leq 4, \\ \quad 2\lambda_1 + 3\lambda_2 \leq 15, \\ \quad 3\lambda_1 + \ \lambda_2 \leq 12, \\ \quad \lambda_1 - \ \lambda_2 \leq 2, \\ \text{where} \quad \lambda_1 \geq 0, \quad \lambda_2 \geq 0. \end{cases}$$

Notice that (LP) is a superconsistent linear program; for example, $x_1 = 0$, $x_2 = 1$, $x_3 = 0$, $x_4 = 0$ yield strict inequalities in both constraints.

The dual program (DLP) can be solved graphically:

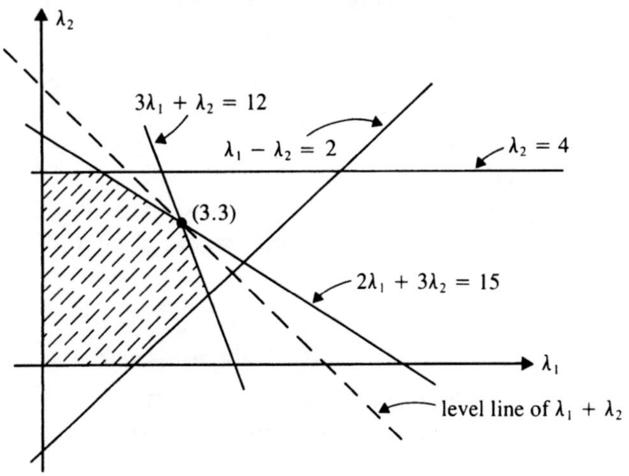

The solution of (DLP) is easily seen to be $\lambda_1^* = 3$, $\lambda_2^* = 3$ and the maximum of the dual objective function is $\lambda_1^* + \lambda_2^* = 6 = MD$.

The Karush–Kuhn–Tucker Complementary Slackness Condition can be written as

$$\begin{aligned} 3(1 \qquad\; - 2x_2 - 3x_3 - x_4) &= 0, \\ 3(1 - x_1 - 3x_2 - \;\; x_3 + x_4) &= 0, \end{aligned}$$

and the Duality Theorem for Linear Programming tells us that

$$4x_1 + 15x_2 + 12x_3 + 2x_4 = 6.$$

This leads us to the system of equations

$$\begin{aligned} x_1 + \;\; 3x_2 + \quad x_3 - \quad x_4 &= 1, \\ 2x_2 + \;\; 3x_3 + \;\; x_4 &= 1, \\ 4x_1 + 15x_2 + 12x_3 + 2x_4 &= 6, \end{aligned}$$

which has the solution

$$x_1 = -2x_4, \qquad x_2 = \tfrac{2}{7} + \tfrac{10}{7}x_4, \qquad x_3 = \tfrac{1}{7} - \tfrac{9}{7}x_4.$$

Since any solution of (LP) must have nonnegative components, we conclude that

$$x_1^* = 0, \qquad x_2^* = \tfrac{2}{7}, \qquad x_3^* = \tfrac{1}{7}, \qquad x_4^* = 0$$

is the unique solution to the given linear program (LP).

Next, we consider a class of quadratic programs for which the duality approach is both straightforward and fruitful.

(5.4.5) Example. Suppose that Q is an $n \times n$-positive definite matrix and that $\mathbf{0} \neq \mathbf{a} \in R^n$, $c \in R$. Consider the quadratic program

$$(P) \quad \begin{cases} \text{Minimize} & f(\mathbf{x}) = \tfrac{1}{2}\mathbf{x} \cdot Q\mathbf{x} \\ \text{subject to} & \mathbf{a} \cdot \mathbf{x} \leq c. \end{cases}$$

If $g(\mathbf{x}) = \mathbf{a} \cdot \mathbf{x} - c$, then $g(\mathbf{x})$ is convex on R^n and the constraint requires that $g(\mathbf{x}) \leq 0$. Also, $f(\mathbf{x})$ is convex on R^n because Q is positive definite. Therefore, (P) is a convex program; moreover, (P) is superconsistent since $\mathbf{a} \neq \mathbf{0}$ allows us to find $\mathbf{x}$ with $\mathbf{a} \cdot \mathbf{x} - c < 0$.

Of course, if $c \geq 0$, then $\mathbf{x}^* = \mathbf{0}$ is feasible for (P) and $\mathbf{x}^* = \mathbf{0}$ is the unique solution to (P) because it is the global minimizer of $f(\mathbf{x})$ on R^n. For this reason, we restrict further consideration of (P) to the case when $c < 0$.

The Lagrangian of (P) is given by

$$L(\mathbf{x}, \lambda) = \tfrac{1}{2}\mathbf{x} \cdot Q\mathbf{x} + \lambda(\mathbf{a} \cdot \mathbf{x} - c),$$

where $\mathbf{x} \in R^n$, $\lambda \geq 0$. Notice that for any fixed $\lambda \geq 0$, the Lagrangian $L(\mathbf{x}, \lambda)$ is a strictly convex function of $\mathbf{x}$. Hence, the unique minimizer $\mathbf{x}^*$ of $L(\mathbf{x}, \lambda)$

is determined by equating the gradient of $L(\mathbf{x}, \lambda)$ to zero, that is, $\mathbf{x}^*$ satisfies

$$\mathbf{0} = Q\mathbf{x}^* + \lambda\mathbf{a}.$$

Therefore,

$$(*) \qquad \mathbf{x}^* = -\lambda Q^{-1}\mathbf{a}$$

is the minimizer of $L(\mathbf{x}, \lambda)$ for fixed $\lambda \geq 0$. We conclude that all $\lambda \geq 0$ are feasible for (DP). Also, because

$$\begin{aligned} L(\mathbf{x}^*, \lambda) &= L(-\lambda Q^{-1}\mathbf{a}, \lambda) \\ &= \tfrac{1}{2}(-\lambda Q^{-1}\mathbf{a})\cdot Q(-\lambda Q^{-1}\mathbf{a}) + \lambda[\mathbf{a}\cdot(-\lambda Q^{-1}\mathbf{a}) - c] \\ &= \frac{\lambda^2}{2}\mathbf{a}\cdot Q^{-1}\mathbf{a} - \lambda^2\mathbf{a}\cdot Q^{-1}\mathbf{a} - \lambda c \\ &= -\left(\frac{\lambda^2}{2}\mathbf{a}\cdot Q^{-1}\mathbf{a} + \lambda c\right), \end{aligned}$$

we see that

$$MD = \sup_{\lambda \geq 0}\left\{\inf_{x \in R^n} L(\mathbf{x}, \lambda)\right\} = \max_{\lambda \geq 0}\left[-\left(\frac{\lambda^2}{2}(\mathbf{a}\cdot Q^{-1}\mathbf{a}) + \lambda c\right)\right].$$

To maximize

$$\varphi(\lambda) = -\left[\frac{\lambda^2}{2}(\mathbf{a}\cdot Q^{-1}\mathbf{a}) + \lambda c\right].$$

we compute

$$\begin{aligned} \varphi'(\lambda) &= -((\lambda\mathbf{a}\cdot Q^{-1}\mathbf{a}) + c), \\ \varphi''(\lambda) &= -\mathbf{a}\cdot Q^{-1}\mathbf{a}. \end{aligned}$$

Note that $\lambda^* = -c/(\mathbf{a}\cdot Q^{-1}\mathbf{a}) > 0$ is the critical point of $\varphi(\lambda)$ and this critical point is a strict maximizer of $\varphi(\lambda)$ since $\varphi''(\lambda^*) < 0$. Therefore, λ^* is a solution of the dual program. The solution $\mathbf{x}^*$ of the primal program (P) is the minimizer of $L(\mathbf{x}, \lambda^*)$ for $\mathbf{x} \in R^n$. Equation $(*)$ shows that this minimizer is given by

$$\mathbf{x}^* = -\lambda^* Q^{-1}\mathbf{a} = \frac{cQ^{-1}\mathbf{a}}{\mathbf{a}\cdot Q^{-1}\mathbf{a}}.$$

This completes the solution of the given class of quadratic programs.

In (5.4.2), we discussed duality in linear programming and its relation to the Karush–Kuhn–Tucker conditions. We shall now derive the corresponding duality result for general convex programs.

(5.4.6) Theorem (The Duality Theorem). *Suppose that $f(\mathbf{x}), g_1(\mathbf{x}), \ldots, g_m(\mathbf{x})$ are convex functions with continuous first partial derivatives defined on a convex* *subset C of* R^n. *If* $\mathbf{y}$ *is a feasible vector for the program (P) and* $\boldsymbol{\lambda}$ *is a feasible vector for the dual program (DP), then*

$$f(\mathbf{y}) \geq h(\boldsymbol{\lambda}) = \inf_{\mathbf{x} \in C} L(\mathbf{x}, \boldsymbol{\lambda}).$$

Consequently, if (P) and (DP) are both consistent, then MP and MD are both finite and

$$MP \geq MD \qquad (\textit{the Primal–Dual Inequality}).$$

PROOF. Because $\mathbf{y}$ is feasible for (P) and $\boldsymbol{\lambda}$ is feasible for (DP), it follows that

$$f(\mathbf{y}) \geq f(\mathbf{y}) + \sum_{i=1}^{m} \lambda_i g_i(\mathbf{y}) = L(\mathbf{y}, \boldsymbol{\lambda}),$$

since $\lambda_i \geq 0$ and $g_i(\mathbf{y}) \leq 0$ for $i = 1, 2, \ldots, m$. But then it is certainly true that

$$f(\mathbf{y}) \geq h(\boldsymbol{\lambda}) = \inf_{\mathbf{x} \in C} L(\mathbf{x}, \boldsymbol{\lambda})$$

which is precisely the first assertion. This inequality also shows that

$$MD = \sup_{\boldsymbol{\lambda} \geq \mathbf{0}} \left\{ \inf_{\mathbf{x} \in C} L(\mathbf{x}, \boldsymbol{\lambda}) \right\} \leq f(\mathbf{y})$$

and that

$$MP = \inf\{f(\mathbf{x}) \colon \mathbf{x} \in C, g_i(\mathbf{x}) \leq 0, i = 1, \ldots, m)\} \geq h(\boldsymbol{\lambda}),$$

whenever $\mathbf{y}$ is feasible for (P) and $\boldsymbol{\lambda}$ is feasible for (DP). This shows that, if (P) and (DP) are both consistent programs, then MP and MD are finite and $MP \geq MD$. This completes the proof.

(5.4.7) Corollary. *Suppose that* $\mathbf{x}^*$ *is a feasible vector for a convex program (P) and that* $\boldsymbol{\lambda}^*$ *is a feasible vector for the dual program (DP). If*

$$f(\mathbf{x}^*) = h(\boldsymbol{\lambda}^*)$$

then $\mathbf{x}^*$ *is a solution of (P) and* $\boldsymbol{\lambda}^*$ *is a solution of (DP).*

PROOF. According to the Primal–Dual Inequality and the definitions of MP and MD, we know that

$$f(\mathbf{x}^*) \geq MP \geq MD \geq h(\boldsymbol{\lambda}^*).$$

By our hypothesis, equality holds in each of these inequalities, so

$$f(\mathbf{x}^*) = MP, \qquad h(\boldsymbol{\lambda}^*) = MD$$

which implies that $\mathbf{x}^*$ and $\boldsymbol{\lambda}^*$ are solutions to (P) and (DP), respectively.

The discussion following Definition (5.4.1) shows that the duality approach to the solution of a convex program (P) works only if MP and MD are finite and $MP = MD$. Therefore, it is of some interest to know when this condition holds. The following example shows that it does not hold in general.

(5.4.8) Example (Duffin's Duality Gap). Consider the convex program

$$(P) \quad \begin{cases} \text{Minimize} \quad f(x, y) = e^{-y} \quad \text{subject to} \\ \quad g(x, y) = \sqrt{x^2 + y^2} - x \le 0; \quad (x, y) \in R^2. \end{cases}$$

This program was discussed in (5.2.9)(c) where it was shown that $MP(z)$ is a discontinuous function of z at $z = 0$. The corresponding dual program is

$$(DP) \quad \begin{cases} \text{Maximize} \quad h(\lambda) = \inf\{e^{-y} + \lambda g(x, y) : (x, y) \in R^2\}, \\ \text{subject to} \quad \lambda \ge 0. \end{cases}$$

Note that $\sqrt{x^2 + y^2} \ge x$ for all (x, y) in R^2 so the constraint $g(x, y) \le 0$ in (P) is satisfied if and only if $x \ge 0$ and $y = 0$. Therefore

$$MP = \inf\{e^{-y} : x \ge 0, y = 0\} = e^{-0} = 1.$$

On the other hand,

$$MD = \sup_{\lambda \ge 0} [\inf\{e^{-y} + \lambda[\sqrt{x^2 + y^2} - x] : (x, y) \in R^2\}].$$

For a fixed $\lambda \ge 0$,

$$e^{-y} + \lambda[\sqrt{x^2 + y^2} - x] \ge 0$$

for any $(x, y) \in R^2$; moreover, since

$$\lim_{x \to +\infty} [\sqrt{x^2 + y^2} - x] = 0$$

for any fixed y, it follows that

$$\lim_{y \to +\infty} \lim_{x \to +\infty} \{e^{-y} + \lambda[\sqrt{x^2 + y^2} - x]\}$$

$$= \lim_{y \to +\infty} \left\{e^{-y} + \lambda \lim_{x \to +\infty} [\sqrt{x^2 + y^2} - x]\right\} = \lim_{y \to +\infty} e^{-y} = 0.$$

Hence, for any fixed $\lambda \ge 0$,

$$h(\lambda) = \inf\{e^{-y} + \lambda g(x, y) : (x, y) \in R^2\} = 0,$$

and so

$$MD = \sup_{\lambda \ge 0} h(\lambda) = 0.$$

Thus, $MP = 1 > 0 = MD$.

(5.4.9) Definition. If (P) is a convex program with dual (DP) and if $MP > MD$, then we say that (P) has a *duality gap*.

We have shown earlier (see the remarks after (5.4.1)) that if a convex program (P) is superconsistent and has a solution, then (P) does not have a duality gap, and therefore the duality approach might be useful for finding the solution of (P). In Chapter 6, we will find other conditions under which we can be certain that a convex program (P) has no duality gap.

*5.5. Trust Regions

This section is meant to be read in conjunction with Section 3.3 on methods of minimization of functions on R^n. In that section, we discuss iterative methods for finding minimizers of a given function $f(\mathbf{x})$ with continuous second partial derivatives on R^n. In particular, the following iterative scheme was studied:

(1) Given $\mathbf{x}^{(k)}$.
(2) Compute $\mu_k > 0$ such that

$$Hf(\mathbf{x}^{(k)}) + \mu_k I$$

is positive definite.
(3) Solve for $\mathbf{p}^{(k)}$:

$$(Hf(\mathbf{x}^{(k)}) + \mu_k I)\mathbf{p}^{(k)} = -\nabla f(\mathbf{x}^{(k)}).$$

(4) Set $t_k > 0$ by backtracking.
(5) Update

$$\mathbf{x}^{(k+1)} = \mathbf{x}^{(k)} + t_k \mathbf{p}^{(k)}.$$

(6) Iterate.

This iterative procedure is based on the following approximation: At the point $\mathbf{x}^{(k)}$, we approximate $f(\mathbf{x})$ by the quadratic function

$$Q_k(\mathbf{x}) = f(\mathbf{x}^{(k)}) + \nabla f(\mathbf{x}^{(k)}) \cdot (\mathbf{x} - \mathbf{x}^{(k)}) + \tfrac{1}{2}(\mathbf{x} - \mathbf{x}^{(k)}) \cdot A_k(\mathbf{x} - \mathbf{x}^{(k)}),$$

where $A_k = Hf(\mathbf{x}^{(k)}) + \mu_k I$ where $\mu_k > 0$ is chosen so that A_k is positive definite. The direction $\mathbf{p}^{(k)}$ pointing from $\mathbf{x}^{(k)}$ toward $\mathbf{x}^{(k+1)}$ is just the Newton direction

$$-A_k^{-1}(\nabla f(\mathbf{x}^{(k)}))$$

for the approximation $Q_k(\mathbf{x})$ of $f(\mathbf{x})$. For most well-behaved functions this is quite satisfactory. However, situations occasionally arise in which

$$\mathbf{p}^{(k)} = -A_k^{-1}(\nabla f(\mathbf{x}^{(k)}))$$

might be very large and therefore not numerically helpful in the search for a minimizer of $f(\mathbf{x})$. If this happens, then we keep the step-length

$$l_k = \|t_k \mathbf{p}^{(k)}\|$$

computed by backtracking, and modify A_k by adding an appropriate multiple λ_k of the identity. Then we compute a new $\mathbf{x}^{(k+1)}$ by solving

$$\mathbf{x}^{(k+1)} - \mathbf{x}^{(k)} = (A_k + \lambda_k I)^{-1}(\nabla f(\mathbf{x}^{(k)})),$$

where $\|\mathbf{x}^{(k+1)} - \mathbf{x}^{(k)}\| = l_k$, that is, we keep the same step-length.

That this can be done is a consequence of the Karush–Kuhn–Tucker Theorem.

(5.5.1) Theorem. *Suppose that $f(\mathbf{x})$ has continuous second partial derivatives on R^n, that $\mathbf{x}^{(k)} \in R^n$ and that*

$$Q_k(\mathbf{x}) = f(\mathbf{x}^{(k)}) + \nabla f(\mathbf{x}^{(k)})\cdot(\mathbf{x} - \mathbf{x}^{(k)}) + \tfrac{1}{2}(\mathbf{x} - \mathbf{x}^{(k)})\cdot A_k(\mathbf{x} - \mathbf{x}^{(k)}).$$

There exists a nonnegative number λ_k such that the minimizer $\mathbf{x}^{(k+1)}$ of $Q_k(\mathbf{x})$ subject to the constraint

$$\|\mathbf{x} - \mathbf{x}^{(k)}\| \le l_k$$

is a solution $\mathbf{x}^{(k+1)}$ of the system

$$(A_k + \lambda_k I)(\mathbf{x} - \mathbf{x}^{(k)}) = -\nabla f(\mathbf{x}^{(k)}).$$

If the Newton direction

$$\mathbf{p}^{(k)} = -A_k^{-1}(\nabla f(\mathbf{x}^{(k)}))$$

satisfies $\|\mathbf{p}^{(k)}\| \le l_k$, then we can take $\lambda_k = 0$; otherwise, λ_k is a positive number such that $\|\mathbf{x}^{(k+1)} - \mathbf{x}^{(k)}\| = l_k$.

PROOF. Let μ_k be a Karush–Kuhn–Tucker multiplier for the constrained minimization (superconsistent) program

$$\text{Minimize} \quad Q_k(\mathbf{x}) \quad \text{subject to} \quad \|\mathbf{x} - \mathbf{x}^{(k)}\| \le l_k.$$

The associated Lagrangian is

$$L(\mathbf{x}, \mu_k) = f(\mathbf{x}^{(k)}) + \nabla f(\mathbf{x}^{(k)})\cdot(\mathbf{x} - \mathbf{x}^{(k)}) + \tfrac{1}{2}(\mathbf{x} - \mathbf{x}^{(k)})\cdot A_k(\mathbf{x} - \mathbf{x}^{(k)})$$
$$+ \mu_k(\|\mathbf{x} - \mathbf{x}^{(k)}\|^2 - l_k^2).$$

By the Karush–Kuhn–Tucker Theorem (5.2.13), $L(\mathbf{x}, \mu_k)$ has a global minimizer at the solution $\mathbf{x}^{(k+1)}$ of the constrained problem. Hence,

$$\begin{aligned} \mathbf{0} &= \nabla_{\mathbf{x}} L(\mathbf{x}^{(k+1)}, \mu_k) \\ &= \nabla f(\mathbf{x}^{(k)}) + A_k(\mathbf{x}^{(k+1)} - \mathbf{x}^{(k)}) + 2\mu_k(\mathbf{x}^{(k+1)} - \mathbf{x}^{(k)}). \end{aligned}$$

Therefore,

$$(A_k + 2\mu_k)(\mathbf{x}^{(k+1)} - \mathbf{x}^{(k)}) = -\nabla f(\mathbf{x}^{(k)}).$$

Set $\lambda_k = 2\mu_k$ and observe that

$$(A_k + \lambda_k I)(\mathbf{x}^{(k+1)} - \mathbf{x}^{(k)}) = -\nabla f(\mathbf{x}^{(k)}).$$

Now, if the constraint is inactive, then $\lambda_k = 0$; otherwise, $\lambda_k = 2\mu_k > 0$, which is just what the statement of the theorem promises.

Now let us see how this theorem is to be interpreted: Let $A_k = Hf(\mathbf{x}^{(k)}) + \mu_k I$ where $\mu_k > 0$ is computed to make A_k positive definite. If

$$\mathbf{p}^{(k)} = A_k^{-1}(\nabla f(\mathbf{x}^{(k)}))$$

is too large, we compute $l_k = \|t_k \mathbf{p}^{(k)}\|$ and compute a different $\mathbf{x}^{(k+1)}$ by solving

$$(Hf(\mathbf{x}^{(k+1)}) + \mu_k I + \lambda_k I)(\mathbf{x}^{(k+1)} - \mathbf{x}^{(k)}) = -\nabla f(\mathbf{x}^{(k)}).$$

In this case we have the extra assurance that the new $\mathbf{x}^{(k+1)}$ is the constrained minimizer of the quadratic approximation $Q_k(\mathbf{x})$. Also note that if $\lambda_k = 0$, then we do not make the modification. If l_k is small, then λ_k is large and $\mathbf{x}^{(k+1)} - \mathbf{x}^{(k)}$ is close to the steepest descent direction $-\nabla f(\mathbf{x}^{(k)})$.

Also, this theorem says that if somehow we know the desired step-length

$$l_k = \|\mathbf{x}^{(k+1)} - \mathbf{x}^{(k)}\|,$$

then we can solve directly for λ_k and the new $\mathbf{x}^{(k+1)}$ without backtracking.

Exercises

1. Prove that if M is a subspace of R^n such that $M \neq R^n$, then the interior M^0 of M is empty.

2. Let C be a closed convex subset of R^n. If $\mathbf{y}$ is not in C, show $\mathbf{x}^* \in C$ is the closest vector to $\mathbf{y}$ in C if and only if $(\mathbf{x} - \mathbf{y})\cdot(\mathbf{x}^* - \mathbf{y}) \geq \|\mathbf{x}^* - \mathbf{y}\|^2$ for all $\mathbf{x} \in C$.

3. Suppose that C_1 and C_2 are convex sets in R^n such that C_1 has interior points and C_2 does not contain any interior points of C_1. Prove that there is a hyperplane H in R^n such that C_1 and C_2 lie in the opposite closed half-spaces determined by H, that is, there exist an $\mathbf{a} \neq \mathbf{0}$ in R^n and an $\alpha \in R$ such that

$$\mathbf{x}\cdot\mathbf{a} \leq \alpha \leq \mathbf{y}\cdot\mathbf{a}$$

for all $\mathbf{x} \in C_1$ and all $\mathbf{y} \in C_2$. (Hint: Consider the set $C = C_1^0 - C_2$. Apply (5.1.9) to obtain the desired result when $C_1 \cap C_2 \neq \varnothing$. Use (5.1.5) to handle the case when $C_1 \cap C_2 = \varnothing$.)

4. Suppose that A is an $m \times n$-matrix and that $\mathbf{b} \in R^n$. Prove that the system

$$A^{\mathrm{T}}\mathbf{x} = \mathbf{b}$$

has a solution $\mathbf{x} \geq \mathbf{0}$ if and only if

$$\mathbf{b}\cdot\mathbf{y} \geq 0 \quad \text{whenever} \quad A\mathbf{y} \geq \mathbf{0}.$$

(This result, which is known as the Farkas Lemma, has a number of interesting and important consequences including the Karush–Kuhn–Tucker Theorem.) The Farkas Lemma can be proved by completing the following steps:

(a) Show that the statement of the Farkas Lemma is equivalent to the following: The system

$$A\mathbf{y} \geq 0, \qquad \mathbf{b}\cdot\mathbf{y} < 0$$

has a solution if and only if **b** does not belong to the closed convex set

$$C = \{A^{\mathrm{T}}\mathbf{x}: \mathbf{x} \geq 0\}.$$

(b) Apply the Basic Separation Theorem (5.1.5) to conclude that if **b** does not belong to C then there exist an $\mathbf{a} \neq \mathbf{0}$ in R^n and an $\alpha \in R$ such that

$$\mathbf{a}\cdot\mathbf{b} < \alpha < \mathbf{a}\cdot A^{\mathrm{T}}\mathbf{x}$$

for all $\mathbf{x} \geq 0$ in R^m.

(c) Show that $\alpha < 0$ by making a special choice of **x** in the inequality in (b). Then show that $A\mathbf{a} \geq 0$ by making special choices of **x** in the same inequality. Conclude that if **b** does not belong to the closed convex set C in (a), then the system

$$A\mathbf{y} \geq \mathbf{0}, \qquad \mathbf{b}\cdot\mathbf{y} < 0$$

has a solution.

(d) Show that if $A^{\mathrm{T}}\mathbf{x} = \mathbf{b}$ has a solution **x** such that $\mathbf{x} \geq 0$ and if $A\mathbf{y} \geq 0$, then $\mathbf{b}\cdot\mathbf{y} \geq 0$.

5. Apply the Karush–Kuhn–Tucker Theorem to locate all solutions of the following convex programs:

(a)
$$\begin{cases} \text{Minimize} \quad f(x_1, x_2) = e^{-(x_1+x_2)} \\ \text{subject to} \\ \quad e^{x_1} + e^{x_2} \leq 20, \\ \qquad\quad x_1 \geq 0. \end{cases}$$

(b)
$$\begin{cases} \text{Minimize} \quad f(x_1, x_2) = x_1^2 + x_2^2 - 4x_1 - 4x_2 \\ \text{subject to the constraints} \\ \quad x_1^2 - x_2 \leq 0, \\ \quad x_1 + x_2 \leq 2. \end{cases}$$

6. Consider the geometric program:

$$\begin{cases} \text{Minimize} \quad f(t_1, t_2) = t_1^{-1}t_2^{-1} \\ \text{subject to the constraint} \\ \quad \tfrac{1}{2}t_1 + \tfrac{1}{2}t_2 \leq 1; \quad t_1 > 0, \quad t_2 > 0. \end{cases}$$

(a) Convert this program to an equivalent convex program and solve the resulting program by applying the Karush–Kuhn–Tucker Theorem.

(b) Solve the given geometric program by the method of Section 5.3.

7. Find the dual of the linear program:

$$(L)' \quad \text{Minimize} \quad \mathbf{b}\cdot\mathbf{x} \quad \text{subject to} \quad A\mathbf{x} \geq \mathbf{c}.$$

8. Solve the following constrained geometric programming problems:

(a)
$$\begin{cases} \text{Minimize} \quad x^{1/2} + y^{-2}z^{-1} \\ \text{subject to the constraint} \\ \quad x^{-1}y^2 + x^{-1}z^2 \le 1, \\ \text{where} \quad x > 0, \quad y > 0, \quad z > 0. \end{cases}$$

(b)
$$\begin{cases} \text{Minimize} \quad x^{1/2} + y^{-2} \\ \text{subject to the constraints} \\ \quad x^{-1}z + x^{-1}w \le 1, \\ \quad yz^{-1} + wz^{-1} \le 1, \\ \text{where} \quad x > 0, \quad y > 0, \quad z > 0, \quad w > 0. \end{cases}$$

9. Let M be a subspace of R^n. From the definitions, it is clear that $M \subseteq (M^{\perp})^{\perp}$. Use the Basic Separation Theorem to show $(M^{\perp})^{\perp} \subseteq M$, thus giving another proof that $M = (M^{\perp})^{\perp}$.

10. For a convex program P, show that $MP(\mathbf{z}_1) \le MP(\mathbf{z}_2)$ whenever $\mathbf{z}_1 \ge \mathbf{z}_2$.

11. Let A, B, C be nonempty closed convex sets in R^n such that
$$A + C = B + C,$$
prove that $A = B$.

12. Let $f(x)$ be a differentiable function on R^1. Suppose $x^{(0)}$ is fixed and there is a number α such that
$$f(x) \ge f(x^{(0)}) + \alpha(x - x^{(0)})$$
for all $x \in R^1$. Show that $\alpha = f'(x_0)$.

13. Recall that a cone C in R^n is a convex set such that $t\mathbf{x} \in C$ provided $\mathbf{x} \in C$ and $t \ge 0$ (see p. 44). For a cone C in R^n, define
$$C^* = \{\mathbf{y} \in R^n : \mathbf{x} \cdot \mathbf{y} \ge 0 \text{ for all } \mathbf{x} \in C\}.$$
Show that if C is a closed cone in R^n, then C^* is a cone in R^n and $(C^*)^* = C$.

14. Let A be an $m \times n$ matrix and let $\mathbf{b} \in R^m$ be a fixed vector. Suppose the convex program
$$\begin{aligned} &\text{Minimize} \quad \|\mathbf{x}\|^2 \\ &\text{subject to} \quad A\mathbf{x} \le \mathbf{b} \end{aligned}$$
is superconsistent and has solution $\mathbf{x}^*$. Use the Karush–Kuhn–Tucker Theorem to show that there is a vector $\mathbf{y}$ in R^m such that $\mathbf{x}^* = A^T\mathbf{y}$. Compare this result with Theorem 4.3.2.

15. Say why the gradient form of the Karush–Kuhn–Tucker Theorem is not applicable to finding solutions of all superconsistent linear programs in minimal standard form while the saddlepoint form is applicable.

CHAPTER 6

Penalty Methods

6.1. Penalty Functions

One way to solve the inequality-constrained minimization problem

$$(P)\quad \begin{cases} \text{Minimize} \quad f(\mathbf{x}) \quad \text{subject to} \\ g_1(\mathbf{x}) \le 0, \quad g_2(\mathbf{x}) \le 0, \ldots, \quad g_m(\mathbf{x}) \le 0; \qquad \mathbf{x} \in R^n, \end{cases}$$

is to approximate this problem with an unconstrained minimization problem

$$(P') \quad \text{Minimize} \quad F(\mathbf{x}) \quad \text{for } \mathbf{x} \in R^n,$$

where the objective function $F(\mathbf{x})$ for the unconstrained problem is constructed from the objective function $f(\mathbf{x})$ and the constraints for the given constrained problem in such a way that:

(1) $F(\mathbf{x})$ includes a "penalty" term which increases the value of $F(\mathbf{x})$ whenever a constraint $g_i(\mathbf{x}) \le 0$ is violated with larger violations resulting in larger increases.
(2) The unconstrained minimizer $\mathbf{x}_F^*$ of $F(\mathbf{x})$ is near the feasibility region and $\mathbf{x}_F^*$ is near a constrained minimizer for the given constrained problem.

Using this approach, we hope that, as the size of the penalty term in $F(\mathbf{x})$ increases, the minimizer $\mathbf{x}_F^*$ of $F(\mathbf{x})$ will approach a point $\mathbf{x}^*$ that is feasible and a minimizer for the given constrained problem.

In this chapter, we will show that this strategy works in very general circumstances. We will begin by discussing the construction of a penalty function for the given constrained minimization problem in this section.

For a given constraint $g(\mathbf{x}) \le 0$, note that the function $g^+(\mathbf{x})$ defined by

$$g^+(\mathbf{x}) = \begin{cases} 0 & \text{if } g(\mathbf{x}) \le 0, \\ g(\mathbf{x}) & \text{if } g(\mathbf{x}) > 0, \end{cases}$$

is zero for all $\mathbf{x}$ that satisfy the constraint and that it has a positive value whenever this constraint is violated. Moreover, large violations in the constraint $g(\mathbf{x}) \le 0$ result in large values for $g^+(\mathbf{x})$. Thus, $g^+(\mathbf{x})$ has the penalty features we want relative to the single constraint $g(\mathbf{x}) \le 0$. Before we go on, let us look at the graphs of $g^+(\mathbf{x})$ for some simple choices of the given constraint.

(6.1.1) Examples

(a) If $g(x) = x - 1 \le 0$ for $x \in R^1$ is the given constraint, then $g^+(x)$ has the graph displayed below:

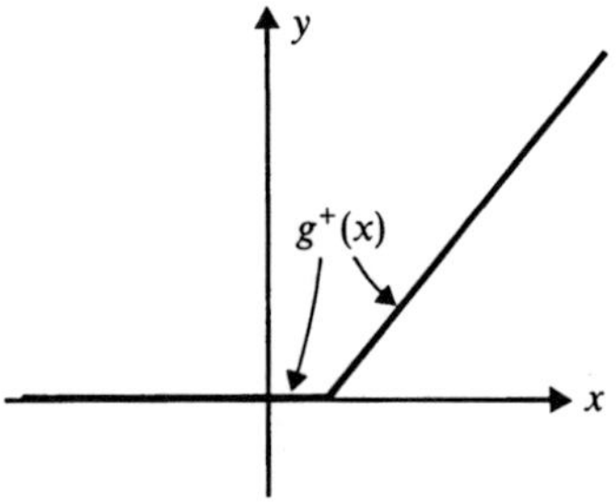

$$g^+(x) = \begin{cases} x - 1 & \text{if } x \ge 1, \\ 0 & \text{if } x < 1. \end{cases}$$

(b) If $g(x) = x^3$ for $x \in R^1$, then $g^+(x)$ has the following description:

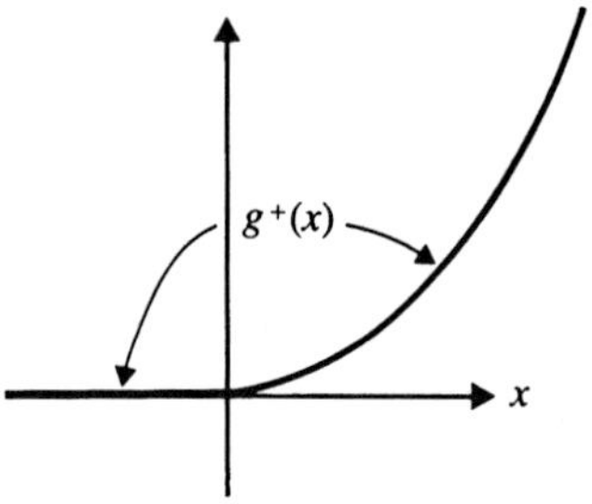

$$g^+(x) = \begin{cases} 0 & \text{if } x < 0, \\ x^3 & \text{if } x \ge 0. \end{cases}$$

(c) If $g(x, y) = x^2 + y^2 - 1$, then the graph of $g^+(x, y)$ is the "truncated paraboloid" depicted below:

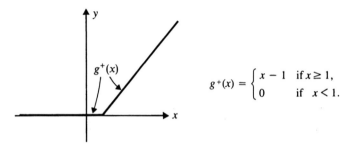

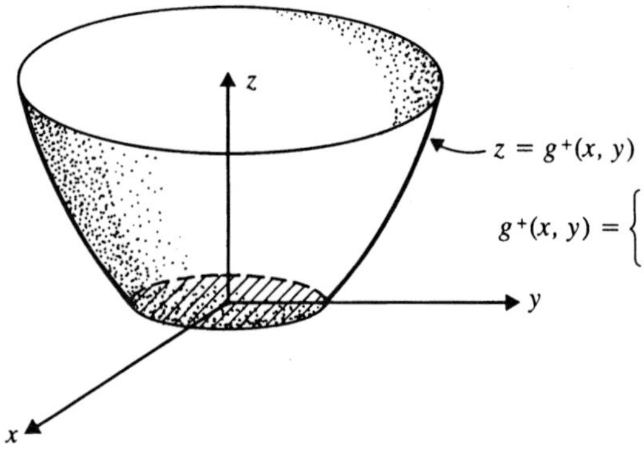

$$g^+(x, y) = \begin{cases} 0 & \text{if } x^2 + y^2 \le 1, \\ x^2 + y^2 - 1 & \text{otherwise.} \end{cases}$$

Examples (a) and (c) show that $g^+(\mathbf{x})$ need not have continuous derivatives even when $g(\mathbf{x})$ is a very smooth function.

If we now return to the original constrained minimization problem

$$(P)\quad \begin{cases} \text{Minimize} \quad f(\mathbf{x}) \quad \text{subject to} \\ g_1(\mathbf{x}) \le 0, \quad g_2(\mathbf{x}) \le 0, \ldots, \quad g_m(\mathbf{x}) \le 0; \qquad \mathbf{x} \in R^n, \end{cases}$$

we see from our discussion of the basic features of the function $g^+(\mathbf{x})$ that one reasonable definition for the objective function for an approximating unconstrained program (P') for (P) is

$$F_k(\mathbf{x}) = f(\mathbf{x}) + k \sum_{i=1}^{m} g_i^+(\mathbf{x}),$$

where k is a positive integer. The penalty term $\sum_{i=1}^{m} g_i^+(\mathbf{x})$ is often called the *Absolute Value Penalty Function* because it is equal to $\sum |g_i(\mathbf{x})|$ where the summation extends over all constraints violated at $\mathbf{x}$.

The role of the positive integer k is obvious: As k increases, so does the penalty associated with a given choice of $\mathbf{x}$ that violates one or more of the constraints $g_i(\mathbf{x}) \le 0$ for $i = 1, 2, \ldots, m$. For this reason, we call k the *penalty parameter*.

Our hope is that, for large k, the value of

$$k \sum_{i=1}^{m} g_i^+(\mathbf{x}_k^*)$$

at a minimizer $\mathbf{x}_k^*$ for $F_k(\mathbf{x})$ should be small, $\mathbf{x}_k^*$ should be near the feasibility region for (P), and $F_k(\mathbf{x}_k^*)$ should be near a minimum for (P). This leads us to hope that there might be at least a subsequence of $\{\mathbf{x}_k^*\}$ that converges to a minimizer $\mathbf{x}^*$ for (P).

One might feel that this penalty function approach to the solution of (P) is too naive to be successful and that the hopes and expectations expressed in the preceding paragraph will simply not be fulfilled in most realistic problems. However, it turns out that this method or one of its close relatives can be effective for the solution of constrained minimization problems and that it is often the method of choice because of its simplicity.

As we have already observed in Example (6.1.1), the function $g^+(\mathbf{x})$ does not in general inherit differentiability properties from $g(\mathbf{x})$. Thus, even if the objective function $f(\mathbf{x})$ and the constraint functions $g_1(\mathbf{x}), g_2(\mathbf{x}), \ldots, g_m(\mathbf{x})$ in the constrained problem (P) have continuous first partial derivatives on R^n, the same may not be true of

$$F_k(\mathbf{x}) = f(\mathbf{x}) + k \sum_{i=1}^{m} g_i^+(\mathbf{x}).$$

Thus, to locate the minimizer $\mathbf{x}_k^*$ of $F_k(\mathbf{x})$, we would be limited to methods that do not require the objective function to be smooth. This is certainly not sufficient reason to abandon penalty functions such as the Absolute Value

Penalty Function used in the definition of $F_k(\mathbf{x})$; in fact, this penalty function can be quite useful in spite of this apparent disadvantage (see (6.2.2)). However, it is also very useful from the standpoint of the development of the Penalty Function Method to know that continuity of the first partial derivatives can be maintained through a suitable modification of the penalty term. To see how this might be accomplished, let us take a closer look at the functions in (6.1.1)(a), (c).

(6.1.2) Examples

(a) If $g(x) = x - 1$, then

$$g^+(x) = \begin{cases} x - 1 & \text{if } x \geq 1, \\ 0 & \text{if } x < 1, \end{cases}$$

fails to have a derivative at $x = 1$. However,

$$h(x) = [g^+(x)]^2 = \begin{cases} (x - 1)^2 & \text{if } x \geq 1, \\ 0 & \text{if } x < 1, \end{cases}$$

has a continuous derivative everywhere; in particular, $h'(1) = 0$.

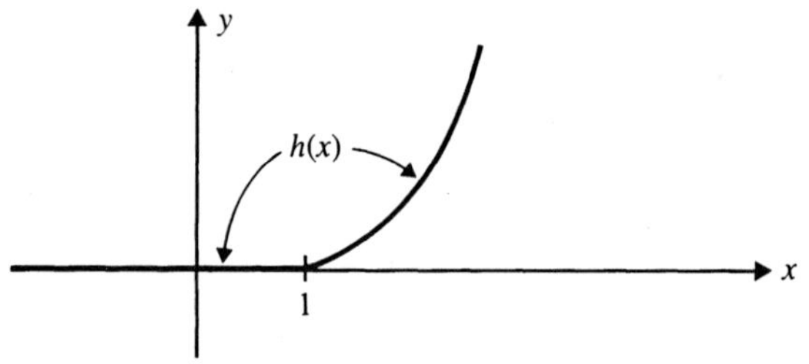

(b) If $g(x, y) = x^2 + y^2 - 1$, then the first partial derivatives of

$$g^+(x, y) = \begin{cases} x^2 + y^2 - 1 & \text{if } x^2 + y^2 > 1, \\ 0 & \text{if } x^2 + y^2 \leq 1, \end{cases}$$

do not exist for (x, y) on the unit circle $x^2 + y^2 = 1$. However,

$$h(x, y) = [g^+(x, y)]^2$$

has the property that

$$\lim_{\substack{x \to a \\ y \to b}} \frac{\partial h}{\partial x}(x, y) = 0 = \lim_{\substack{x \to a \\ y \to b}} \frac{\partial h}{\partial y}(x, y)$$

for any (a, b) on the unit circle $x^2 + y^2 = 1$. It is then a routine matter to verify that $h(x, y)$ has continuous first partial derivatives on R^2.

The following result shows that the situation described in the preceding example holds quite generally.

(6.1.3) Lemma. *If* $g(\mathbf{x})$ *has continuous first partial derivatives on* R^n, *the same is true of* $h(\mathbf{x}) = [g^+(\mathbf{x})]^2$. *Moreover,*

$$(*) \qquad \frac{\partial h}{\partial x_i}(\mathbf{x}) = 2g^+(\mathbf{x})\frac{\partial g}{\partial x_i}(\mathbf{x}), \qquad i = 1, 2, \ldots, n,$$

for all $\mathbf{x} \in R^n$.

PROOF. If $g(\mathbf{z}) > 0$, then $g^+(\mathbf{x}) = g(\mathbf{x})$ for all $\mathbf{x}$ in some ball $B(\mathbf{z}, r)$ centered at $\mathbf{z}$. Hence $h(\mathbf{x})$ has continuous first partial derivatives throughout $B(\mathbf{z}, r)$ and the formula

$$\frac{\partial h}{\partial x_i}(\mathbf{z}) = 2g^+(\mathbf{z})\frac{\partial g}{\partial x_i}(\mathbf{z})$$

holds for $i = 1, 2, \ldots, n$.

If $g(\mathbf{z}) < 0$, then $g^+(\mathbf{x}) = 0$ for all $\mathbf{x}$ in some ball $B(\mathbf{z}, r)$ centered at $\mathbf{z}$ and so

$$\frac{\partial h}{\partial x_i}(\mathbf{z}) = 0$$

for $i = 1, \ldots, n$. But $2g^+(\mathbf{z})(\partial g/\partial x_i)(\mathbf{z}) = 0$ for $i = 1, 2, \ldots, n$ since $g^+(\mathbf{z}) = 0$. Consequently, the formula $(*)$ is valid if either $g(\mathbf{z}) > 0$ or $g(\mathbf{z}) < 0$.

If $g(\mathbf{z}) = 0$, then a careful limit argument shows that

$$\frac{\partial h}{\partial x_i}(\mathbf{z}) = 2\left[\frac{\partial g}{\partial x_i}(\mathbf{z})\right]g^+(\mathbf{z}) = 0$$

for $i = 1, 2, \ldots, n$ to complete the proof.

It is now evident how the penalty term should be altered to preserve smoothness. If the objective function $f(\mathbf{x})$ and the constraint functions $g_1(\mathbf{x})$, $\ldots, g_m(\mathbf{x})$ in (P) have continuous first partial derivatives, then the same is true of

$$P_k(\mathbf{x}) = f(\mathbf{x}) + k \sum_{i=1}^{m} [g_i^+(\mathbf{x})]^2$$

and $P_k(\mathbf{x})$ serves as a suitable objective function for the penalty approach to the solution of (P). The penalty term in $P_k(\mathbf{x})$ is sometimes called the *Courant–Beltrami Penalty Function.*

6.2. The Penalty Method

We are now ready to provide a more precise description of the penalty approach to constrained minimization problems, first for the Courant–Beltrami Penalty Function.

(6.2.1) The Penalty Function Method. Suppose that $f(\mathbf{x}), g_1(\mathbf{x}), g_2(\mathbf{x}), \ldots, g_m(\mathbf{x})$ have continuous first partial derivatives on R^n. To solve the constrained minimization problem

$$(P) \quad \begin{cases} \text{Minimize} \quad f(\mathbf{x}) \quad \text{subject to} \\ g_1(\mathbf{x}) \le 0, \quad g_2(\mathbf{x}) \le 0, \ldots, \quad g_m(\mathbf{x}) \le 0; \qquad \mathbf{x} \in R^n, \end{cases}$$

we proceed as follows:

(1) For each positive integer k, suppose $\mathbf{x}_k^*$ is a global minimizer of

$$P_k(\mathbf{x}) = f(\mathbf{x}) + k \sum_{i=1}^{m} [g_i^+(\mathbf{x})]^2.$$

(2) Show that some subsequence of $\{\mathbf{x}_k^*\}$ converges to a solution $\mathbf{x}^*$ for (P).

We will show later that the Penalty Function Method is guaranteed to produce a solution $\mathbf{x}^*$ of (P) under relatively mild additional restrictions on (P). However, before we do this, let us try this method on a concrete problem. The program in the following example is quite simple but yet it is general enough to reveal most of the features of the method.

(6.2.2) Examples. Consider the program

$$(P) \quad \begin{cases} \text{Minimize } f(x) = x^2 \quad \text{subject to} \\ g(x) = 1 - x \le 0; \qquad x \in R. \end{cases}$$

Since this program simply asks us to minimize x^2 for $x \ge 1$, it is obvious that the solution to (P) is $x^* = 1$ and the minimum value of (P) is $MP = 1$. Let us see how the Penalty Function Method produces this result. In this case,

$$\begin{aligned} P_k(x) &= x^2 + k[(1 - x)^+]^2 \\ &= \begin{cases} x^2 + k(1 - x)^2 & \text{for } x \le 1, \\ x^2 & \text{for } x > 1. \end{cases} \end{aligned}$$

The graph of $P_k(x)$ is pictured below

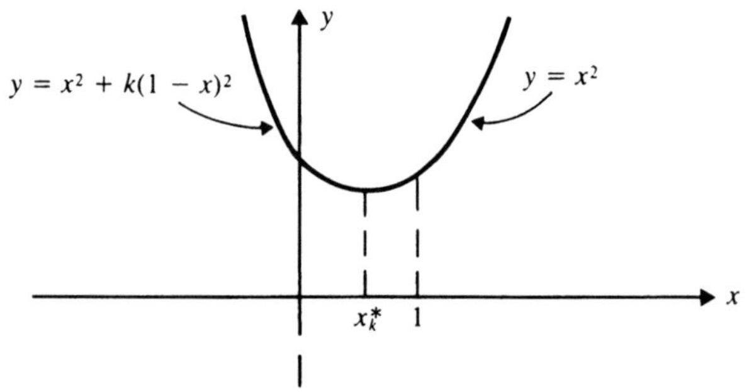

We know from (6.1.3) that $P_k(x)$ is continuously differentiable everywhere. It is an increasing function at $x = 1$ and it has a unique minimizer x_k^* to the left of $x = 1$; in fact, x_k^* is obviously the unique solution of the equation

$$0 = P_k'(x) = 2x - 2k(1 - x) = (2 + 2k)x - 2k$$

for $x < 1$. Thus,

$$x_k^* = \frac{k}{1+k}.$$

Notice the following features of the sequence $\{x_k^*\}$:

(1) $x_k^* < 1$ for all k so the sequence $\{x_k^*\}$ consists of points that are not feasible for (P).

(2) $\lim_k x_k^* = 1 = x^*$, that is, the sequence $\{x_k^*\}$ converges to the solution of (P); moreover, the higher the value of the penalty parameter k, the closer $x_k^* = k/(k+1)$ is to being feasible for (P).

(3)
$$\begin{aligned} P_k(x_k^*) &= \left(\frac{k}{k+1}\right)^2 + k\left(1 - \left(\frac{k}{k+1}\right)\right)^2 \\ &= \left(\frac{k}{k+1}\right)^2 + k\left(\frac{1}{k+1}\right)^2 \\ &= \frac{k}{(k+1)^2}(k+1) = \frac{k}{k+1} < 1 = MP. \end{aligned}$$

Thus $P_k(x_k^*) \leq MP$ for all k and

$$\lim_k P_k(x_k^*) = MP.$$

It turns out that all of the features (1), (2), and (3) of the Penalty Function Method in the preceding example are present in most well-behaved programs. The following theorem and its corollaries tell us why.

(6.2.3) Theorem. *Suppose that $f(\mathbf{x}), g_1(\mathbf{x}), \ldots, g_m(\mathbf{x})$ are continuous on R^n and that $f(\mathbf{x})$ is bounded from below in R^n (that is, there is a constant c such that $c \leq f(\mathbf{x})$ for all $\mathbf{x} \in R^n$). If $\mathbf{x}^*$ is a solution of the program*

$$(P) \quad \begin{cases} \text{Minimize} \quad f(\mathbf{x}) \quad \text{subject to} \\ \quad g_1(\mathbf{x}) \leq 0, \quad g_2(\mathbf{x}) \leq 0, \ldots, \quad g_m(\mathbf{x}) \leq 0, \end{cases}$$

and if, for each positive integer k, there is an $\mathbf{x}_k \in R^n$ such that

$$P_k(\mathbf{x}_k) = \min_{\mathbf{x} \in R^n} P_k(\mathbf{x}),$$

then:

(1) $P_k(\mathbf{x}_k) \leq P_{k+1}(\mathbf{x}_{k+1}) \leq f(\mathbf{x}^*) = MP$

for each positive integer k, and

(2) $\lim_{k\to\infty} \sum_{i=1}^{m} [g_i^+(\mathbf{x}_k)]^2 = 0.$

Consequently, if $\{\mathbf{x}_{k_p}\}$ *is any convergent subsequence of* $\{\mathbf{x}_k\}$ *and if*

$$\lim_{p} \mathbf{x}_{k_p} = \mathbf{x}^{**},$$

then $\mathbf{x}^{**}$ *is a solution of* (P).

PROOF. To prove that $P_k(\mathbf{x}_k) \le P_{k+1}(\mathbf{x}_{k+1})$, simply note that

$$\begin{aligned} P_k(\mathbf{x}_k) &= \min_{\mathbf{x}\in R^n} P_k(\mathbf{x}) \le P_k(\mathbf{x}_{k+1}) = f(\mathbf{x}_{k+1}) + k \sum_{i=1}^{m} [g_i^+(\mathbf{x}_{k+1})]^2 \\ &\le f(\mathbf{x}_{k+1}) + (k+1) \sum_{i=1}^{m} [g_i^+(\mathbf{x}_{k+1})]^2 = P_{k+1}(\mathbf{x}_{k+1}). \end{aligned}$$

Since $\mathbf{x}^*$ is a solution of (P), we know that $\mathbf{x}^*$ is feasible, so that $g_i^+(\mathbf{x}^*) = 0$ for $i = 1, \ldots, m$. Therefore,

$$\begin{aligned} P_{k+1}(\mathbf{x}_{k+1}) &= \min_{\mathbf{x}\in R^n} P_{k+1}(\mathbf{x}) \le P_{k+1}(\mathbf{x}^*) \\ &= f(\mathbf{x}^*) + (k+1) \sum_{i=1}^{m} [g_i^+(\mathbf{x}^*)]^2 = f(\mathbf{x}^*) = MP. \end{aligned}$$

This proves statement (1).

Next, we shall prove that

$$\lim_{k} \sum_{i=1}^{m} (g_i^+(\mathbf{x}_k))^2 = 0.$$

To this end, choose a number c such that $f(\mathbf{x}) \ge c$ for all $\mathbf{x} \in R^n$. Then, for each positive integer k,

$$\begin{aligned} c + k \sum_{i=1}^{m} [g_i^+(\mathbf{x}_k)]^2 &\le f(\mathbf{x}_k) + k \sum_{i=1}^{m} [g_i^+(\mathbf{x}_k)]^2 \\ &= P_k(\mathbf{x}_k) \le MP \end{aligned}$$

so that $k \sum_{i=1}^{m} [g_i^+(\mathbf{x}_k)]^2 \le MP - c$ for each positive integer k. It follows that

$$0 \le \sum_{i=1}^{m} [g_i^+(x_k)]^2 \le \frac{MP - c}{k}$$

and so $\lim_k \sum_{i=1}^{m} [g_i^+(\mathbf{x}_k)]^2 = 0$, which completes the proof of statement (2).

Finally, assume that $\mathbf{x}^{**}$ is the limit of some subsequence $\{\mathbf{x}_{k_p}\}$ of $\{\mathbf{x}_k\}$. Then since each $g_i(\mathbf{x})$ is continuous on R^n, it follows that

$$0 \le \sum_{i=1}^{m} [g_i^+(\mathbf{x}^{**})]^2 = \lim_{p} \sum_{i=1}^{m} [g_i^+(\mathbf{x}_{k_p})]^2 = 0$$

by virtue of the conclusion obtained in the preceding paragraph of the proof. We conclude that $g_i(\mathbf{x}^{**}) = 0$ for $i = 1, 2, \ldots, m$; in particular, $\mathbf{x}^{**}$ is feasible for (P).

To complete the proof that $\mathbf{x}^{**}$ is a solution to (P), it is only necessary to show that $f(\mathbf{x}^{**}) \leq MP$, since MP is the constrained minimum of $f(\mathbf{x})$ and $\mathbf{x}^{**}$ is feasible for (P). But this is an immediate consequence of the continuity of $f(\mathbf{x})$ and $g_1(\mathbf{x}), \ldots, g_m(\mathbf{x})$ since

$$\begin{aligned} f(\mathbf{x}^{**}) &= f(\mathbf{x}^{**}) + \sum_{i=1}^{m} [g_i^+(\mathbf{x}^{**})]^2 \\ &= \lim_p \, [f(\mathbf{x}_{k_p}) + \sum_{i=1}^{m} [g_i^+(\mathbf{x}_{k_p})]^2] \\ &\leq \lim_p \, [f(\mathbf{x}_{k_p}) + k_p \sum_{i=1}^{m} [g_i^+(\mathbf{x}_{k_p})]^2] \\ &= \lim_p P_{k_p}(\mathbf{x}_{k_p}) \leq MP. \end{aligned}$$

A careful review of the proof of the preceding theorem shows that it remains valid if the Courant–Beltrami penalty term is replaced by the Absolute Value Penalty Function

$$\sum_{i=1}^{m} g_i^+(\mathbf{x}).$$

that is, if $P_k(\mathbf{x})$ is replaced by $F_k(\mathbf{x})$. More generally, if

$$Q_k(\mathbf{x}) = f(\mathbf{x}) + k \sum_{i=1}^{m} G_i(\mathbf{x})$$

where

(1) $G_i(\mathbf{x})$ is continuous on R^n provided $g_i(\mathbf{x})$ is continuous on R^n for $i = 1, \ldots, m$;
(2) $G_i(\mathbf{x}) \geq 0$ for all $\mathbf{x} \in R^n$ and $i = 1, \ldots, m$;
(3) $G_i(\mathbf{x}) = 0$ for $i = 1, 2, \ldots, m$ if and only if $\mathbf{x}$ is feasible for (P);

then the proof of (6.2.3) is valid when $P_k(\mathbf{x})$ is replaced by $Q_k(\mathbf{x})$.

Given a constrained minimization problem

$$(P) \quad \begin{cases} \text{Minimize} \quad f(\mathbf{x}) \quad \text{subject to} \\ \quad g_1(\mathbf{x}) \leq 0, \quad g_2(\mathbf{x}) \leq 0, \ldots, \quad g_m(\mathbf{x}) \leq 0; \qquad \mathbf{x} \in R^n, \end{cases}$$

we call any function

$$\sum_{i=1}^{m} G_i(\mathbf{x}),$$

where the $G_i(\mathbf{x})$ for $i = 1, \ldots, m$ have properties (1), (2), (3), a *generalized penalty function* for (P) and the function $Q_k(\mathbf{x})$ is called the *generalized penalty method objective function* for (P). Not only does (6.2.3) remain valid when $P_k(\mathbf{x})$ is replaced by $Q_k(\mathbf{x})$, but the same is also true of the following useful corollary. In particular, this corollary holds for the Absolute Value Penalty Function.

(6.2.4) Corollary. *Suppose that $f(\mathbf{x}), g_1(\mathbf{x}), \ldots, g_m(\mathbf{x})$ are continuous functions on R^n and suppose that*

$$(P) \quad \begin{cases} \textit{Minimize} \quad f(\mathbf{x}) \quad \textit{subject to} \\ \quad g_1(\mathbf{x}) \leq 0, \ldots, \quad g_m(\mathbf{x}) \leq 0; \qquad \mathbf{x} \in R^n, \end{cases}$$

has a solution $\mathbf{x}^*$. If $f(\mathbf{x})$ is coercive, and if

$$P_k(\mathbf{x}) = f(\mathbf{x}) + k \sum_{i=1}^{n} [g_i^+(\mathbf{x})]^2,$$

then:

(1) *For each k, there is a point $\mathbf{x}_k$ in R^n such that*

$$P_k(\mathbf{x}_k) = \min_{\mathbf{x} \in R^n} P_k(\mathbf{x}).$$

(2) *The sequence $\{\mathbf{x}_k\}$ is bounded and has convergent subsequences, all of which converge to solutions of (P).*

PROOF. Since

$$\lim_{\|\mathbf{x}\| \to +\infty} f(\mathbf{x}) = +\infty$$

and since $f(\mathbf{x})$ is continuous on R^n, it follows that $f(\mathbf{x})$ is bounded from below on R^n by (1.4.4).

Next, we shall show that each $P_k(\mathbf{x})$ has an (unconstrained) minimizer on R^n. To this end, simply note that for each positive integer k,

$$P_k(\mathbf{x}) = f(\mathbf{x}) + k \sum_{i=1}^{m} [g_i^+(\mathbf{x})]^2 \geq f(\mathbf{x})$$

for all $\mathbf{x} \in R^n$, so that

$$\lim_{\|\mathbf{x}\| \to +\infty} P_k(\mathbf{x}) = +\infty.$$

Therefore, by (1.4.4), $P_k(\mathbf{x})$ has an unconstrained minimizer $\mathbf{x}_k$.

We will now show that the sequence $\{\mathbf{x}_k\}$ is bounded. Suppose, to the contrary, that $\{\mathbf{x}_k\}$ is not bounded. Then $\{f(\mathbf{x}_k)\}$ contains arbitrarily large terms since

$$\lim_{\|\mathbf{x}\| \to +\infty} f(\mathbf{x}) = +\infty.$$

But $f(\mathbf{x}_k) \leq P_k(\mathbf{x}_k) \leq f(\mathbf{x}^*)$ for all k, which contradicts the statement that $\{f(\mathbf{x}_k)\}$ contains arbitrarily large terms. Consequently, $\{\mathbf{x}_k\}$ must be a bounded sequence.

The Bolzano–Weierstrass Property guarantees that $\{\mathbf{x}_k\}$ has convergent subsequences and (6.2.3) implies that the limit of any convergent subsequence of $\{\mathbf{x}_k\}$ must be a solution of (P). This completes the proof.

The following simple examples illustrate not only the Penalty Function Method but also some practical problems related to the application of this method.

(6.2.5) Example. Consider the program

$$(P)\quad \begin{cases} \text{Minimize} \quad f(x, y) = x^2 + y^2 & \text{subject to} \\ g(x, y) = 1 - x - y \leq 0; & (x, y) \in R^2. \end{cases}$$

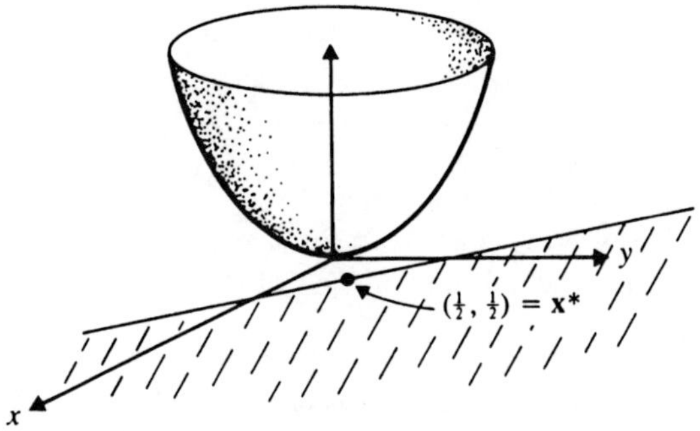

It is quite easy to see from the graph of $f(x, y)$ and the feasibility region for (P) that the unique solution of this program is $(x^*, y^*) = (\frac{1}{2}, \frac{1}{2})$. However, let us apply the Penalty Function Method to see what happens. In this case,

$$P_k(x, y) = \begin{cases} x^2 + y^2 & \text{if } x + y \geq 1, \\ x^2 + y^2 + k(1 - x - y)^2 & \text{if } x + y < 1, \end{cases}$$

has its minimizer in the region where $x + y < 1$. In fact, in this region

$$0 = \frac{\partial P_k}{\partial x} = 2x - 2k(1 - x - y),$$

$$0 = \frac{\partial P_k}{\partial y} = 2y - 2k(1 - x - y).$$

If we solve these equations for the critical points, we find that

$$x_k = \frac{k}{1 + 2k}, \qquad y_k = \frac{k}{1 + 2k}.$$

Note that the sequence $\{(x_k, y_k)\}$ is outside of the feasibility region for (P) and that it converges to the solution $(\frac{1}{2}, \frac{1}{2})$ for (P).

Because of problems related to computational accuracy, it is not in general practical to attempt to compute an approximate value of the solution $\mathbf{x}^*$ to (P) by computing the minimizer $\mathbf{x}_k$ of P_k to within a preassigned accuracy for a single large value of the penalty parameter k. (See Section 6.2.1 in *Practical Optimization* by P. E. Gill, W. Murray and M. H. Wright (Academic Press,

New York, 1981).) Rather, a sequence of computations of approximations to $\mathbf{x}_k$ with increasing values of k is required to implement the Penalty Function Method for the Courant–Beltrami Penalty term.

(6.2.6) Example. Let us look at the Penalty Function Method again applied to the simple example in (6.2.2)

$$(P) \quad \begin{cases} \text{Minimize} \quad f(x) = x^2 \quad \text{subject to} \\ \quad g(x) = 1 - x \le 0; \qquad x \in R, \end{cases}$$

this time using the objective function $F_k(x)$ with the Absolute Value Penalty term

$$F_k(x) = \begin{cases} x^2 + k(1 - x) & \text{for } x < 1, \\ x^2 & \text{for } x \ge 1. \end{cases}$$

The minimizer x_k^* of $F_k(x)$ is at $x = \frac{1}{2}$ for $k = 1$ and at $x = 1$ for all positive integers $k \ge 2$. Thus the solution $x^* = 1$ of (P) is actually the minimizer of $F_k(x)$ for all $k \ge 2$.

Surprisingly enough, the apparent very special feature of Example (6.2.6) obtains under quite general conditions when the Absolute Value Penalty Function is used, that is, the solution x^* of a constrained program (P) is the unconstrained minimizer x_k^* of F_k for all sufficiently large k. Penalty functions with this property are referred to as *exact* penalty functions in the literature. Example (6.2.5) shows that the Courant–Beltrami Penalty Function is not exact.

6.3. Applications of the Penalty Function Method to Convex Programs

We will now use the Penalty Function Method to study duality in convex programming. In particular, we will obtain a new derivation of the Karush–Kuhn–Tucker Theorem that is independent of the abstract methods of Chapter 5 and identify conditions under which convex programs do not have duality gaps.

Consider the convex program

$$(P) \quad \begin{cases} \text{Minimize} \quad f(\mathbf{x}) \quad \text{subject to} \\ \quad g_1(\mathbf{x}) \le 0, \ldots, \quad g_m(\mathbf{x}) \le 0; \qquad \mathbf{x} \in R^n, \end{cases}$$

where $f(\mathbf{x}), g_1(\mathbf{x}), \ldots, g_m(\mathbf{x})$ are convex functions on R^n. In Chapter 5, we defined the Lagrangian $L(\mathbf{x}, \boldsymbol{\lambda})$ for (P) by

$$L(\mathbf{x}, \boldsymbol{\lambda}) = f(\mathbf{x}) + \sum_{i=1}^{m} \lambda_i g_i(\mathbf{x}),$$

where $\mathbf{x} \in R^n$ and $\boldsymbol{\lambda} \geq \mathbf{0}$ in R^m and the dual program

$$(DP) \quad \begin{cases} \text{Maximize} & h(\boldsymbol{\lambda}) = \inf\{L(\mathbf{x}, \boldsymbol{\lambda}) \colon \mathbf{x} \in R^n\} \\ \text{subject to} & \boldsymbol{\lambda} \geq \mathbf{0}. \end{cases}$$

We showed that the quantities

$$MP = \inf\{f(\mathbf{x}) \colon g_i(\mathbf{x}) \leq 0, i = 1, \ldots, m; \mathbf{x} \in R^n\},$$

and

$$MD = \sup\{h(\boldsymbol{\lambda}) \colon \boldsymbol{\lambda} \geq \mathbf{0}; \boldsymbol{\lambda} \in R^m\}$$

are related by the Primal–Dual Inequality

$$MP \geq MD$$

whenever both (P) and (DP) are consistent. We also gave an example of a convex program (P) for which strict inequality holds in the Primal–Dual Inequality, that is,

$$MP > MD.$$

Such programs are said to have a *duality gap*. Convex programs with a duality gap are intractable by the Duality Method discussed in Chapter 5. Consequently it is useful to find conditions on a convex program that assure that it does not have a duality gap, that is, that $MP = MD$. The next theorem uses the Penalty Function Method to identify a class of convex programs with this desirable feature.

(6.3.1) Theorem. *Suppose that $f(\mathbf{x}), g_1(\mathbf{x}), \ldots, g_m(\mathbf{x})$ are convex functions with continuous first partial derivatives on R^n and suppose that $f(\mathbf{x})$ is coercive; that is,*

$$\lim_{\|\mathbf{x}\| \to +\infty} f(\mathbf{x}) = +\infty.$$

If the convex program

$$(P) \quad \begin{cases} \text{Minimize} \quad f(\mathbf{x}) \quad \text{subject to} \\ \quad g_1(\mathbf{x}) \leq 0, \quad g_2(\mathbf{x}) \leq 0, \ldots, \quad g_m(\mathbf{x}) \leq 0; \qquad \mathbf{x} \in R^n, \end{cases}$$

is consistent, then its dual program (DP) is consistent and $MP = MD$.

Proof. Use the objective function

$$P_k(\mathbf{x}) = f(\mathbf{x}) + k \sum_{i=1}^{m} [g_i^+(\mathbf{x})]^2$$

with the Courant–Beltrami Penalty term. According to (6.2.4), there is a vector $\mathbf{x}_k$ such that

$$P_k(\mathbf{x}_k) = \min\{P_k(\mathbf{x}) \colon \mathbf{x} \in R^n\}$$

for each positive integer k; moreover, $\{\mathbf{x}_k\}$ is a bounded sequence and all of its convergent subsequences have limits that are solutions of (P).

Suppose that $\{\mathbf{x}_{k_j}\}$ is a convergent subsequence of $\{\mathbf{x}_k\}$. By virtue of Lemma (6.1.3),

$$\nabla P_k(\mathbf{x}) = \nabla f(\mathbf{x}) + k \sum_{i=1}^{m} 2g_i^+(\mathbf{x})\nabla g_i(\mathbf{x}).$$

Since $\mathbf{x}_{k_j}$ is a minimizer for $P_{k_j}(\mathbf{x})$, it follows that

$$(*) \qquad 0 = \nabla P_{k_j}(\mathbf{x}_{k_j}) = \nabla f(\mathbf{x}_{k_j}) + \sum_{i=1}^{m} 2k_j g_i^+(\mathbf{x}_{k_j})\nabla g_i(\mathbf{x}_{k_j}).$$

Let $\lambda_i^{(j)} = 2k_j g_i^+(\mathbf{x}_{k_j})$ for $i = 1, \ldots, m$ and all j, and let

$$\boldsymbol{\lambda}^{(j)} = (\lambda_1^{(j)}, \lambda_2^{(j)}, \ldots, \lambda_m^{(j)})$$

for all j. Then $\boldsymbol{\lambda}^{(j)} \geq \mathbf{0}$ and $(*)$ shows that

$$\nabla L(\mathbf{x}_{k_j}, \boldsymbol{\lambda}^{(j)}) = 0$$

for each positive integer j. But, for each j, $L(\mathbf{x}, \boldsymbol{\lambda}^{(j)})$ is a convex function on R^n since $f(\mathbf{x}), g_1(\mathbf{x}), \ldots, g_m(\mathbf{x})$ are convex functions and $\boldsymbol{\lambda}^{(j)} \geq \mathbf{0}$; so $\mathbf{x}_{k_j}$ is an unconstrained global minimizer of $L(\mathbf{x}, \boldsymbol{\lambda}^{(j)})$ on R^n. Thus,

$$L(\mathbf{x}_{k_j}, \boldsymbol{\lambda}^{(j)}) = \min\{L(\mathbf{x}, \boldsymbol{\lambda}^{(j)}) : \mathbf{x} \in R^n\} > -\infty.$$

This shows that the vector $\boldsymbol{\lambda}^{(j)}$ is feasible for the dual program (DP).

Since (P) is consistent and $f(\mathbf{x})$ is coercive, that is,

$$\lim_{\|\mathbf{x}\| \to +\infty} f(\mathbf{x}) = +\infty$$

it follows that (P) has solutions and that $MP > -\infty$. Theorem (6.2.3) shows that the limit $\mathbf{x}^{**}$ of the convergent subsequence $\{\mathbf{x}_{k_j}\}$ is a solution of (P).

Observe that for each j,

$$\begin{aligned} f(\mathbf{x}_{k_j}) \leq P_{k_j}(\mathbf{x}_{k_j}) &= f(\mathbf{x}_{k_j}) + \sum_{i=1}^{m} k_j[g_j^+(\mathbf{x}_{k_j})]^2 \leq f(\mathbf{x}_{k_j}) + \sum_{i=1}^{m} 2k_j[g_i^+(\mathbf{x}_{k_j})]^2 \\ &= f(\mathbf{x}_{k_j}) + \sum_{i=1}^{m} 2k_j g_i^+(\mathbf{x}_{k_j})g_i(\mathbf{x}_{k_j}) \qquad \text{(Why?)} \\ &= f(\mathbf{x}_{k_j}) + \sum_{i=1}^{m} \lambda_i^{(j)} g_i(\mathbf{x}_{k_j}) \\ &= L(\mathbf{x}_{k_j}, \boldsymbol{\lambda}^{(j)}) = \min\{L(\mathbf{x}, \boldsymbol{\lambda}^{(j)}) : \mathbf{x} \in R^n\} \leq MD. \end{aligned}$$

Hence, $f(\mathbf{x}_{k_j}) \leq MD$ for all j. Since $f(\mathbf{x})$ is a continuous function on R^n and $\{\mathbf{x}_{k_j}\}$ converges to $\mathbf{x}^{**}$, it follows that

$$MP = f(\mathbf{x}^{**}) = \lim_j f(\mathbf{x}_{k_j}) \leq MD.$$

Since $MP \geq MD$ by the Primal–Dual Inequality, it follows that $MP = MD$, which completes the proof.

Note that in the preceding proof, the assumption that (P) is a consistent program is not needed to establish the consistency of the dual program (DP); the consistency of (DP) follows from the assumption that $f(\mathbf{x})$ is a coercive function (apply (1.4.4) instead of (6.2.4)) and the smoothness condition on $f(\mathbf{x})$ and the constraint functions.

If the objective function $f(\mathbf{x})$ in the convex program

$$(P) \quad \begin{cases} \text{Minimize} \quad f(\mathbf{x}) \quad \text{subject to} \\ g_1(\mathbf{x}) \le 0, \quad g_2(\mathbf{x}) \le 0, \ldots, \quad g_m(\mathbf{x}) \le 0; \qquad \mathbf{x} \in R^n, \end{cases}$$

is not coercive, it is always possible to perturb $f(\mathbf{x})$ so that this condition is satisfied. More specifically, for each $\varepsilon > 0$, define

$$f^\varepsilon(\mathbf{x}) = f(\mathbf{x}) + \varepsilon\|\mathbf{x}\|^2.$$

Then $f^\varepsilon(\mathbf{x})$ is a convex function because $f(\mathbf{x})$ and $\|\mathbf{x}\|^2$ are convex and $\varepsilon > 0$.

We will now show that $f^\varepsilon(\mathbf{x})$ is also coercive; that is, that

$$\lim_{\|\mathbf{x}\| \to +\infty} f^\varepsilon(\mathbf{x}) = +\infty.$$

First, note that there is a vector $\mathbf{d} \in R^n$ such that

$$f(\mathbf{x}) \ge f(\mathbf{0}) + d \cdot \mathbf{x}$$

for all $\mathbf{x} \in R^n$. In fact, if $f(\mathbf{x})$ has continuous first partial derivatives, we can take $\mathbf{d} = \nabla f(\mathbf{0})$ by (2.3.5). In the general case, $\mathbf{d}$ can be taken to be the subgradient of $f(\mathbf{x})$ at $\mathbf{0}$. (See (5.1.10) and the discussion following that result.) Next, observe that

$$\begin{aligned} f^\varepsilon(\mathbf{x}) = f(\mathbf{x}) + \varepsilon\|\mathbf{x}\|^2 &\ge f(\mathbf{0}) + \mathbf{d}\cdot\mathbf{x} + \varepsilon\|\mathbf{x}\|^2 \\ &= f(\mathbf{0}) - (-\mathbf{d}\cdot\mathbf{x}) + \varepsilon\|\mathbf{x}\|^2. \end{aligned}$$

By the Cauchy–Schwarz Inequality,

$$-\mathbf{d}\cdot\mathbf{x} \le \|\mathbf{d}\|\,\|\mathbf{x}\|$$

so

$$\begin{aligned} f^\varepsilon(\mathbf{x}) &\ge f(\mathbf{0}) - \|\mathbf{d}\|\,\|\mathbf{x}\| + \varepsilon\|\mathbf{x}\|^2 \\ &= f(\mathbf{0}) + \|\mathbf{x}\|(\varepsilon\|\mathbf{x}\| - \|\mathbf{d}\|). \end{aligned}$$

As $\|\mathbf{x}\| \to +\infty$, it is clear that $(\varepsilon\|\mathbf{x}\| - \|\mathbf{d}\|) \to +\infty$ so

$$\lim_{\|\mathbf{x}\| \to +\infty} f^\varepsilon(\mathbf{x}) = +\infty.$$

Hence, $f^\varepsilon(\mathbf{x})$ is coercive.

Now suppose that we are given a program

$$(P) \quad \begin{cases} \text{Minimize} \quad f(\mathbf{x}) \quad \text{subject to} \\ g_1(\mathbf{x}) \le 0, \quad g_2(\mathbf{x}) \le 0, \ldots, \quad g_m(\mathbf{x}) \le 0; \qquad \mathbf{x} \in R^n, \end{cases}$$

where $f(\mathbf{x}), g_1(\mathbf{x}), \ldots, g_m(\mathbf{x})$ are convex functions with continuous first partial derivatives on R^n but $f(\mathbf{x})$ is not coercive. Then, for each $\varepsilon > 0$, the program

$$(P^\varepsilon) \quad \begin{cases} \text{Minimize} \quad f^\varepsilon(\mathbf{x}) \quad \text{subject to} \\ g_1(\mathbf{x}) \le 0, \quad g_2(\mathbf{x}) \le 0, \ldots, \quad g_m(\mathbf{x}) \le 0; \qquad \mathbf{x} \in R^n, \end{cases}$$

is convex, its objective and constraint functions have continuous first partial derivatives on R^n and $f^\varepsilon(\mathbf{x})$ is coercive. Obviously, (P_ε) is consistent if and only if (P) is consistent because both programs have the same constraints. Therefore, the program (P^ε) satisfies the hypotheses of (6.3.1) whenever (P) is consistent.

The Lagrangian $L^\varepsilon(\mathbf{x}, \boldsymbol{\lambda})$ for (P^ε) is related to the Lagrangian $L(\mathbf{x}, \boldsymbol{\lambda})$ for (P) as follows:

$$\begin{aligned} L^\varepsilon(\mathbf{x}, \boldsymbol{\lambda}) &= f(\mathbf{x}) + \varepsilon\|\mathbf{x}\|^2 + \sum_{i=1}^{m} \lambda_i g_i(\mathbf{x}) \\ &= L(\mathbf{x}, \boldsymbol{\lambda}) + \varepsilon\|\mathbf{x}\|^2. \end{aligned}$$

Thus, the dual (DP^ε) of (P^ε) is

$$(DP^\varepsilon) \quad \begin{cases} \text{Maximize} \quad h^\varepsilon(\lambda) = \inf\{L(\mathbf{x}, \boldsymbol{\lambda}) + \varepsilon\|\mathbf{x}\|^2 : \mathbf{x} \in R^n\} \\ \text{subject to} \quad \boldsymbol{\lambda} \ge \mathbf{0} \quad \text{in } R^m. \end{cases}$$

In keeping with our notation for the given program (P) and its dual (DP), we define

$$MP^\varepsilon = \inf\{f^\varepsilon(\mathbf{x}) : g_1(\mathbf{x}) \le 0, \ldots, g_m(\mathbf{x}) \le 0; \mathbf{x} \in R^n\},$$

$$MD^\varepsilon = \sup\{h^\varepsilon(\boldsymbol{\lambda}) : \mathbf{0} \le \boldsymbol{\lambda} \in R^m\}.$$

Note that if $0 < \varepsilon \le \delta$, then $f^\varepsilon(\mathbf{x}) \le f^\delta(\mathbf{x})$ for all $\mathbf{x} \in R^n$ and also $L^\varepsilon(\mathbf{x}, \boldsymbol{\lambda}) \le L^\delta(\mathbf{x}, \boldsymbol{\lambda})$ for all $\mathbf{x} \in R^n$ and $\mathbf{0} \le \boldsymbol{\lambda} \in R^m$, so $MP^\varepsilon \le MP^\delta$ and $MD^\varepsilon \le MP^\delta$.

The preceding considerations and (6.3.1) now yield the following result.

(6.3.2) Lemma. *Suppose that $f(\mathbf{x}), g_1(\mathbf{x}), \ldots, g_m(\mathbf{x})$ have continuous first partial derivatives on R^n and that the program*

$$(P) \quad \begin{cases} \text{Minimize} \quad f(\mathbf{x}) \quad \text{subject to} \\ g_1(\mathbf{x}) \le 0, \quad g_2(\mathbf{x}) \le 0, \ldots, \quad g_m(\mathbf{x}) \le 0; \qquad \mathbf{x} \in R^n, \end{cases}$$

is consistent. Then for each $\varepsilon > 0$, the programs (P^ε) and (DP^ε) are consistent and $MD^\varepsilon = MP^\varepsilon$.

Before we proceed to use the programs (P^ε) and (DP^ε) to obtain a new proof of the Karush–Kuhn–Tucker Theorem by way of the Penalty Function Method, let us look at these programs in a simple example.

(6.3.3) Example. Note that the objective function in the program

$$(P) \quad \begin{cases} \text{Minimize} \quad f(x, y) = x + y \quad \text{subject to} \\ g(x, y) = x^2 + y^2 - 2 \le 0; \qquad (x, y) \in R^2, \end{cases}$$

is not coercive; for example, the value of $f(x, y)$ approaches $-\infty$ along the negative x-axis or negative y-axis.

A glance at the level curves of $f(x, y)$ and the feasibility region for (P) shows that (P) has the unique solution $(x^*, y^*) = (-1, -1)$ and that $MP = -2$. The program (P) is obviously superconsistent and $MD = MP = -2$. (See the remarks following (5.4.1).)

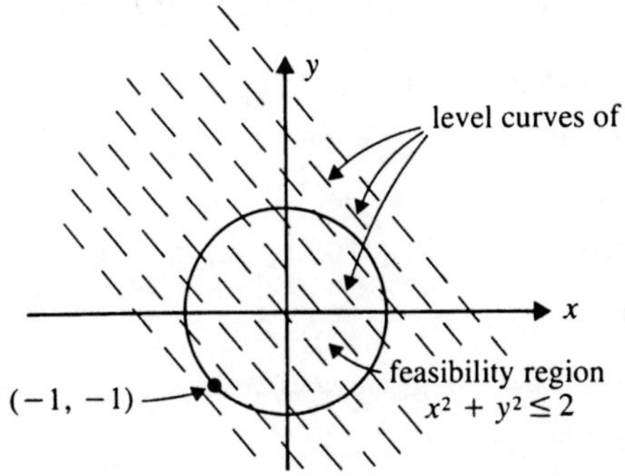

For each $\varepsilon > 0$, the objective function $f^\varepsilon(x, y)$ can be expressed as follows:

$$f^\varepsilon(x, y) = x + y + \varepsilon(x^2 + y^2) = \varepsilon\left(x^2 + \frac{1}{\varepsilon}x\right) + \varepsilon\left(y^2 + \frac{1}{\varepsilon}y\right)$$

$$= \varepsilon\left[\left(x + \frac{1}{2\varepsilon}\right)^2 + \left(y + \frac{1}{2\varepsilon}\right)^2\right] - \frac{1}{2\varepsilon}.$$

Thus, the level curves for $f^\varepsilon(x, y)$ are circles centered at $(-1/2\varepsilon, -1/2\varepsilon)$. Consequently, we see that, for $\varepsilon \geq \frac{1}{2}$, the constrained minimum value of $f^\varepsilon(x, y)$ is

$$MP^\varepsilon = -\frac{1}{2\varepsilon}$$

and that this value is assumed at the point $(-1/2\varepsilon, -1/2\varepsilon)$. On the other hand, if $0 < \varepsilon \leq \frac{1}{2}$, the constrained minimum value of $f^\varepsilon(x, y)$ is

$$MP^\varepsilon = 2\varepsilon - 2$$

and that this value is assumed at $(-1, -1)$. The graph of MP^ε as a function of ε is pictured below.

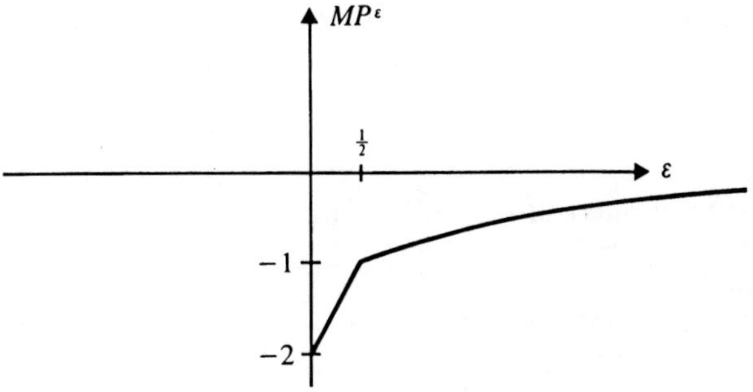

Of course, since (P) satisfies the hypotheses of (6.3.2), we see that $MD^\varepsilon = MP^\varepsilon$ for all $\varepsilon > 0$.

Note that, in the preceding example, the infimum of MD^ε over all $\varepsilon > 0$ is -2 which in turn is equal to MP for that example. The following lemma shows that this equality holds under rather general circumstances.

(6.3.4) Lemma. *Suppose that $f(\mathbf{x}), g_1(\mathbf{x}), \ldots, g_m(\mathbf{x})$ have continuous first partial derivatives on R^n. If the program*

$$(P)\quad \begin{cases} \text{Minimize} \quad f(\mathbf{x}) \quad \text{subject to} \\ \quad g_1(\mathbf{x}) \le 0, \quad g_2(\mathbf{x}) \le 0, \ldots, \quad g_m(\mathbf{x}) \le 0; \qquad \mathbf{x} \in R^n, \end{cases}$$

is consistent and if $MP > \infty$, then:

(1) *The program (DP^ε) is consistent for all $\varepsilon > 0$.*
(2) $MP = \inf\{MD^\varepsilon: \varepsilon > 0\}$.

Proof. For a given $\varepsilon > 0$, $f(\mathbf{x}) \le f^\varepsilon(\mathbf{x})$ for all $\mathbf{x} \in R^n$ so $MP \le MP^\varepsilon$ because (P) and (P^ε) have the same constraints. But $MP^\varepsilon = MD^\varepsilon$ and (D^ε) is consistent for each $\varepsilon > 0$ by Lemma (6.3.2). Therefore

$$\begin{aligned} MP \le \inf\{MD^\varepsilon: \varepsilon > 0\} &= \inf\{MP^\varepsilon: \varepsilon > 0\} \\ &= \inf_{\varepsilon > 0} \left[\inf\{f(\mathbf{x}) + \varepsilon\|\mathbf{x}\|^2: g_i(\mathbf{x}) \le 0, i = 1, \ldots, m\}\right] \\ &= \inf\left\{\inf_{\varepsilon > 0} \left[f(\mathbf{x}) + \varepsilon\|\mathbf{x}\|^2\right]: g_i(\mathbf{x}) \le 0, i = 1, \ldots, m\right\} \\ &= \inf\{f(\mathbf{x}): g_i(\mathbf{x}) \le 0, i = 1, 2, \ldots, m\} = MP \end{aligned}$$

which completes the proof.

Remark. It is not true, in general, under the hypotheses of (6.3.4) that

$$(*) \qquad MD = \inf\{MD^\varepsilon: \varepsilon > 0\}.$$

In fact, if $(*)$ holds, then (6.3.4) implies that $MP = MD$, that is, the program (P) does not have a duality gap. The example in (5.3.5) shows that programs satisfying the hypotheses of (5.4.8) can have duality gaps.

The next theorem allows us to mesh the work of this section with that of Section 5.2. This theorem should be studied together with Theorems (5.2.13) and (5.2.16).

(6.3.5) Theorem. *Suppose that $f(\mathbf{x}), g_1(\mathbf{x}), \ldots, g_m(\mathbf{x})$ are convex and have continuous first partial derivatives on R^n. If the program*

$$(P)\quad \begin{cases} \text{Minimize} \quad f(\mathbf{x}) \quad \text{subject to} \\ \quad g_1(\mathbf{x}) \le 0, \quad g_2(\mathbf{x}) \le 0, \ldots, \quad g_m(\mathbf{x}) \le 0; \qquad \mathbf{x} \in R^n, \end{cases}$$

is superconsistent and $MP > -\infty$, *then:*

(1) *the dual program* (*D*) *is consistent;*
(2) $MP = MD$, *that is,* (*P*) *does not have a duality gap;*
(3) *there is a vector* $\boldsymbol{\lambda}^* \in R^m$ *which is a solution of the dual program*

$$(P) \quad \begin{cases} \text{Maximize} & h(\boldsymbol{\lambda}) = \inf\{L(\mathbf{x}, \boldsymbol{\lambda}) : \mathbf{x} \in R^n\} \\ \text{subject to} & \boldsymbol{\lambda} \geq 0. \end{cases}$$

If there is a solution $\mathbf{x}^*$ *of* (*P*), *then*

$$f(\mathbf{x}^*) = L(\mathbf{x}^*, \boldsymbol{\lambda}^*) = h(\boldsymbol{\lambda}^*).$$

Moreover,

$$\lambda_i^* g_i(\mathbf{x}^*) = 0 \quad \textit{for } i = 1, 2, \ldots, m,$$

and $\boldsymbol{\lambda}^*$ *is a sensitivity vector for* (*P*).

PROOF. According to Lemma (6.3.4), the dual program (DP^ε) of (P^ε) is consistent for all $\varepsilon > 0$ and

$$MP = \inf\{MD^\varepsilon : \varepsilon > 0\}.$$

Because

$$MD^\varepsilon = \sup_{\boldsymbol{\lambda} \geq \mathbf{0}} \inf_{\mathbf{x} \in R^n} (L(\mathbf{x}, \boldsymbol{\lambda}) + \varepsilon \|\mathbf{x}\|^2),$$

and because MD^ε decreases as $\varepsilon > 0$ decreases, it follows that for each positive integer k there is a positive integer m_k such that

$$MP \leq \sup_{\boldsymbol{\lambda} \geq \mathbf{0}} \left(\inf_{\mathbf{x} \in R^n} \left(L(\mathbf{x}, \boldsymbol{\lambda}) + \frac{1}{m_k} \|\mathbf{x}\|^2 \right) \right) \leq MP + \frac{1}{k},$$

and $m_k > m_p$ when $k > p$. But then, for each positive integer k, we can choose $\boldsymbol{\lambda}^{(k)} \in R^m$ with $\boldsymbol{\lambda}^{(k)} \geq \mathbf{0}$ such that

$$(*) \qquad MP \leq \inf_{\mathbf{x} \in R^n} \left(L(\mathbf{x}, \boldsymbol{\lambda}^{(k)}) + \frac{1}{m_k} \|\mathbf{x}\|^2 \right) + \frac{1}{k} \leq MP + \frac{2}{k}.$$

We shall now show that $\{\boldsymbol{\lambda}^{(k)}\}$ is a bounded sequence. Let $\mathbf{y}$ be a Slater point for (*P*); that is, $g_i(\mathbf{y}) < 0$ for $i = 1, 2, \ldots, m$. Choose $\alpha > 0$ so that $g_i(\mathbf{y}) < -\alpha$ for $i = 1, 2, \ldots, m$. Then by virtue of (∗),

$$\begin{aligned} MP &\leq L(\mathbf{y}, \boldsymbol{\lambda}^{(k)}) + \frac{1}{m_k} \|\mathbf{y}\|^2 + \frac{1}{k} \\ &\leq f(\mathbf{y}) + \sum_{i=1}^m \lambda_i^{(k)} g_i(\mathbf{y}) + \frac{1}{m_k} \|\mathbf{y}\|^2 + \frac{1}{k} \\ &\leq f(\mathbf{y}) - \alpha \sum_{i=1}^m \lambda_i^{(k)} + \|\mathbf{y}\|^2 + 1, \end{aligned}$$

and so

$$\sum_{i=1}^{m} \lambda_i^{(k)} \leq \frac{1}{\alpha}(f(\mathbf{y}) + \|\mathbf{y}\|^2 + 1 - MP)$$

for all k. This latter inequality, together with the fact that $\lambda_i^{(k)} \geq 0$ for all k and all $i = 1, 2, \ldots, m$, yields the boundedness of the sequence $\{\boldsymbol{\lambda}^{(k)}\}$ in R^m.

The Bolzano–Weierstrass Property assures us that some subsequence of $\{\boldsymbol{\lambda}^{(k)}\}$ converges to a vector $\boldsymbol{\lambda}^* \in R^m$. Clearly, $\boldsymbol{\lambda}^* \geq 0$ and $(*)$ implies that

$$MP \leq L(\mathbf{x}, \boldsymbol{\lambda}^*)$$

for all $\mathbf{x} \in R^n$. Therefore, $\boldsymbol{\lambda}^*$ is a feasible vector for (D) and

$$MP \leq \inf_{\mathbf{x} \in R^n} \{L(\mathbf{x}, \boldsymbol{\lambda}^*)\} \leq \sup_{\lambda \geq 0} \inf_{\mathbf{x} \in R^n} \{L(\mathbf{x}, \boldsymbol{\lambda})\} = MD.$$

Since the Primal–Dual Inequality assures us that $MP \geq MD$, we conclude that

$$MP = MD = \inf_{\mathbf{x} \in R^n} \{L(\mathbf{x}, \boldsymbol{\lambda}^*)\} = h(\boldsymbol{\lambda}^*),$$

which completes the proof of assertions (1), (2), and (3) of the theorem.

If $\mathbf{x}^*$ is a solution to (D), then by (2),

$$f(\mathbf{x}^*) = MP = MD = \inf_{\mathbf{x} \in R^n} \{L(\mathbf{x}, \boldsymbol{\lambda}^*)\}.$$

But, because $\boldsymbol{\lambda}^* \geq \mathbf{0}$ and $g_i(\mathbf{x}^*) \leq 0$ for $i = 1, 2, \ldots, m$, it follows that

$$f(\mathbf{x}^*) \geq f(\mathbf{x}^*) + \sum_{i=1}^{m} \lambda_i^* g_i(\mathbf{x}^*) = L(\mathbf{x}^*, \boldsymbol{\lambda}^*) \geq h(\boldsymbol{\lambda}^*) = MD.$$

Consequently,

$$f(\mathbf{x}^*) = L(\mathbf{x}^*, \boldsymbol{\lambda}^*) = h(\boldsymbol{\lambda}^*)$$

and

$$\sum_{i=1}^{m} \lambda_i^* g_i(\mathbf{x}^*) = 0.$$

The last equation implies that

$$\lambda_i^* g_i(\mathbf{x}^*) = 0, \qquad i = 1, 2, \ldots, m,$$

because $\lambda_i \geq 0$, $g_i(\mathbf{x}^*) \leq 0$ for $i = 1, 2, \ldots, m$.

Finally, to show that $\boldsymbol{\lambda}^*$ is a sensitivity vector for the program (P), we note from (5.2.16) that we need only show that

$$L(\mathbf{x}^*, \boldsymbol{\lambda}) \leq L(\mathbf{x}^*, \boldsymbol{\lambda}^*)$$

for all $\boldsymbol{\lambda} \geq \mathbf{0}$ in R^m. To this end, we observe that

$$\begin{aligned} L(\mathbf{x}^*, \boldsymbol{\lambda}^*) - L(\mathbf{x}^*, \boldsymbol{\lambda}) &= \sum_{i=1}^{m} (\lambda_i^* - \lambda_i) g_i(\mathbf{x}^*) \\ &= -\sum_{i=1}^{m} \lambda_i g_i(\mathbf{x}^*) \geq 0, \end{aligned}$$

because $\boldsymbol{\lambda} \geq \mathbf{0}$, $g_i(\mathbf{x}^*) \leq 0$, and $\lambda_i^* g_i(\mathbf{x}^*) = 0$ for $i = 1, 2, \ldots, m$. This yields the desired inequality and we conclude that $\boldsymbol{\lambda}^*$ is a sensitivity vector for (P). The proof of the theorem is now complete.

The preceding result actually includes the Karush–Kuhn–Tucker Theorem (5.2.13). For if $\mathbf{x}^*$ is a solution to a program (P) satisfying the hypotheses of (5.2.13), then $MP = f(\mathbf{x}^*) > -\infty$ and so (6.3.5) asserts that there is a vector $\boldsymbol{\lambda}^*$ in R^m such that:

(1) $\lambda_i^* \geq 0$ for $i = 1, 2, \ldots, m$;
(2) $\lambda_i^* g_i(\mathbf{x}^*) = 0$ for $i = 1, 2, \ldots, m$.

Moreover, because $L(\mathbf{x}^*, \boldsymbol{\lambda}^*) = f(\mathbf{x}^*)$ and (2) holds, we see that

$$L(\mathbf{x}^*, \boldsymbol{\lambda}^*) \leq L(\mathbf{x}, \boldsymbol{\lambda}^*)$$

for all $\mathbf{x} \in R^n$. Thus, $\mathbf{x}^*$ is a global minimizer of $L(\mathbf{x}, \boldsymbol{\lambda}^*)$ on R^n and so $\nabla L(\mathbf{x}^*, \boldsymbol{\lambda}^*) = 0$. Consequently,

(3) $$\nabla f(\mathbf{x}^*) + \sum_{i=1}^{m} \lambda_i^* \nabla g_i(\mathbf{x}^*) = 0.$$

Conversely, if $\mathbf{x}^*$ is feasible for (P) and if $\boldsymbol{\lambda}^* \in R^m$ satisfies (1), (2), and (3), then the calculation in the second part of the proof of (5.2.14) shows that $\mathbf{x}^*$ is a solution to (P).

The point of these observations is that (6.3.5) provides a new way to establish the Karush–Kuhn–Tucker necessary conditions (1), (2), and (3) that is independent of the abstract methods of Chapter 5. The sufficiency of these conditions, which is quite elementary and completely independent of these abstract methods, is established just as in (5.2.14).

One final comment: All of this chapter has dealt with programs of the form

$$(P) \quad \begin{cases} \text{Minimize} \quad f(\mathbf{x}) \quad \text{subject to} \\ \quad g_1(\mathbf{x}) \leq 0, \ldots, \quad g_m(\mathbf{x}) \leq 0; \qquad \mathbf{x} \in R^n, \end{cases}$$

We can replace the domain R^n for (P) by any closed convex subset of R^n and establish all of the results of this chapter by the same methods with a little extra care at appropriate places.

Exercises

1. Consider the following program:

$$(P) \quad \begin{cases} \text{Minimize} \quad f(x) = x^2 - 2x \\ \text{subject to} \quad 0 \leq x \leq 1. \end{cases}$$

   (a) Sketch the graphs of the Absolute Value and Courant–Beltrami Penalty Terms for (P).
   (b) For each positive integer k, compute the minimizer x_k of the corresponding unconstrained objective function $P_k(x)$ with the Courant–Beltrami Penalty Term.
   (c) For each positive integer k, compute the minimizer x_k of the corresponding unconstrained objective function $F_k(x)$ with the Absolute Value Penalty Term.

2. (a) Use the Penalty Function Method with the Courant–Beltrami Penalty Term to solve the problem

$$\begin{cases} \text{Minimize} & f(x_1, x_2) = x_1 + x_2 \\ \text{subject to} & x_1^2 - x_2 \leq 2. \end{cases}$$

(b) Show that the objective function $F_k(\mathbf{x})$ corresponding to the Absolute Value Penalty Term has no critical points off the parabola

$$x_1^2 - x_2 = 2$$

for $k > 1$ and compute the minimizer of $F_k(\mathbf{x})$.

3. Use the Penalty Function Method with the Courant–Beltrami Penalty Term to minimize

$$f(x, y) = x^2 + y^2$$

subject to the constraint $x + y \geq 1$.

4. Consider the program:

$$(P) \quad \text{Minimize} \quad f(\mathbf{x}) \quad \text{subject to} \quad g(\mathbf{x}) \leq 0,$$

where $f(\mathbf{x})$ and $g(\mathbf{x})$ have continuous first partial derivatives on R^n and $f(\mathbf{x})$ is convex and coercive.

(a) Prove that the associated unconstrained program

$$\text{Minimize} \quad F_k(\mathbf{x}) = f(\mathbf{x}) + kg^+(\mathbf{x})$$

has a minimizer $\mathbf{x}_k$ for each positive integer k.

(b) Prove that if the gradient of

$$\varphi_k(\mathbf{x}) = f(\mathbf{x}) + kg(\mathbf{x})$$

is nonzero for all nonfeasible points for (P), then $\mathbf{x}_k$ must be feasible for (P).

(c) Show by example that $\{\mathbf{x}_k\}$ may converge to a point $\mathbf{x}^*$ that is not a solution of (P). (Hint: Try a simple inconsistent program (P).)

5. Suppose $f(\mathbf{x}), g_1(\mathbf{x}), \ldots, g_m(\mathbf{x})$ are continuous functions on R^n. Suppose the Penalty Function Method is used to minimize $f(\mathbf{x})$ subject to $g_1(\mathbf{x}) \leq 0, \ldots, g_m(\mathbf{x}) \leq 0$. If $\mathbf{x}_k$ is the global minimizer of $P_k(\mathbf{x})$ on R^n and $g_i(\mathbf{x}_k) \leq 0$ for all $i = 1, \ldots, m$, then show that $\mathbf{x}_k$ also minimizes $f(\mathbf{x})$ subject to $g_1(\mathbf{x}) \leq 0, \ldots, g_m(\mathbf{x}) \leq 0$.

6. Let $g(x)$ be a differentiable function on R^1 and suppose $g(x_0) = 0$.

(a) Show $g^+(x)$ is differentiable at x_0 if and only if $g'(x_0) = 0$.

(b) Show carefully that $(g^+(x))^2$ is differentiable at x_0 and that its derivative at x_0 is zero.

7. Let $f(\mathbf{x}), g_1(\mathbf{x}), \ldots, g_m(\mathbf{x})$ be continuous functions on R^n. Suppose there is a vector $\mathbf{y} \in R^n$ such that $g_i(\mathbf{y}) \leq 0$ for all $i = 1, \ldots, m$. Also suppose there is an i_0 with $1 \leq i_0 \leq m$ such that $g_{i_0}(\mathbf{x})$ is coercive. Prove:

(a) For each k there is a point $\mathbf{x}_k$ in R^n such that

$$P_k(\mathbf{x}_k) = \min_{\mathbf{x} \in R^n} P_k(\mathbf{x}).$$

(b) The sequence $(\mathbf{x}_k)$ is bounded.

(c) The sequence $(\mathbf{x}_k)$ has at least one convergent subsequence.

(d) If $\mathbf{x}^{**}$ is the limit of any convergent subsequence of $(\mathbf{x}_k)$ then $\mathbf{x}^{**}$ minimizes $f(\mathbf{x})$ subject to $g_i(\mathbf{x}) \le 0$, $i = 1, \ldots, m$.

8. Suppose $f(\mathbf{x}), g_1(\mathbf{x}), \ldots, g_m(\mathbf{x})$ are all differentiable convex functions defined on R^n. Suppose also that $f(\mathbf{x})$ is coercive. Let $(\mathbf{x}_k)$ be the sequence produced by the Penalty Function Method with the Courant–Beltrami Penalty Term. Prove

$$\lim_{k\to\infty} k \sum_{i=1}^{m} (g_i^+(\mathbf{x}_k))^2 = 0.$$

(Hint: See the proof of Theorem 6.3.1.)

9. Let $\varepsilon > 0$. Show that if a vector $\boldsymbol{\lambda}$ is feasible for the dual (D) of a convex program (P), then $\boldsymbol{\lambda}$ is also feasible for the program (D^ε).

10. Find a convex program (P) that is not superconsistent and yet $MP = MD$.

11. Let $f(\mathbf{x}), g_1(\mathbf{x}), \ldots, g_m(\mathbf{x})$ be differentiable convex functions defined on R^n. Assume $f(x)$ is coercive and assume the dual (D) of the convex program

$$(P)\quad \begin{cases} \text{Minimize} & f(\mathbf{x}) \\ \text{subject to} & g_1(\mathbf{x}) \le 0, \ldots, \quad g_m(\mathbf{x}) \le 0, \end{cases}$$

is consistent and $MD < \infty$. Prove (P) is also consistent and $MD = MP$. In fact, show that there is a vector $\mathbf{x}^*$ feasible for (P) such that $f(\mathbf{x}^*) = MP$.

12. Let (P) be a convex program and suppose that for some $\varepsilon > 0$ the program (DP^ε) is consistent. Prove (P) is also consistent. Show that if $\lim_{\varepsilon\to\infty} MD^\varepsilon < \infty$, then $MP = MD$. (Hint: Refer to the preceding exercise.)

13. Suppose that $f(\mathbf{x}), g_1(\mathbf{x}), \ldots, g_m(\mathbf{x})$ are continuous functions on R^n. Suppose that

$$(P)\quad \begin{cases} \text{Minimize} \quad f(\mathbf{x}) \quad \text{subject to} \\ g_1(\mathbf{x}) \le 0, \ldots, \quad g_m(\mathbf{x}) \le 0; \qquad \mathbf{x} \in R^n \end{cases}$$

has solution $\mathbf{x}^*$. Suppose $f(\mathbf{x})$ is bounded from below on R^n and suppose that $g_{i_0}(\mathbf{x})$ is coercive for some i_0 with $1 \le i_0 \le m$. Show that for each k, $P_k(\mathbf{x})$ has an unconstrained minimizer $\mathbf{x}_k$ and that the sequence $\{\mathbf{x}_k\}$ is bounded. Show any convergent subsequence of $(\mathbf{x}_k)$ converges to a solution of P. (Hint: Mimic the proof of (6.2.4).)

CHAPTER 7

Optimization with Equality Constraints

Many optimization techniques of practical interest include *equality constraints*, that is, constraints of the form

$$g(\mathbf{x}) = 0, \qquad \mathbf{x} \in R^n. \tag{1}$$

For example, the classical Lagrange Multiplier Method for constrained optimization deals exclusively with equality constraints. On the other hand, some very useful methods such as the Wolfe Algorithm for quadratic programming (see Section 7.3) are based on the solution of optimization problems that include both equality and inequality constraints.

The Karush–Kuhn–Tucker theory, as developed for superconsistent convex programs in Chapter 5, is not easily modified to cover the case in which some or all of the constraints are equality constraints. The obvious ploy of replacing an equality constraint (1) by two inequality constraints

$$g(\mathbf{x}) \leq 0, \qquad -g(\mathbf{x}) \leq 0, \qquad \mathbf{x} \in R^n, \tag{2}$$

is not very helpful because both $g(\mathbf{x})$ and $-g(\mathbf{x})$ cannot be convex unless $g(\mathbf{x})$ is linear. Also, the superconsistency hypotheses is obviously not appropriate for programs that include equality constraints.

The development of suitable optimization methods for problems involving some equality constraints will require a rather different approach than that employed for convex programs in Chapter 5—an approach based on the development of the classical Lagrange Multiplier Method. We will see that the Karush–Kuhn–Tucker Theorem (5.2.14) survives in a modified but completely recognizable form and that the new version of this theorem yields methods and insights that are every bit as useful as the earlier version for superconsistent convex (inequality-constrained) programs.

To set the stage for the work in this chapter, let us consider the following simple constrained minimization problem

$$(P)\quad \begin{cases} \text{Minimize} \quad f(x, y) = y \quad \text{subject to} \\ \quad g(x, y) = 3y - 3x^4 - 4x^3 + 12x^2 = 0. \end{cases}$$

Geometrically, the program (P) seeks the point(s) (x, y) on the curve

$$y = x^4 + \tfrac{4}{3}x^3 - 4x^2 \tag{3}$$

for which the ordinate is least, that is, the lowest point(s) on the graph of (3) in the xy-plane.

A little calculus shows that (3) has three critical points $x = 0$, $x = 1$, $x = -2$, and that $x = 0$ is a local maximizer while $x = -2$, $x = 1$ are local minimizers for $g(x) = x^4 + \frac{4}{3}x^3 - 4x^2$. The graph of $y = g(x)$, that is, of $(*)$ is pictured below.

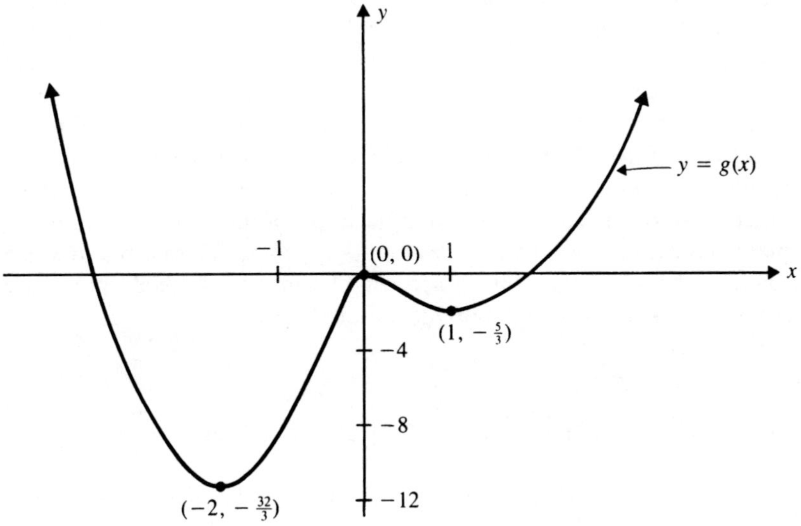

It is evident from this graph that $(x^*, y^*) = (-2, -\frac{32}{3})$ is the solution to (P), that is, (x^*, y^*) is the global minimizer of $f(x, y)$ on the set F of feasible vectors for (P)

$$F = \{(x, y) \in R^2 \colon 3y - 3x^4 - 4x^3 + 12x^2 = 0\}.$$

Note that

$$\nabla f(x, y) = (0, 1), \qquad \nabla g(x, y) = (-12x^3 - 12x^2 + 24x, 3).$$

Consequently, at the solution $(x^*, y^*) = (-2, -\frac{32}{3})$ for (P), the gradient condition

$$\nabla f(x^*, y^*) + \lambda^* \nabla g(x^*, y^*) = (0, 0), \tag{4}$$

familiar from the Karush–Kuhn–Tucker theory, is satisfied if we take

$\lambda^* = -\frac{1}{3}$ since $x^* = -2$ is a solution of

$$-12x^3 - 12x^2 + 24x = 0. \tag{5}$$

However, observe that, unlike the Karush–Kuhn–Tucker theory, λ^* is negative. Note further that $x = 0$ and $x = 1$ are also solutions of (3) and hence that the gradient condition (2) is also satisfied for $\lambda^* = -\frac{1}{3}$ at the points $(x^*, y^*) = (0, 0)$ and $(x^*, y^*) = (1, -\frac{5}{3})$. The point (0, 0) is a local maximizer for (P) while the point $(1, -\frac{5}{3})$ is a local minimizer for (P). These three points (x^*, y^*) and $\lambda^* = -\frac{1}{3}$ are the only solutions to the gradient condition for (P). It is clear from the graph of $y = g(x)$ that (P) has no global maximizer.

The preceding example indicates that the gradient condition familiar from the Karush–Kuhn–Tucker theory can be expected to play a role in the solution of minimization problems in which equality constraints are present. However, it also shows that this condition can only be counted on to identify candidates for local minimizers. This condition may also turn up points that are neither local minimizers nor local maximizers of the objective function on the set of feasible vectors. The example also shows that we cannot expect the gradient condition to determine which local minimizers and maximizers are global, or indeed if global minimizers and maximizers exist. Finally, as we have already noted, the multiplier λ^* associated with an equality constrained program need not be nonnegative as in the case of inequality constraints.

Before we can begin to study minimization of functions subject to some equality constraints, we need to develop some geometric concepts concerning surfaces in R^n. We will do this in Section 7.1 and then proceed to the general development of constrained minimization in Section 7.2.

7.1. Surfaces and Their Tangent Planes

In calculus, a surface S in space (i.e., in R^3) is often described by an equation

$$g(x, y, z) = 0, \tag{1}$$

where g is a function of three variables with continuous first partial derivatives. At any point $\mathbf{x}^{(0)}$, the gradient vector $\nabla g(\mathbf{x}^{(0)})$ is perpendicular to S because the gradient vector points in the direction of maximum increase of g and, according to (1), S is a "level surface" of g. Therefore, the tangent plane to S at $\mathbf{x}^{(0)}$ is given by

$$(\mathbf{x} - \mathbf{x}^{(0)}) \cdot \nabla g(\mathbf{x}^{(0)}) = 0. \tag{2}$$

Put another way, the tangent plane to S at $\mathbf{x}^{(0)}$ consists of all vectors $\mathbf{x}^{(0)} + \mathbf{y}$ where

$$\mathbf{y} \cdot \nabla g(\mathbf{x}^{(0)}) = 0. \tag{3}$$

Although the set of all vectors $\mathbf{y}$ satisfying (3) is not normally singled out for special attention in calculus texts, it is called the *tangent space* $\mathrm{T}(\mathbf{x}^{(0)})$ *of the surface S at* $x^{(0)}$ in more advanced books. Unlike the tangent plane to S at $\mathbf{x}^{(0)}$, the tangent space $T(\mathbf{x}^{(0)})$ is always a subspace of R^3. Moreover, the tangent plane to S at $\mathbf{x}^{(0)}$ is simply the translate $\mathbf{x}^{(0)} + T(x^{(0)})$ of the tangent space at $\mathbf{x}^{(0)}$.

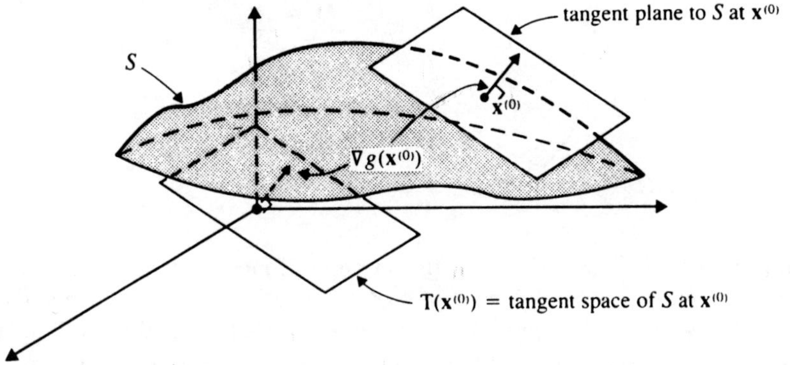

(7.1.1) Definitions. Suppose that $g_1(\mathbf{x}), \ldots, g_p(\mathbf{x})$ are functions with continuous first partial derivatives on some open subset of C of R^n.

(a) A *surface* S in R^n is the set of points in R^n satisfying

$$g_1(\mathbf{x}) = 0, \quad g_2(\mathbf{x}) = 0, \ldots, \quad g_p(\mathbf{x}) = 0; \qquad \mathbf{x} \in C.$$

(b) If $\mathbf{x}^{(0)}$ is a point of a surface S defined in (a), then the *normal space* $N(x^{(0)})$ *to S at* $x^{(0)}$ is the set of all linear combinations of the vectors

$$\nabla g_1(\mathbf{x}^{(0)}), \ldots, \nabla g_p(\mathbf{x}^{(0)}),$$

and the *tangent space* $T(\mathbf{x}^{(0)})$ *to S at* $\mathbf{x}^{(0)}$ is simply the orthogonal complement $N(\mathbf{x}^{(0)})^{\perp}$ of the normal space $N(\mathbf{x}^{(0)})$ to S at $\mathbf{x}^{(0)}$.

(7.1.2) Examples

(a) Note that if a surfaces S in R^3 is described by a single function of three variables as in (1) above and if $\mathbf{x}^{(0)}$ is a point of S for which $\nabla g(\mathbf{x}^{(0)}) \neq \mathbf{0}$, then the normal space $N(\mathbf{x}^{(0)})$ is simply the line L through the origin consisting of all multiples of the vector $\nabla g(\mathbf{x}^{(0)})$, and the tangent space $T(\mathbf{x}^{(0)})$ is the plane through the origin that is perpendicular to L.

(b) It is important to note that the term "surface," as defined in (7.1.1), includes objects that are not normally called surfaces in calculus even when $n = 3$. For example, let us consider the surface (in the sense of (7.1.1)) defined by

$$g_1(\mathbf{x}) = 0, \qquad g_2(\mathbf{x}) = 0, \tag{4}$$

where $g_1(\mathbf{x}), g_2(\mathbf{x})$ have continuous first partial derivatives on R^3. Typically, (4) describes the curve of intersection of the surfaces S_1 given by $g_1(\mathbf{x}) = 0$ and the surface S_2 given by $g_2(\mathbf{x}) = 0$.

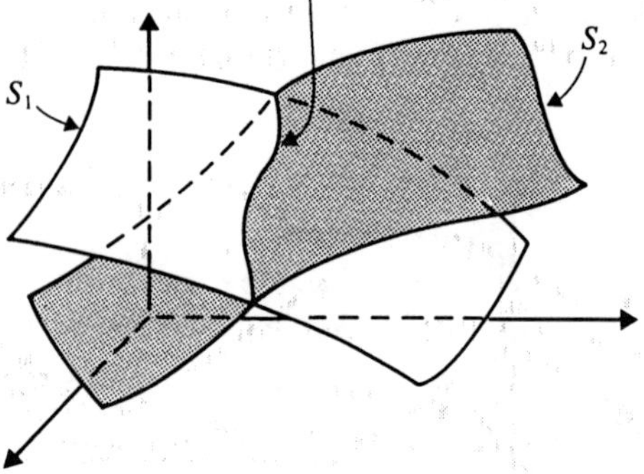

In this case, if $\mathbf{x}^{(0)}$ is a point on this curve S of intersection of S_1, S_2 and if $\nabla g_1(\mathbf{x}^{(0)})$, $\nabla g_2(\mathbf{x}^{(0)})$ are not multiples of one another, then the normal space $\mathrm{N}(\mathbf{x}^{(0)})$ is actually the plane through the origin that is parallel to the plane perpendicular to the curve S at $\mathbf{x}^{(0)}$. In this case, the tangent space $\mathrm{T}(\mathbf{x}^{(0)})$ is the line through the origin that is parallel to the tangent line to the curve S at $\mathbf{x}^{(0)}$.

(c) Suppose that $f(\mathbf{x})$ is a function with continuous first partial derivatives defined on some open subset C of R^n. In Chapter 2 (see the discussion preceding (2.3.5)), we introduced the *tangent hyperplane* $P_{\mathbf{x}^{(0)}}$ *to the graph of* $f(x)$ *at* $x^{(0)}$ to obtain a very useful characterization of convex functions. We defined $P_{\mathbf{x}^{(0)}}(0)$ as follows:

$$P_{\mathbf{x}^{(0)}} = \{(\mathbf{y}, y) \in R^{n+1}: y = f(\mathbf{x}^{(0)}) + \nabla f(\mathbf{x}^{(0)})\cdot(\mathbf{y} - \mathbf{x}^{(0)})\}.$$

To see how this notation fits into our present discussion of surfaces, note that if

$$z = f(\mathbf{x}),$$

then the function $g(\mathbf{x}, z)$ defined on the set

$$C \times R = \{(\mathbf{x}, z): \mathbf{x} \in C, z \in R\}$$

in R^{n+1} by

$$g(\mathbf{x}, z) = f(\mathbf{x}) - z$$

has the property that

$$\nabla g(\mathbf{x}, z) = (\nabla f(\mathbf{x}), -1).$$

Moreover, the graph of $f(\mathbf{x})$ is simply the surface in R^{n+1} defined by

$$g(\mathbf{x}, z).$$

For any $\mathbf{x}^{(0)} \in C$, the point $(\mathbf{x}^{(0)}, f(\mathbf{x}^{(0)}))$ is on the graph of $f(\mathbf{x})$ and the tangent plane to the graph of $f(\mathbf{x})$ at $(\mathbf{x}^{(0)}, f(\mathbf{x}^{(0)}))$ coincides with $P_{\mathbf{x}^{(0)}}$ as can be seen from the following computation:

$$\{(\mathbf{y}, y) \in R^{n+1}: y = f(\mathbf{x}^{(0)}) + \nabla f(\mathbf{x}^{(0)}) \cdot (\mathbf{y} - \mathbf{x}^{(0)})\}$$
$$= \{(\mathbf{y}, y) \in R^{n+1}: (\mathbf{y} - \mathbf{x}^{(0)}) \cdot \nabla f(\mathbf{x}^{(0)}) - (y - f(\mathbf{x}^{(0)})) = 0\}$$
$$= \{(\mathbf{y}, y) \in R^{n+1}: [(\mathbf{y}, y) - (\mathbf{x}^{(0)}, f(\mathbf{x}^{(0)}))] \cdot \nabla g(\mathbf{x}^{(0)}, f(\mathbf{x}^{(0)})) = 0\}.$$

If a surface S in space is described by an equation

$$g_1(x, y, z) = 0, \tag{1}$$

where g_1 is a function of three variables with continuous first partial derivatives and if $\mathbf{x}^{(0)} = (x_0, y_0, z_0)$ is a point of S, then it is intuitively plausible that the tangent space $T(\mathbf{x}^{(0)})$ to S at $\mathbf{x}^{(0)}$ can also be described as the set of all vectors $\mathbf{z}_C$ that are tangent vectors at $\mathbf{x}^{(0)}$ to some curve C that lies on the surface S and passes through $\mathbf{x}^{(0)}$. Once we make precise what we mean by the phrase "a tangent vector to a curve C that lies on the surface S," we will see that this alternate description of the tangent space is available. This alternate description will be very helpful in our study of minimization problems with equality constraints.

(7.1.3) Definition. If S is a surface in R^n, then a function

$$\boldsymbol{\varphi}(t) = (\varphi_1(t), \varphi_1(t), \ldots, \varphi_n(t))$$

of a real variable t with values in R^n is a *path in S* if:

(a) $\boldsymbol{\varphi}(t)$ is defined on an open interval (α, β) of the real line and $\boldsymbol{\varphi}(t) \in S$ for all $t \in (\alpha, \beta)$;
(b) the component function $\varphi_1(t), \ldots, \varphi_n(t)$ are differentiable at each point of the interval (α, β).

(7.1.4) Examples

(a) The surface S_1 in R^3 defined by

$$g_1(x, y, z) = x^2 + y^2 - z = 0$$

is a paraboloid with its vertex at the origin and the positive z-axis as its central axis. The curve in R^3 defined by

$$\boldsymbol{\varphi}(t) = (t \cos t, t \sin t, t^2), \qquad 0 < t < 2\pi,$$

is a path in S since the component functions are clearly differentiable on $(0, 2\pi)$ and

$$(t \cos t)^2 + (t \sin t)^2 - t^2 = 0$$

for all $t \in (0, 2\pi)$.

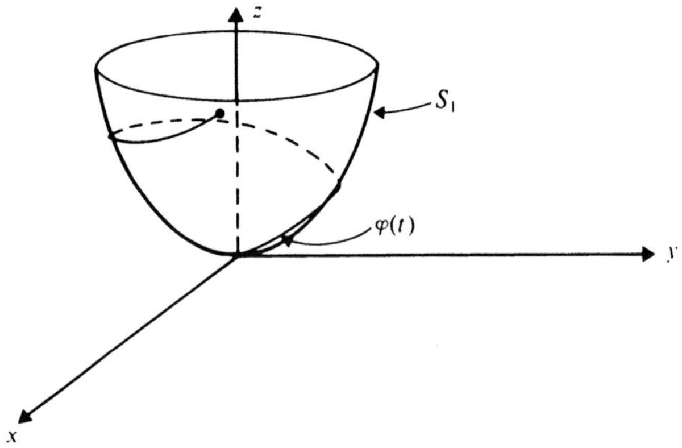

(b) The surface S in R^3 defined by

$$g_1(x, y, z) = x^2 + y^2 - z = 0,$$
$$g_2(x, y, z) = 8 - x^2 - y^2 - z = 0,$$

is the curve of intersection of the paraboloid in part (a) with a paraboloid with vertex at (0, 0, 8), and opening down the z-axis. The curve in R^3 defined by

$$\boldsymbol{\varphi}(t) = (2 \cos t, 2 \sin t, 4), \qquad 0 < t < \frac{\pi}{2},$$

is a path in S which is a quarter circle extending from (2, 0, 4) to (0, 2, 4) in the plane $z = 4$.

The next theorem allows us to identify members of the tangent space $T(\mathbf{x}^{(0)})$ at a given point $\mathbf{x}^{(0)}$ of a given surface S in R^n with tangent vectors to paths in S passing through $\mathbf{x}^{(0)}$.

(7.1.5) Theorem. *Suppose that $g_1(\mathbf{x}), g_2(\mathbf{x}), \ldots, g_p(\mathbf{x})$ are functions with continuous first partial derivatives on some open set C in R^n and that S is the surface in R^n defined by*

$$g_1(\mathbf{x}) = 0, \quad g_2(\mathbf{x}) = 0, \ldots, \quad g_p(\mathbf{x}) = 0.$$

If $\mathbf{x}^{(0)} \in S$ and if

$$\{\nabla g_1(\mathbf{x}^{(0)}), \nabla g_2(\mathbf{x}^{(0)}), \ldots, \nabla g_p(\mathbf{x}^{(0)})\}$$

is a linearly independent set, then a vector $\mathbf{y} \in R^n$ is in the tangent space $T(\mathbf{x}^{(0)})$ to S at $\mathbf{x}^{(0)}$ if and only if there is a path $\boldsymbol{\varphi}(t)$ in S such that $\mathbf{x}^{(0)} = \boldsymbol{\varphi}(0)$ and

$$\mathbf{y} = \boldsymbol{\varphi}'(0) = (\varphi_1'(0), \ldots, \varphi_n'(0)).$$

Discussion of the Proof. It is very easy to show that if $\boldsymbol{\varphi}(t)$ is a path in S such that $\mathbf{x}^{(0)} = \boldsymbol{\varphi}(0)$ and $\mathbf{y} = \boldsymbol{\varphi}'(0)$, then $\mathbf{y}$ is in the tangent space $T(\mathbf{x}^{(0)})$ to S

at $\mathbf{x}^{(0)}$. For, by definition of S,

$$g_i(\varphi_1(t), \ldots, \varphi_n(t)) = 0$$

for $i = 1, \ldots, p$ and all t in the interval of definition of $\boldsymbol{\varphi}$. Therefore, the Chain Rule yields

$$0 = \frac{d}{dt} g_i(\varphi_1(t), \ldots, \varphi_n(t))\bigg|_{t=0} = \nabla g_i(\mathbf{x}^{(0)}) \cdot \boldsymbol{\varphi}'(0)$$

$$= \nabla g_i(\mathbf{x}^{(0)}) \cdot \mathbf{y}$$

for $i = 1, \ldots, p$. Hence, $\mathbf{y}$ is orthogonal to $\nabla g_i(\mathbf{x}^{(0)})$ for $i = 1, \ldots, p$. But then $\mathbf{y}$ is orthogonal to any linear combination of $\nabla g_i(\mathbf{x}^{(0)}), \ldots, \nabla g_p(\mathbf{x}^{(0)})$; that is, $\mathbf{y} \in \mathrm{N}(\mathbf{x}^{(0)})^{\perp} = \mathrm{T}(\mathbf{x}^{(0)})$.

The theorem also asserts that for any $\mathbf{y}$ in the tangent space $\mathrm{T}(\mathbf{x}^{(0)})$, there is a path $\boldsymbol{\varphi}(t)$ in S with $\boldsymbol{\varphi}(0) = \mathbf{x}^{(0)}$ and $\boldsymbol{\varphi}'(0) = \mathbf{y}$. A proof of this statement seems to require the full power of the Implicit Function Theorem, a result whose discussion is beyond the scope of this book. An excellent discussion of the Implicit Function Theorem and the proof of (7.1.5) can be found in Section 4.4 of *Vector Calculus* by G. Marsden and T. Tromba.

Note that the main hypothesis of (7.1.5), namely that the set

$$\{\nabla g_1(\mathbf{x}^{(0)}), \nabla g_2(\mathbf{x}^{(0)}), \ldots, \nabla g_p(\mathbf{x}^{(0)})\} \tag{5}$$

is linearly independent, was not used in our proof that all vectors $\mathbf{y} = \boldsymbol{\varphi}'(0)$, where $\boldsymbol{\varphi}(t)$ is a path in S such that $\boldsymbol{\varphi}(0) = \mathbf{x}^{(0)}$, are members of the tangent space $\mathrm{T}(\mathbf{x}^{(0)})$ to S at $\mathbf{x}^{(0)}$. However, this hypothesis is essential to the proof (omitted above) for the converse statement. For the sake of convenience, we will call a point $\mathbf{x}^{(0)}$ on a surface S a *regular point* if the set (5) is linearly independent.

7.2. Lagrange Multipliers and the Karush–Kuhn–Tucker Theorem for Mixed Constraints

The general problem that we will study in this section is

$$(P) \quad \begin{cases} \text{Minimize} \quad f(\mathbf{x}) \quad \text{subject to the constraints} \\ \quad g_1(\mathbf{x}) = 0, \ldots, \quad g_{m-1}(\mathbf{x}) = 0; \quad g_m(x) \le 0, \ldots, \quad g_p(\mathbf{x}) \le 0, \\ \text{where } f(\mathbf{x}), g_1(\mathbf{x}), \ldots, g_p(\mathbf{x}) \text{ have continuous first partial} \\ \text{derivatives on some open subset } C \text{ of } R^n. \end{cases}$$

Note that if $m = 1$, the program (P) includes only inequality constraints and, if $f(\mathbf{x})$, $g_i(\mathbf{x})$ are convex functions, (P) is a convex program of the sort we have already considered in Chapter 5. On the other hand, if $m - 1 = p$, then (P)

includes only equality constraints and fits the context of the classical Lagrange Multiplier theory as seen in calculus. If $1 < m - 1 < p$, then (P) includes both equality and inequality constraints and we say that the constraints of (P) are *mixed*.

A point $\mathbf{x} \in C$ that satisfies all of the constraints of (P) is *feasible* for (P), and (P) is *consistent* if the set F of feasible points for (P) is not empty. A feasible point $\mathbf{x}^*$ is a *local minimizer* (resp. a *solution*) for (P) if it is a local minimizer (resp. global minimizer) of $f(\mathbf{x})$ on the set F of feasible points for (P).

A feasible point $\mathbf{x}^*$ for (P) is a *regular point* for (P) if the set of vectors

$$\{\nabla g_j(\mathbf{x}^*): j \in J(\mathbf{x})\}$$

is linearly independent where

$$J(\mathbf{x}^*) = \{j: 1 \leq j \leq p, g_j(\mathbf{x}^*) = 0\},$$

that is, $J(\mathbf{x}^*)$ contains all of the indices corresponding to the equality constraints in (P) together with the indices for those inequality constraints for (P) that are active at $\mathbf{x}^*$.

The following theorem provides necessary conditions for a local minimizer for the program (P).

(7.2.1) Theorem. *Suppose that* $\mathbf{x}^*$ *is a regular point for* (P). *If* $\mathbf{x}^*$ *is a local minimizer for* (P), *then there exists a* $\boldsymbol{\lambda}^* \in R^p$ *such that*:

(1) $\lambda_j^* \geq 0$ *for* $j = m, \ldots, p$;
(2) $\lambda_j^* g_j(\mathbf{x}^*) = 0$ *for* $j = m, \ldots, p$;
(3) $\nabla f(\mathbf{x}^*) + \sum_{j=1}^{p} \lambda_j^* \nabla g_j(\mathbf{x}) = 0$.

PROOF. Consider the new program (EP) obtained from (P) by replacing all of the inequality constraints in (P) that are active at $\mathbf{x}^*$ by the corresponding equality constraints and deleting any remaining inequality constraints in (P). Then (EP) can be formulated as follows:

$$(EP) \quad \begin{cases} \text{Minimize} \quad f(\mathbf{x}) \quad \text{subject to} \\ \quad g_j(\mathbf{x}) = 0 \quad \text{for} \ \ j \in J(\mathbf{x}^*), \end{cases}$$

and $\mathbf{x}^*$ is a local minimizer for the program (EP).

Let S be the surface in R^n defined by the constraints for (EP) and let $\varphi(t)$ be a path in S such that $\mathbf{x}^* = \varphi(0)$. Because $\mathbf{x}^*$ is a local minimizer for (EP), there is an $r > 0$ such that $f(\mathbf{x}^*) \leq f(\mathbf{x})$ for all $\mathbf{x} \in B(\mathbf{x}^*, r) \cap S$

$$f(\mathbf{x}^*) = f(\varphi(0)) \leq f(\varphi(t))$$

for all t in some open interval containing $t^* = 0$. But then $t^* = 0$ is a local minimizer of $f(\varphi(t))$ so

$$(*) \qquad 0 = \frac{d}{dt} f(\varphi(t))\bigg|_{t=0} = \nabla f(\varphi(0)) \cdot \varphi'(0) = \nabla f(\mathbf{x}^*) \cdot \varphi'(0).$$

But x^* is a regular point for (P) so it is a regular point for S by definition of $J(\mathbf{x}^*)$. Consequently, by (7.1.5), every vector $\mathbf{y}$ in the tangent space $\mathrm{T}(\mathbf{x}^*)$ to S at $\mathbf{x}^*$ is given by $\mathbf{y} = \boldsymbol{\varphi}'(0)$ for some path $\varphi(t)$ in S. It follows from $(*)$ that $\nabla f(\mathbf{x}^*) \in \mathrm{T}(\mathbf{x}^*)^{\perp}$. Also, by (4.2.7),

$$\mathrm{T}(\mathbf{x}^*)^{\perp} = (\mathrm{N}(\mathbf{x}^*)^{\perp})^{\perp} = \mathrm{N}(\mathbf{x}^*),$$

so we conclude that the gradient vector $\nabla f(\mathbf{x}^*)$ is a linear combination of the gradient vectors $\{\nabla g_j(\mathbf{x}^*): j \in J(\mathbf{x}^*)\}$. This means that there exists a $\boldsymbol{\lambda}^* \in R^p$ such that

$$\nabla f(\mathbf{x}^*) + \sum_{j=1}^{p} \lambda_j^* \nabla g_j(\mathbf{x}^*) = 0,$$

and such that $\lambda_j^* = 0$ for $j \notin J(x^*)$ and $1 \leq j \leq p$. This establishes condition (3), and condition (2) follows from the observation that

$$\lambda_j^* g_j(\mathbf{x}^*) = 0$$

for $j \in J(x^*)$ because $g_j(\mathbf{x}^*) = 0$, and for $j \notin J(\mathbf{x}^*)$, $1 \leq j \leq p$ because $\lambda_j^* = 0$.

Suppose, contrary to (1), that $\lambda_j^* < 0$ for some j such that $m \leq j \leq p$. Then $j \in J(\mathbf{x}^*)$ since $\lambda_i^* = 0$ for $i \notin J(x^*)$, $m \leq i \leq p$. Consider the surface S_j defined by

$$g_i(\mathbf{x}) = 0 \quad \text{for } i \in J(\mathbf{x}^*), i \neq j.$$

Then $\mathbf{x}^* \in S_j$ by definition of $J(\mathbf{x}^*)$. If $\mathrm{T}_j(\mathbf{x}^*)$ is the tangent space to S_j at $\mathbf{x}^*$, then since $\mathbf{x}^*$ is a regular point for (P), there is a $\mathbf{y} \in \mathrm{T}_j(\mathbf{x}^*)$ such that

$$\nabla g_j(\mathbf{x}^*) \cdot \mathbf{y} < 0.$$

Theorem (7.1.5) asserts that there is a path $\boldsymbol{\varphi}(t)$ is S_j such that $\boldsymbol{\varphi}(0) = \mathbf{x}^*$ and $\mathbf{y} = \boldsymbol{\varphi}'(0)$. But then

$$\begin{aligned} \left.\frac{d}{dt} f(\boldsymbol{\varphi}(t))\right|_{t=0} &= \nabla f(\mathbf{x}^*) \cdot \mathbf{y} \\ &= -\left(\sum_{i=1}^{m-1} \lambda_i^* \nabla g_i(\mathbf{x}) \cdot \mathbf{y} + \sum_{i=m}^{p} \lambda_i^* \nabla g_i(\mathbf{x}) \cdot \mathbf{y} \right) \\ &= -\lambda_j^* \nabla g_j(\mathbf{x}^*) \cdot y < 0. \end{aligned}$$

This contradicts the fact that $\mathbf{x}^*$ is a local minimizer for (P) since $\boldsymbol{\varphi}(t)$ is feasible for (P) for all t in some interval $[0, \varepsilon]$ with $\varepsilon > 0$.

As we mentioned in the introductory remarks of this section, the classical Lagrange Multiplier theory is concerned with the minimization of an objective function subject only to equality constraints. For such problems, (7.2.1) asserts that if $\mathbf{x}^*$ is a local minimizer for an equality constrained program (P), then there is vector $\boldsymbol{\lambda}^* \in R^p$ such that (7.2.1)(3) holds:

(3) $\nabla f(\mathbf{x}^*) + \sum_{j=1}^{p} \lambda_j^* \nabla g_j(\mathbf{x}^*) = 0.$

The other conditions (1), (2) of (7.2.1) are vacuous when only equality constraints are present. In this setting, we refer to the components λ_j^* of $\mathbf{x}^*$ as *Lagrange multipliers* and to equation (3) as the *Lagrange Multiplier Condition* for (P).

Note that the Lagrange Multiplier Condition is also a necessary condition for maximizers of an objective function subject only to equality constraints. For if $\mathbf{x}^*$ is a local maximizer of $f(\mathbf{x})$ subject to equality constraints

$$g_1(\mathbf{x}) = 0, \quad g_2(\mathbf{x}) = 0, \ldots, \quad g_p(\mathbf{x}) = 0, \qquad \mathbf{x} \in C,$$

then $\mathbf{x}^*$ is a local minimizer of $-f(\mathbf{x})$ subject to these same constraints. Consequently, by (7.2.1), there is a $\boldsymbol{\mu}^* \in R^p$ such that

$$-\nabla f(\mathbf{x}^*) + \sum_{j=1}^{p} \mu_j^* \nabla g_j(\mathbf{x}^*) = 0,$$

that is,

$$\nabla f(\mathbf{x}^*) + \sum_{j=1}^{p} (-\mu_j^*) \nabla g_j(\mathbf{x}^*) = 0,$$

which reduces to the Lagrange Multiplier Condition (3) if we take $\lambda_j^* = -\mu_j^*$ for $j = 1, 2, \ldots, p$.

The following simple example shows that it may happen that the Lagrange Multiplier Condition

$$\nabla f(\mathbf{x}) + \sum_{j=1}^{p} \lambda_j \nabla g_j(\mathbf{x}) = 0$$

may have a solution $\mathbf{x}^* \in R^n$, $\boldsymbol{\lambda}^* \in R^p$ for which $\mathbf{x}^*$ is neither a local maximizer nor a local minimizer of $f(\mathbf{x})$ subject to given equality constraints.

(7.2.2) Example. Consider the following equality-constrained program:

$$\begin{cases} \text{Minimize} & f(x_1, x_2, x_3) = x_1^2 + x_2^2 + x_3^2 \\ \text{subject to} & g_1(x_1, x_2, x_3) = x_1^2 + \dfrac{x_2^2}{4} + \dfrac{x_3^2}{9} - 1 = 0. \end{cases}$$

In geometric terms, this program asks us to find the points on the ellipsoid

$$x_1^2 + \frac{x_2^2}{4} + \frac{x_3^2}{9} = 1$$

that are closest to the origin. Evidently, the solutions of this geometric problem are $(\pm 1, 0, 0)$, which are the endpoints of the shortest axis of this ellipsoid.

Note that the Lagrange Multiplier Condition for this problem

$$\nabla f(\mathbf{x}) + \lambda_1 \nabla g_1(\mathbf{x}) = 0$$

is satisfied at precisely those points $\mathbf{x}^*$ at which the gradient vector $\nabla f(\mathbf{x}^*)$

is a multiple of the gradient vector $\nabla g_1(\mathbf{x}^*)$, which are the same as those points where the (spherical) level surfaces of $f(\mathbf{x})$ are tangent to the constraint ellipsoid. Thus, without doing any algebra, but rather relying on the geometry, we can see that the points $\mathbf{x}^* \in R^3$ for which there is a Lagrange multiplier λ_1^* are

$$(\pm 1, 0, 0), \quad (0, \pm 2, 0), \quad (0, 0, \pm 3).$$

We have already noted that $(\pm 1, 0, 0)$ are local minimizers for the given problem. It is also readily seen from the geometry that $(0, 0, \pm 3)$ are local maximizers for $f(x)$ subject to $g_1(x) = 0$, while $(0, \pm 2, 0)$ are solutions of the Lagrange Multiplier Condition (along with $\lambda_1^* = -4$) that are neither local maximizers nor local minimizers of $f(x)$ subject to $g_1(x) = 0$. Note that at the local minimizers $(\pm 1, 0, 0)$, the value of λ_1^* is -1, while at the local maximizers $(0, 0, \pm 3)$, the value λ_1^* is -9.

Although the Lagrange Multiplier Condition is only a necessary condition for a minimizer of an equality-constrained program, it does provide a powerful method for the solution of such problems. In practice, it is usually known from the context or the mathematical nature of the constraints that a given problem has a solution, that is, that the given objective function has a global minimizer on the set of feasible points for the problem. In such cases, it is then only necessary to identify the solution(s) among the feasible points that satisfy the Lagrange Multiplier Condition provided that all these points are regular for the given problem.

The main catch in this approach is finding the feasible points that satisfy the Lagrange Multiplier Condition. In general, this requires the solution of a system of nonlinear equations by some iterative scheme such as Newton's Method or Broyden's Method. Nevertheless, the use of Lagrange multipliers can lead to some interesting and important results as the following examples show.

Our first example shows that Lagrange multipliers can be used to give a new proof of the Arithmetic–Geometric Mean Inequality (2.4.1).

(7.2.3) Example. We will show that if $\delta_1, \ldots, \delta_n$ are positive numbers such that $\delta_1 + \delta_2 + \cdots + \delta_n = 1$, then for any positive numbers $x_1, \ldots, x_n$,

$$\text{(A–G)} \qquad \prod_{i=1}^{n} x_i^{\delta_i} \leq \sum_{i=1}^{n} \delta_i x_i$$

with equality in (A–G) if and only if $x_1 = x_2 = \cdots = x_n$. To this end, we consider, for a given positive number L, the program

$$(P_L) \qquad \begin{cases} \text{Minimize} & f(\mathbf{x}) = \sum_{i=1}^{n} \delta_i x_i \\ \text{subject to} & g_1(\mathbf{x}) = \prod_{i=1}^{n} x_i^{\delta_i} - L = 0. \end{cases}$$

The Lagrange Multiplier Condition for (P_L) reduces to

$$\delta_i + \lambda\delta_i(x_i)^{\delta_i - 1} \prod_{j \neq i} x_j^{\delta_j} = 0, \qquad i = 1, \ldots, n.$$

Since $\delta_i \neq 0$, we can multiply the ith equation in this system by x_i/δ_i to obtain the equivalent system

$$x_i + \lambda L = 0, \qquad i = 1, \ldots, n.$$

Thus, we conclude that if $\mathbf{x}^*$ is a global minimizer of (P_L), then $x_1^* = x_2^* = \cdots = x_n^* = -\lambda L$ for an appropriate choice of λ. (Why does λ^* exist?) But

$$L = \prod_{i=1}^{n} (x_i)^{\delta_i} = \prod_{i=1}^{n} (-\lambda L)^{\delta_i} = (-\lambda L)^{\delta_1 + \cdots + \delta_n} = -\lambda L,$$

so $\lambda = -1$ and $x_1^* = x_2^* = \cdots = x_n^* = L$.

We conclude that

$$\sum_{i=1}^{n} \delta_i x_i = f(\mathbf{x}) \geq f(\mathbf{x}^*) = \sum_{i=1}^{n} \delta_i L = L = \prod_{i=1}^{n} (x_i)^{\delta_i}.$$

Since L is an arbitrary positive number, this establishes the inequality (A–G) for all positive $x_1, \ldots, x_n$ and it shows that equality holds in (A–G) if all of the x_i's are equal. On the other hand, if equality holds in the inequality (A–G) for given positive numbers $x_1, \ldots, x_n$ and if L is the common value of the two sides of this inequality, then $x_1, \ldots, x_n$ is a global minimizer for (P_L), so all of the x_i's have the same value.

On a number of occasions, we have used the fact that any symmetric $n \times n$-matrix has n mutually orthogonal eigenvectors of unit length. For example, this result made it possible to diagonalize any symmetric matrix with an orthogonal change of variables, and this in turn was basic to our development of eigenvalue criteria for positive and negative definiteness, and to the computation of the square root of a positive definite matrix. Our next example shows that this special eigenvector structure of symmetric matrices can be derived easily and clearly by means of Lagrange multipliers.

(7.2.4) Example. Suppose that A is an $n \times n$-symmetric matrix. We will show inductively that A has n mutually orthogonal eigenvectors of unit length.

To produce the first such eigenvector, we consider the program

$$(P_1) \quad \begin{cases} \text{Maximize} & f(\mathbf{x}) = \mathbf{x} \cdot A\mathbf{x} \\ \text{subject to} & g_1(\mathbf{x}) = \|\mathbf{x}\|^2 - 1 = 0. \end{cases}$$

Since $f(\mathbf{x})$ is continuous and the set of feasible points for (P_1) is not empty, closed and bounded, (P_1) has a local maximizer $\mathbf{x}^{(1)}$, so (7.2.1) implies that there is a constant λ_1 such that

$$\nabla f(\mathbf{x}^{(1)}) + \lambda_1 \nabla g_1(\mathbf{x}^{(1)}) = \mathbf{0}.$$

That gives

$$2A\mathbf{x}^{(1)} + \lambda_1 2\mathbf{x}^{(1)} = 0.$$

But then $A\mathbf{x}^{(1)} = -\lambda_1\mathbf{x}^{(1)}$ so $\mathbf{x}^{(1)}$ is an eigenvector of A of unit length.

Given k mutually orthogonal unit eigenvectors $\mathbf{x}^{(1)}, \ldots, \mathbf{x}^{(k)}$, consider the program

$$(P_k) \quad \begin{cases} \text{Maximize} \quad f(\mathbf{x}) = \mathbf{x}\cdot A\mathbf{x} \quad \text{subject to} \\ g_1(\mathbf{x}) = \|\mathbf{x}\|^2 - 1 = 0 \quad \text{and} \\ g_2(\mathbf{x}) = \mathbf{x}\cdot\mathbf{x}^{(1)} = 0, \ldots, \quad g_{k+1}(\mathbf{x}) = \mathbf{x}\cdot\mathbf{x}^{(k)} = 0. \end{cases}$$

Again, because $f(\mathbf{x})$ is a continuous function and the set of feasible points is not empty, closed and bounded, the program (P_k) must have a local maximizer $\mathbf{x}^{(k+1)}$. But then (7.2.1) asserts that there exist $\lambda_1, \lambda_2, \ldots, \lambda_{k+1}$ such that

$$\nabla f(\mathbf{x}^{(k+1)}) + \lambda_1\nabla g_1(\mathbf{x}^{(k+1)}) + \cdots + \lambda_{k+1}\nabla g_{k+1}(\mathbf{x}^{(k+1)}) = 0.$$

Therefore,

$$2A\mathbf{x}^{(k+1)} + 2\lambda_1\mathbf{x}^{(k+1)} + \lambda_2\mathbf{x}^{(1)} + \cdots + \lambda_{k+1}\mathbf{x}^{(k)} = 0.$$

It follows from the fact that $\mathbf{x}^{(1)}, \ldots, \mathbf{x}^{(k+1)}$ are mutually orthogonal unit vectors and the last equation that

$$2\mathbf{x}^{(i)}\cdot A\mathbf{x}^{(k+1)} + \lambda_i = 0, \qquad i = 1, \ldots, k.$$

Also $\mathbf{x}^{(i)}$ is an eigenvector of A for $1 \leq i \leq k$ so there is a μ_i such that

$$A\mathbf{x}^{(i)} = \mu_i\mathbf{x}^{(i)}.$$

Therefore

$$\begin{aligned} -\frac{\lambda_i}{2} &= \mathbf{x}^{(i)}\cdot A\mathbf{x}^{(k+1)} = (A^{\mathrm{T}}\mathbf{x}^{(i)})\cdot\mathbf{x}^{(k+1)} = (A\mathbf{x}^{(i)})\cdot\mathbf{x}^{(k+1)} \\ &= \mu_i\mathbf{x}^{(i)}\cdot\mathbf{x}^{(k+1)} = 0, \end{aligned}$$

so that $\lambda_i = 0$ for $2 \leq i \leq k+1$. It follows that

$$2A\mathbf{x}^{(k+1)} + 2\lambda_1\mathbf{x}^{(k+1)} = 0,$$

so $\mathbf{x}^{(k+1)}$ is an eigenvector of A. Since $\mathbf{x}^{(k+1)}$ is of unit length and $\mathbf{x}^{(1)}, \ldots, \mathbf{x}^{(k+1)}$ are mutually orthogonal, this completes the proof.

Two comments are in order in regard to the proof in the preceding example.

(1) The set of feasible points for the program (P_k) is not empty only as long as $1 \leq k \leq n-1$, so the inductive process stops with $\mathbf{x}^{(n)}$ as expected.
(2) Although we did not explicitly check the hypothesis that $\{\nabla g_1(\mathbf{x}^{(k+1)}), \ldots, \nabla g_{k+1}(\mathbf{x}^{(k+1)})\}$ is a linearly independent set when we applied (7.2.1) to the program (P_k), note that this hypothesis is indeed satisfied because $\{\mathbf{x}^{(1)}, \ldots, \mathbf{x}^{(k+1)}\}$ are mutually orthogonal unit vectors.

(7.2.5) Example. Before the advent of the Space Shuttle, our astronauts were carried into space aboard three-stage rocket vehicles. The main advantages and disadvantages of multistage rocket vehicles are fairly obvious. A multistage rocket jettisons its stages as they become useless for further propulsion. As a result, the engines of the upper stages are not burdened with this useless weight so they can provide more acceleration for a given amount of fuel to the remainder of the vehicle. On the other hand, the complexity and cost of a rocket vehicle increases with the number of stages. But why are rockets usually built with precisely three stages? Why not two stages, or four stages? We will now show why three-stage rockets are the optimal choice by applying Lagrange multipliers to solve a constrained minimization problem arising from a simple mathematical model of multistage rocket propulsion. We begin by developing a simple model for single-stage rocket propulsion.

For this purpose, we assume that all outside forces on the rocket (such as gravity and aerodynamic drag) are so small in comparison with the thrust of the rocket engine that these outside forces can be neglected. We will also assume that the exhaust gases are expelled straight out of the back of the rocket at a constant speed c relative to the rocket, and that the rocket consumes fuel at a constant rate. Under these circumstances, the motion of the rocket is governed by the so-called *rocket equation*

$$\frac{dv}{dt} = -\frac{c}{M}\frac{dM}{dt}, \tag{1}$$

where $v = v(t)$ and $M = M(t)$ denote the velocity and mass of the rocket at time t.

We will now introduce two structural parameters related to the design of the rocket. Suppose that the mass of the rocket payload is P, that the mass of the rocket vehicle without payload or fuel is M_v, and that the mass of the fully fueled rocket vehicle without payload is M_0. Then the *mass ratio* R of the rocket is defined by

$$R = \frac{P}{M_0}, \tag{2}$$

and the *structural factor* S is defined by

$$S = \frac{M_v}{M_0}. \tag{3}$$

The values of R and S depend on the design characteristics of a particular rocket. However, typical values are $R = 0.01$ and $S = 0.2$.

Since our model assumes that the rocket is consuming fuel at a constant rate $-k$, we can rewrite the rocket equation as

$$\frac{dv}{dt} = \frac{ck}{P + M_0 - kt}, \tag{4}$$

where t is the length of time that the rocket engine has been operating. If we separate the variables in (4) and integrate, we obtain

$$v(t) = -c\ln(P + M_0 - kt) + B. \tag{5}$$

The constant of integration B can be expressed in terms of the initial velocity v_0 of the rocket by setting $t = 0$ in (5) to obtain

$$v_0 = -c\ln(P + M_0) + B.$$

If we use the preceding equation to eliminate B from (5), we obtain

$$\begin{aligned} v(t) &= -c\ln(P + M_0 - kt) + v_0 + c\ln(P + M_0) \\ &= v_0 - c\ln\left[\frac{P + M_0 - kt}{P + M_0}\right] \\ &= v_0 - c\ln\left[1 - \frac{kt}{P + M_0}\right]. \end{aligned} \tag{6}$$

Since the rocket is burning fuel at a rate $-k$ and since the mass of fuel on the rocket initially is $M_0 - M_v$, the fuel will burn out completely at time

$$t_b = \frac{M_0 - M_v}{k} = \frac{(1 - S)M_0}{k}.$$

Therefore, the velocity increment ΔV imparted to the rocket by burning the full fuel load is

$$\begin{aligned} \Delta V = v(t_b) - v_0 &= -c\ln\left(1 - \frac{(1 - S)M_0}{P + M_0}\right) \\ &= -c\ln\left\{\frac{R + S}{R + 1}\right\}. \end{aligned} \tag{7}$$

Now let us consider a rocket vehicle with n stages. Let the ith stage have total mass M_i, engine exhaust speed c_i, and structural factor S_i. Our objective is to minimize the total mass

$$M_1 + M_2 + \cdots + M_n \tag{8}$$

subject to the constraint that the final velocity of the rocket after burnout of the last stage is a prescribed value v_f.

If P is the mass of the payload of the n-stage rocket vehicle, then the ith stage can be thought of as a single stage rocket with a total mass

$$M_i + M_{i+1} + \cdots + M_n + P.$$

Therefore, by equation (7), the velocity increment ΔV_i provided by the operation of the ith stage is

$$\begin{aligned} \Delta V_i &= -c_i\ln\left[1 - \frac{(1 - S_i)M_i}{M_i + M_{i+1} + \cdots + M_n + P}\right] \\ &= c_i\ln\left[\frac{M_i + M_{i+1} + \cdots + M_n + P}{S_iM_i + M_{i+1} + \cdots + M_n + P}\right]. \end{aligned}$$

Our objective is to minimize the total mass (8) of the rocket vehicle subject to the constraint

$$v_{\mathrm{f}} - \sum_{i=1}^{n} c_i \ln\left[\frac{M_i + M_{i+1} + \cdots + M_n + P}{S_i M_i + M_{i+1} + \cdots + M_n + P}\right] = 0. \tag{9}$$

The complexity of the constraint (9) makes it difficult to apply Lagrange multipliers to this minimization problem in its present form. Instead, we first make the change of variables

$$N_i = \frac{M_i + M_{i+1} + \cdots + M_N + P}{S_i M_i + M_{i+1} + \cdots + M_n + P}, \qquad i = 1, \ldots, n.$$

For these variables, the constraint (9) assumes the simple form

$$v_{\mathrm{f}} - \sum_{i=1}^{n} c_i \ln N_i = 0. \tag{10}$$

The simple objective function (8) becomes quite complicated when expressed in terms of the variables $N_1, \ldots, N_n$. However, if we can find a simpler objective function of $N_1, \ldots, N_n$ that has its minimum value at the same place as the one determined by (8), we can use it in place of the given one. To this end, we note that by definition of N_i

$$\frac{M_1 + \cdots + M_n + P}{P} = \left[\frac{M_1 + \cdots + M_n + P}{M_2 + \cdots + M_n + P}\right]\left[\frac{M_2 + \cdots + M_n + P}{M_3 + \cdots + M_n + P}\right] \cdots \left[\frac{M_{n-1} + M_n + P}{M_n + P}\right]\left[\frac{M_n + P}{P}\right].$$

But for each i,

$$\begin{aligned}
&\frac{M_i + M_{i+1} + \cdots + M_n + P}{M_{i+1} + \cdots + M_n + P} \\
&\qquad = \frac{(1 - S_i)(M_i + M_{i+1} + \cdots + M_n + P)}{S_i M_i + (1 - S_i)M_{i+1} + \cdots + (1 - S_i)P - S_i M_i} \\
&\qquad = \frac{(1 - S_i)(M_i + \cdots + M_n + P)}{S_i M_i + M_{i+1} + \cdots + M_n + P - S_i(M_i + \cdots + M_n + P)} \\
&\qquad = \frac{(1 - S_i)N_i}{1 - S_i N_i}.
\end{aligned}$$

Therefore,

$$\frac{M_1 + \cdots + M_n + P}{P} = \left[\frac{(1 - S_1)N_1}{1 - S_1 N_1}\right]\left[\frac{(1 - S_2)N_2}{1 - S_2 N_2}\right] \cdots \left[\frac{(1 - S_n)N_n}{1 - S_n N_n}\right]. \tag{11}$$

Since P is fixed, minimizing the function (8) is equivalent to minimizing the function

$$\frac{M_1 + \cdots + M_n + P}{P},$$

which in turn is equivalent to minimizing the function

$$\ln\left[\frac{M_1 + \cdots + M_n + P}{P}\right] = \sum_{i=1}^{n} [\ln N_i + \ln(1 - S_i) - \ln(1 - S_i N_i)].$$

Thus, we are led to consider the following equivalent minimization problem:

$$\begin{cases} \text{Minimize} \quad f(N_1, \ldots, N_n) = \sum_{i=1}^{n} [\ln N_i + \ln(1 - S_i) - \ln(1 - S_i N_i)] \\ \text{subject to} \quad g(N_1, \ldots, N_n) = v_f - \sum_{i=1}^{n} c_i \ln N_i = 0. \end{cases} \tag{12}$$

If $N_1^*, \ldots, N_n^*$ is a solution to the problem (12) there is a λ^* that satisfies the Lagrange Multiplier Condition

$$\frac{1}{N_i^*} + \frac{S_i}{1 - S_i N_i^*} - \lambda^* \frac{c_i}{N_i^*} = 0, \qquad i = 1, \ldots, n. \tag{13}$$

The equations (13) can be solved for N_i^* in terms of λ^* to obtain

$$N_i^* = \frac{\lambda^* c_i - 1}{\lambda^* c_i S_i}, \qquad i = 1, \ldots, n. \tag{14}$$

Substitute the values (14) into the constraint equation in (12) to obtain

$$v_f = \sum_{i=1}^{n} c_i \ln\left[\frac{\lambda^* c_i - 1}{\lambda^* c_i S_i}\right]. \tag{15}$$

In the special case in which the structural factors S_i all have the same value S and the exhaust speeds c_i all have the same value c, we obtain from (14) and (15) that

$$N_i^* = \frac{\lambda^* c - 1}{\lambda^* c S}, \qquad i = 1, \ldots, n, \tag{14$'$}$$

$$v_f = nc \ln\left[\frac{\lambda^* c - 1}{\lambda^* c S}\right]. \tag{15$'$}$$

Equation (15)′ can be used to compute the common value N^* of N_i^* from (14)′ as

$$N^* = e^{v_f/nc}. \tag{16}$$

If we substitute N^* into (11), we obtain

$$\frac{M_1 + \cdots + M_n + P}{P} = \left[\frac{(1 - S)e^{v_f/nc}}{1 - Se^{v_f/nc}}\right]^n. \tag{17}$$

Equation (17) can be solved for the minimum total mass of the rocket vehicle without payload

$$M_1 + \cdots + M_n = \left[\frac{(1-S)^n e^{v_f/c}}{(1-Se^{v_f/nc})^n} - 1\right]P. \tag{18}$$

Let us now interpret the preceding results in terms of a concrete problem. Suppose that we wish to place a space capsule of mass P in a circular orbit at an altitude of 100 miles above the earth's surface by using a multistage rocket for which each stage has the same structural factor $S = 0.2$, mass ratio $R = 0.01$, and the same exhaust speed $c = 6{,}000$ miles per hour. Let us compute the minimum total mass of the n-stage rocket vehicle required for this orbital insertion for $n = 1, 2, 3, 4$.

The final velocity required for the prescribed orbit is approximately 17,500 miles per hour, that is,

$$v_f = 17{,}500 \text{ miles per hour.}$$

For a single-stage rocket, the final velocity can be computed from (7) as

$$\Delta V = -6{,}000 \ln\left(\frac{0.21}{1.01}\right) = 9{,}424 \text{ miles per hour.}$$

Consequently, we see that it is not possible to achieve the required orbital insertion with a single-stage rocket!

According to (18) the minimum total mass for an n-stage rocket vehicle to achieve this orbital insertion is

$$M_1 + M_2 + \cdots + M_n = \left[\frac{(0.8)^n e^{17{,}500/(6000)}}{(1-(0.2)e^{17{,}500/6000n})^n} - 1\right]P.$$

In particular, this yields (to two decimal places)

$$\text{for two stages:} \quad M_1 + M_2 = 600.33P;$$

$$\text{for three stages:} \quad M_1 + M_2 + M_3 = 89.42P;$$

$$\text{for four stages:} \quad M_1 + M_2 + M_3 + M_4 = 63.48P;$$

$$\text{for five stages:} \quad M_1 + M_2 + M_3 + M_4 + M_5 = 54.70P.$$

Of course, as the number of stages increases, the value of the minimum total mass of the rocket vehicle decreases and approaches a positive value (which value?), but we can see from the above computations that it is at least reasonable that the cost–benefit trade-off occurs at three stages. Beyond three stages, the decrease in total mass is offset by the increased cost, complexity, and potential for failure.

Note that the optimal masses of the three rocket stages can be computed by applying equation (17). These turn out to be

$$M_3 = 3.49P, \qquad M_2 = 15.67P, \qquad M_1 = 70.36P.$$

We conclude this section with some important observations about convex programs and constraint qualifications.

(7.2.6) Remarks

(a) Suppose that $\mathbf{x}^*$ is a regular point for a *convex* program

$$(P) \quad \begin{cases} \text{Minimize} \quad f(\mathbf{x}) \quad \text{subject to} \\ g_1(\mathbf{x}) \leq 0, \ldots, \quad g_p(\mathbf{x}) \leq 0; \qquad \mathbf{x} \in C, \end{cases}$$

where $f(\mathbf{x}), g_1(\mathbf{x}), \ldots, g_p(\mathbf{x})$ are convex functions with continuous first partial derivatives on an open convex set C in R^n. Then (7.2.1) asserts that a necessary condition for $\mathbf{x}^*$ to be a local minimizer for (P) is that there exists a $\boldsymbol{\lambda}^* \in R^p$ such that

(1) $\lambda_j^* \geq 0$ for $j = 1, \ldots, p$;
(2) $\lambda_j^* g_j(\mathbf{x}^*) = 0$ for $j = 1, \ldots, p$;
(3) $\nabla f(\mathbf{x}^*) + \sum_{j=1}^{p} \lambda_j^* \nabla g_j(\mathbf{x}^*) = 0$.

However, because the objective function $f(\mathbf{x})$ and the constraint $g_1(\mathbf{x}), \ldots, g_p(\mathbf{x})$ are all convex, we can apply Theorem (2.3.4) to $f(\mathbf{x})$ on the convex set F of feasible points for (P) to conclude that any local minimizer $\mathbf{x}^*$ for a convex (P) is actually a global minimizer, that is, a *solution* for (P). Thus, we see that the necessary conditions for $\mathbf{x}^*$ to be a solution for (P) given by (7.2.1) are exactly the same as those given in the gradient form of the Karush–Kuhn–Tucker Theorem (5.2.14) even though the hypotheses of these two results are different (that is, (P) is superconsistent in (5.2.14) and $\mathbf{x}^*$ is a regular point for (P) in (7.2.1)). Moreover, although (7.2.1) states that (1), (2), and (3) are only *necessary* conditions for $\mathbf{x}^*$ to be a solution to (P), they are also *sufficient* conditions when (P) is a convex program. The sufficiency of these conditions follows exactly as in the proof of (5.2.14) because the superconsistency hypothesis is not used in that part of the proof. Thus, *for a convex program* (P), *conditions* (1), (2), and (3) *are both necessary and sufficient for a point* $\mathbf{x}^*$ *to be a solution under either the hypothesis that* (P) *is superconsistent or the hypothesis that* $\mathbf{x}^*$ *is a regular point for* (P).

(b) In this book, we have seen reasonably coherent theories emerge under the hypotheses of superconsistency or regularity. These two conditions are called *constraint qualifications*. Additional theories corresponding to other constraint qualifications also appear in the literature. For instance, the problem

$$(P) \quad \begin{cases} \text{Minimize} \quad f(\mathbf{x}) \quad \text{subject to} \\ g_1(\mathbf{x}) = 0, \ldots, \quad g_{m-1}(\mathbf{x}) = 0; \quad g_m(\mathbf{x}) \leq 0, \ldots, \quad g_p(\mathbf{x}) \leq 0, \\ \text{where } f(\mathbf{x}), g_1(\mathbf{x}), \ldots, g_p(\mathbf{x}) \text{ have continuous first partial} \\ \text{derivatives on some open subset of } R^n, \end{cases}$$

can be studied successfully by using the Mangasarian–Fromowitz constraint qualification. This constraint qualification at a feasible point $\mathbf{x}$ can be stated as follows:

(a) $\{\nabla g_i(\mathbf{x}); i = 1, \ldots, m-1\}$ is linearly independent; and

$$\text{(b)} \quad \begin{cases} \text{There exists a vector } \mathbf{y} \text{ in } R^n \text{ such that} \\ \nabla g_i(\mathbf{x})\cdot\mathbf{y} = 0 \quad \text{for } i = 1, \ldots, m-1, \\ \nabla g_i(\mathbf{x})\cdot\mathbf{y} < 0 \quad \text{if } g_i(\mathbf{x}) = 0 \text{ and } i = m, \ldots, p. \end{cases}$$

In fact, in their original derivation of the Karush–Kuhn–Tucker Theorem in 1951, Kuhn and Tucker imposed a constraint qualification similar to the Mangasarian–Fromowitz constraint qualification. For more on this, see *Constrained Optimization* by R. Fletcher (Wiley, New York, 1981).

An excellent, although somewhat advanced, discussion of constraint qualifications can be found in *Optimization and Non-Smooth Analysis* by Frank H. Clarke (Wiley-Interscience, New York, 1983). This book details the study of how constraint qualifications can be used to determine the sensitivity of (P) to perturbations in the constraints, a topic that we discussed briefly in Chapter 5. (Recall that in Chapter 5 the constraint qualification of superconsistency actually led to the existence of sensitivity vectors in the convex programming case.)

7.3. Quadratic Programming

Nonlinear optimization problems in which a quadratic function is maximized or minimized subject to linear constraints and nonnegativity restrictions on the variables, arise in a wide variety of applications including regression analysis in statistics, economic models of optimal sales revenues, and investment portfolio analysis. Several efficient algorithms have been developed to take full advantage of the linearity of the constraints and the quadratic character of the objective function in such problems. This section will discuss one such algorithm, called Wolfe's Algorithm, which is a variant of the Simplex Algorithm for linear programming. As we shall see, the Karush–Kuhn–Tucker Theorem in the form (7.2.1) provides the theoretical basis for the algorithm.

(7.3.1) Definition. The *standard quadratic programming* problem can be formulated as follows:

$$(QP) \quad \begin{cases} \text{Minimize} \quad f(\mathbf{x}) = a + \mathbf{c}\cdot\mathbf{x} + \tfrac{1}{2}\mathbf{x}\cdot Q\mathbf{x} \\ \text{subject to the constraints} \quad A\mathbf{x} \le \mathbf{b}, \quad \mathbf{x} \ge \mathbf{0}, \\ \text{where } Q \text{ is a positive definite } n \times n\text{-matrix, } \mathbf{b} \in R^m, \mathbf{c} \in R^n, a \in R, \\ \text{and } A \text{ is an } m \times n\text{-matrix of rank } m. \end{cases}$$

It is convenient to allow for equality constraints in (QP). For this purpose we will assume $(A\mathbf{x})_i \le b_i$ for $i = 1, \ldots, k$ and that $(A\mathbf{x})_i = b_i$ for $i = k+1, \ldots, m$.

The nonnegativity restrictions $\mathbf{x} \geq 0$ on the variables can be incorporated as additional inequality constraints. Thus, the constraints in the standard quadratic program (QP) can be listed in the format of Section 7.2 as follows:

$$
\begin{aligned}
a_{11}x_1 + \cdots + a_{1n}x_n - b_1 &\leq 0, \\
a_{k1}x_1 + \cdots + a_{kn}x_n - b_k &\leq 0, \\
a_{k+1,1}x_1 + \cdots + a_{k+1,n}x_n - b_{k+1} &= 0, \\
&\vdots \\
a_{m1}x_1 + \cdots + a_{mn}x_n - b_m &= 0, \\
-x_1 \leq 0, \ldots, \quad -x_n &\leq 0.
\end{aligned}
$$

Thus, (QP) is a constrained minimization problem (in the broad context that allows both equality and inequality constraints), so the Karush–Kuhn–Tucker theory can be applied to conclude that if a regular point $\mathbf{x}^*$ minimizes (QP), then there are $\boldsymbol{\mu}^* \in R^m$, $\mathbf{v}^* \in R^n$ such that:

(1) $\mu_i^* \geq 0$ for $i = 1, \ldots, k$, $v_j^* \geq 0$ for $j = 1, \ldots, n$;
(2) $\mathbf{c} + Q\mathbf{x}^* + A^{\mathrm{T}}\boldsymbol{\mu}^* - \mathbf{v}^* = \mathbf{0}$;
(3) $\mu_i^*[(A\mathbf{x}^*)_i - b_i] = 0$ for $i = 1, \ldots, m$.

Some explanation of the formulation of these conditions is in order. Note that

$$\nabla(a_{p1}x_1 + \cdots + a_{pn}x_n - b_p) = (a_{p1}, \ldots, a_{pn}),$$

so that

$$\sum_{i=1}^{n} \mu_i^* \nabla((A\mathbf{x})_i - b_i) = A^{\mathrm{T}}\boldsymbol{\mu}^*.$$

Also, the gradient of the inequality constraints $-x_j \leq 0$ are simply the negative of the jth unit vector in R^n, so that the term $-\mathbf{v}^*$ in (2) simply accounts for these constraints.

Note that if $x_j^* = 0$, then certainly $v_j^* x_j^* = 0$. On the other hand, if $x_j^* \geq 0$, then $v_j^* = 0$ because v_j^* is the Karush–Kuhn–Tucker multiplier corresponding to the constraint $-x_j \leq 0$. Therefore, the following additional restriction on $\mathbf{x}^*$, $\mathbf{v}^*$ is imposed:

(4) $x_j^* v_j^* = 0$ for $j = 1, \ldots, n$.

The inequality constraints determined by the matrix A can be replaced by introducing "slack" variables x_{n+i} such that $x_{n+i} \geq 0$ and

$$\sum_{j=1}^{n} a_{ij}x_j + x_{n+i} = b_i, \qquad i = 1, \ldots, k.$$

With this modification, the linear constraints in (QP) assume the form

$$\text{(I)} \qquad \begin{cases} \sum_{j=1}^{n} a_{ij}x_j + x_{n+i} = b_i, & i = 1, \ldots, k, \\ \sum_{j=1}^{n} a_{ij}x_j = b_i, & i = k+1, \ldots, m. \end{cases}$$

Conditions (2), (3), and (4) can be expressed as follows:

$$\text{(II)} \qquad \sum_{j=1}^{n} q_{ij}x_j + \sum_{j=1}^{m} a_{ji}\mu_j - v_i = -c_i, \qquad i = 1, \ldots, n,$$

$$\text{(III)} \qquad \mu_i x_{n+i} = 0, \qquad i = 1, \ldots, k,$$

$$\text{(IV)} \qquad x_j v_j = 0 \quad \text{for } j = 1, \ldots, n.$$

The variables in (I) and (II) are $\{x_1, \ldots, x_n, x_{n+1}, \ldots, x_{n+k}, \mu_1, \ldots, \mu_m, v_1, \ldots, v_n\}$. $2n + m + k$ variables altogether. However, conditions (III) and (IV) imply that $k + n$ of these variables must take the value 0, so that at most $n + m$ of these variables can have nonzero values. Thus, any solution of (I), (II), (III), and (IV) must be a basic solution of (I) and (II) in the sense of the simplex algorithm for linear programming. This suggests that the simplex algorithm can be suitably modified to solve quadratic programming problems. Before we discuss these modifications, we will present a brief summary of the simplex method.

(7.3.2) A Short Course on the Simplex Method. The *equality standard form* of a linear program is

$$(ELP) \qquad \begin{cases} \text{Minimize} \quad b_1x_1 + b_2x_2 + \cdots + b_nx_n \\ \text{subject to} \\ \quad a_{11}x_1 + \cdots + a_{1n}x_n = c_1, \\ \quad \vdots \qquad\qquad\qquad \vdots \\ \quad a_{m1}x_1 + \cdots + a_{mn}x_n = c_m, \\ \text{and } x_1 \geq 0, \quad x_2 \geq 0, \ldots, \quad x_n \geq 0. \end{cases}$$

By possibly multiplying equality constraints by -1, we can arrange to have $c_i \geq 0$ for $i = 1, \ldots, m$. We will always assume that this is the case.

(a) *Matrix Form.* If

$$A = \begin{pmatrix} a_{11} & \cdots & a_{1n} \\ \vdots & & \vdots \\ a_{m1} & \cdots & a_{mn} \end{pmatrix}, \qquad \mathbf{x} = \begin{pmatrix} x_1 \\ \vdots \\ x_n \end{pmatrix}, \qquad \mathbf{b} = \begin{pmatrix} b_1 \\ \vdots \\ b_n \end{pmatrix}, \qquad \mathbf{c} = \begin{pmatrix} c_1 \\ \vdots \\ c_m \end{pmatrix},$$

we can rewrite (ELP) as follows:

$$(ELP)\quad \text{Minimize}\quad \mathbf{b}\cdot\mathbf{x}\quad \text{subject to}\quad A\mathbf{x} = \mathbf{c},\quad \mathbf{x} \geq \mathbf{0}.$$

We will assume that $n \geq m$ and that the rank of A is m.

(b) *Conversion of Linear Programs to Equality Standard Form.* Other forms of linear programs can be converted to (ELP) as follows:

(a) Inequality constraints such as $a_{i1}x_1 + \cdots + a_{in}x_n \leq c_i$ with $c_i \geq 0$ can be replaced by $a_{i1}x_1 + \cdots + a_{in}x_n + x_{n+i} = c_i$ where $x_{n+i} \geq 0$ is a new variable called a *slack variable.*
(b) Inequality constraints such as $a_{i1}x_1 + \cdots + a_{in}x_n \geq c_i$ with $c_i \geq 0$ can be replaced by $a_{i1}x_1 + \cdots + a_{in}x_n - x_{n+i} = c_i$ where $x_{n+i} \geq 0$ is a new variable called a *surplus variable.*
(c) Maximization problems for $b_1x_1 + \cdots + b_nx_n$ can be replaced by minimization problems for $(-b_1)x_1 + (-b_2)x_2 + \cdots + (-b_n)x_n$.

Using these techniques, we can convert any linear program to (ELP).

(c) *Basic Solutions.* Since A has rank m, A contains at least one $m \times m$-submatrix B of rank m. For simplicity, assume that B consists of the first m columns of A. Then there is a unique vector $\mathbf{x}_B \in R^m$ such that $B\mathbf{x}_B = \mathbf{b}$. If we augment $\mathbf{x}_B$ with $n - m$ zero entries, we obtain a vector $\mathbf{x} \in R^n$ such that $A\mathbf{x} = \mathbf{b}$. Such a solution of $A\mathbf{x} = \mathbf{b}$ is called a *basic solution.* In addition, if $\mathbf{x}$ satisfies the nonnegativity constraints, $\mathbf{x}$ is called a *basic feasible solution.*

(d) *The Geometry of* (ELP). Each of the equality constraints defines a hyperplane in R^n (that is, a translate of an $(n-1)$-dimensional subspace of R^n) and each of the nonnegativity constraints defines a half-space in R^n (see (2.1.2)(d)). Consequently, the set F of feasible points for (ELP) is the convex set obtained by intersecting all of these hyperplanes and half-spaces. Of course, F may be empty, in which case there are no feasible points for (ELP) and hence no solution to (ELP).

If F is not empty, then a point $\mathbf{e} \in F$ is an *extreme point of* F if $\mathbf{e} = \lambda\mathbf{x} + (1-\lambda)\mathbf{y}$ for $\mathbf{x}, \mathbf{y} \in F$ and $0 < \lambda < 1$ imply that $\mathbf{e} = \mathbf{x} = \mathbf{y}$; that is, $\mathbf{e}$ is not an any open interval $(\mathbf{x}, \mathbf{y}) = \{\lambda\mathbf{x} + (1-\lambda)\mathbf{y}: 0 < \lambda < 1\}$ joining two distinct points $\mathbf{x}, \mathbf{y}$ of F. It can be shown that the *extreme points of* F *are precisely the basic feasible solutions of* (ELP).

The level sets of the objective function $\mathbf{b}\cdot\mathbf{x}$ for (ELP) are hyperplanes in R^n which may or may not intersect the set F of feasible points for (ELP). Solving (ELP) amounts to finding an $\mathbf{x}^{(0)}$ in F that is on a level set for the objective function and that satisfies $\mathbf{b}\cdot\mathbf{x} \geq \mathbf{b}\cdot\mathbf{x}^{(0)}$ for all $\mathbf{x} \in F$, that is, F is "supported below" by the level set through $\mathbf{x}^{(0)}$. Such a supporting hyperplane for F may contain many points of F but the important thing is that it must contain at least one extreme point of F. In view of the identification of extreme points of F and basic feasible solutions of (ELP), we obtain the so-called *Fundamental Theorem of Linear Programming*:

Given a linear program (ELP), *it is true that*:

(a) *if there is a feasible solution of the constraints, there is also a basic feasible solution of the constraints*;

(b) *if there is a feasible solution of the constraints that minimizes the objective function, there is also a basic feasible solution that minimizes this function.*

Thus, if (ELP) has a minimizer, one can be found among the finitely many extreme points of the convex set F of feasible points.

The *Simplex Method* provides an algorithm for deciding if a minimizer for (ELP) exists and for finding it when it does exist. It begins with a basic feasible solution (that is, an extreme point of F) and it decides if it is a minimizer. If it is not, it moves to an adjacent extreme point where the objective function value is smaller and repeats the procedure until it finds a minimizer or determines that no minimizer exists.

(e) *The Simplex Table.* Given a basic feasible solution to $A\mathbf{x} = \mathbf{c}$, $\mathbf{x} \geq 0$, let $\mathbf{a}^{(i_1)}, \ldots, \mathbf{a}^{(i_m)}$ be the columns of A in the $m \times m$-submatrix B of A corresponding to the given basic feasible solution. Let $\mathbf{a}^{(j)}$ for $j \neq i_k$ denote the vectors of coefficients needed to express the jth column of A as a linear combination of $\mathbf{a}^{(i_1)}, \ldots, \mathbf{a}^{(i_m)}$. The simplex table for (ELP) and the given basic feasible solution has the following form:

			b_1	$\cdots$	b_n
		$\mathbf{c}$	$\mathbf{a}^{(1)}$	$\cdots$	$\mathbf{a}^{(n)}$
	$\mathbf{z} - \mathbf{b}$				
b_{i_1}	$\mathbf{a}^{(i_1)}$				
$\vdots$	$\vdots$		$\vdots$	$\vdots$ $\vdots$	$\vdots$
b_{i_m}	$\mathbf{a}^{(i_m)}$				

← coefficient of the objective function

Table entries t_{ij}, $i = 0, \ldots, m$; $j = 0, \ldots, n$,

$$z_j = \sum_{k=1}^{m} b_{i_k} t_{i_k j}.$$

↑ Columns of B

↑ coefficient of objective function for variables corresponding to B

$$t_{0j} = \begin{cases} z_0 & \text{if } j = 0, \\ z_j - b_j & \text{if } j = 1, \ldots, n. \end{cases}$$

(f) *Pivoting Rules*

(1) Find any positive entry in the $\mathbf{z} - \mathbf{b}$ row and an $a^{(j)}$ column (that is, find $t_{0j} > 0$ for $j \neq 0$). Call the corresponding column the *pivot column*. Suppose that $a^{(j)}$ is the selected pivot column.

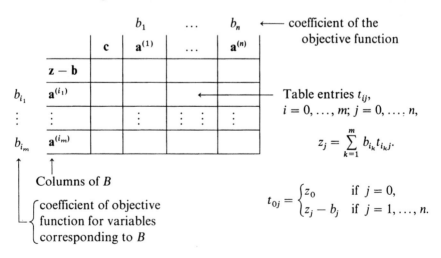

(2) Form all ratio t_{i0}/t_{ij} for which $t_{ij} > 0$, $i \neq 0$, and let the minimum of these ratios be t_{p0}/t_{pj}. Then t_{pj} is called the *pivot element*.

(g) *Constructing the Next Table.* (Just like Gaussian Elimination!)

(1) Divide the numbers in the row containing the pivot element by the pivot element to produce a new row with $+1$ in the pivot position.
(2) Add multiples of the new pivot row to the other rows to make all other entries in the pivot column equal to 0.
(3) Change the labels $a^{(i_p)}$ and b_{i_p} on the left-hand side of the pivot row to the labels $a^{(j)}$ and b_j at the top of the pivot column.
(4) If possible, select a new pivot element in this table in accordance with the Pivoting Rules.

(h) *Stopping Condition.* There are basic conditions in the simplex table that can cause the iteration described above to stop.

(1) There are no positive t_{0j} for $j \neq 0$. In this case a minimizing basic feasible solution has been reached. It can be obtained by setting the variables corresponding to the left-hand side labels of the table equal to the corresponding entries in the **c** column, and setting the remaining variables equal to zero.
(2) There are positive t_{0j} for $j \neq 0$ but no corresponding t_{ij} for $i \neq 0$ is positive. In this case, the objective function has no minimum.

In case (1), the minimum value of the objective function is the upper left-hand entry in the simplex table for which the iteration stopped.

(i) *Obtaining the Initial Simplex Table.* If all of the constraints in the given problem are inequality constraints, then m new slack or surplus variables must be introduced to put the problem in the form (*ELP*). Then the columns of A corresponding to these new variables form the identity matrix I. If we set these new variables equal to the corresponding c_i and set the remaining (old) variables equal to zero, we have an initial basic feasible solution. The simplex table for this solution consists of $t_{ij} = a_{ij}$ for $i = 1, \ldots, m$; $j = 1, \ldots, n$, $t_{00} = 0$, $t_{0j} = b_j$.

If some or all of the given constraints are equality constraints, we can obtain an initial basic feasible solution and the corresponding initial simplex table by applying the Simplex Method to the problem

$$(*)\quad \begin{cases} \text{Minimize} \quad y_1 + y_2 + \cdots + y_m \quad \text{subject to} \\ a_{11}\, x_1 + \cdots + a_{1n} x_n + y_1 \qquad\qquad = c_2, \\ a_{21}\, x_1 + \cdots + a_{2n} x_n \qquad + y_2 \qquad = c_2, \\ \quad \vdots \qquad\qquad \vdots \qquad\qquad \ddots \qquad \vdots \\ a_{m1}\, x_1 + \cdots + a_{mn} x_n \qquad\qquad + y_m = c_m. \end{cases}$$

The variables y_i introduced in this way are called *artificial variables*. It can be shown that (ELP) has a feasible solution if and only if the program $(*)$ has minimum value equal to zero. If $(*)$ has minimum value equal to zero, the values of x_i in the minimizing basic solution for $(*)$ are a basic feasible solution for (ELP), and the part of the final simplex table for $(*)$ corresponding to the x_i's is the initial simplex table for (ELP) corresponding to this solution.

We are now ready to describe a modification of the simplex algorithm, called *Wolfe's Algorithm*, for solving the quadratic programming problem (QP). The algorithm is based on the initial simplex table for the linear programming problem

$$\begin{cases} \text{Minimize} \quad y_1 + y_2 + \cdots + y_{m-k+n} \\ \text{subject to the constraints} \\ \displaystyle\sum_{j=1}^{n} a_{ij}x_j + x_{n+i} = b_i, \qquad i = 1, \ldots, k, \\ \displaystyle\sum_{j=1}^{n} a_{ij}x_j + y_i = b_i, \qquad i = k+1, \ldots, m, \\ \displaystyle\sum_{j=1}^{n} q_{ij}x_j + \sum_{j=1}^{n} a_{ji}\mu_j - v_i + y_{m-k+i} = -c_i, \qquad i = 1, 2, \ldots, n. \end{cases}$$

The original variables $x_1, \ldots, x_n$, the slack variables $x_{n+1}, \ldots, x_{n+k}$, the Karush–Kuhn–Tucker multipliers $\mu_1, \ldots, \mu_k$, $v_1, \ldots, v_n$, and the artificial variables $y_1, \ldots, y_m$ are all nonnegative variables, but the variables $\mu_{k+1}, \ldots, \mu_m$ are unrestricted since they correspond to equality constraints. Consequently, since the simplex algorithm is based on nonnegativity of the variables, we write the latter μ's as differences of nonnegative variables

$$\mu_i = \mu_i^+ - \mu_i^-, \qquad \mu_i^+ \geq 0, \qquad \mu_i^- \geq 0, \qquad i = k+1, \ldots, m.$$

Once the initial simplex table is established, we proceed to apply the simplex pivoting rules with the following modifications which result from restrictions (III) and (IV) on (QP):

(III)′ μ_i and x_{n+i} must never be allowed to appear together as nonzero basic variables for $i = 1, \ldots, k$.

(IV)′ x_j and v_j must never be allowed to appear together as nonzero basic variables for $j = 1, \ldots, n$.

The Wolfe Algorithm terminates when no further pivots are possible according to the simplex pivoting rules with the additional restrictions (III)′, (IV)′. The minimizer $\mathbf{x}^*$ for (QP) is then given by the current values for $x_1, \ldots, x_n$ in the final simplex table.

Before we proceed to a concrete example illustrating the application of the Wolfe Algorithm, we will describe how the above procedure is modified if inequality constraints of the form

$$a_{i1}x_1 + \cdots + a_{in}x_n \geq b_i \quad \text{where } b_i > 0$$

are present in a given problem. Since the simplex algorithm requires that $b_i \geq 0$, the trivial change

$$-a_{i1}x_1 - \cdots - a_{in}x_n \leq -b_i$$

will not suffice to incorporate such an inequality into the Wolfe Algorithm format. Instead, we simply use a *surplus* variable x_{n+i} to rewrite the given inequality constraint as an equality constraint

$$a_{i1}x_1 + \cdots + a_{in}x_n - x_{n+i} = b_i.$$

The effect of this change on the construction of the initial simplex table for Wolfe's Algorithm is simply that the coefficient of x_{n+i} is -1 instead of $+1$, that the corresponding Karush–Kuhn–Tucker multiplier μ_i is replaced by $\mu_i' = -\mu_i$, and that the corresponding artificial variable y_i is not necessary.

(7.3.3) Example. We will apply Wolfe's Algorithm to minimize the quadratic function

$$f(x_1, x_2) = x_1^2 - x_1x_2 + 2x_2^2 - x_1 - x_2$$

subject to the constraints

$$x_1 - x_2 \geq 3, \qquad x_1 + x_2 = 4, \qquad x_1 \geq 0, \qquad x_2 \geq 0.$$

In this case, we introduce a surplus variable x_3 to convert the first inequality constraint to an equality constraint. We then introduce artificial variables y_1, y_2, y_3, y_4, the Kuhn–Tucker multipliers μ_1 for the inequality constraint and $\mu_2 = \mu_2^+ - \mu_2^-$ for the equality constraint, and ν_1, ν_2 for the nonnegativity constraints on x_1, x_2, and construct the initial simplex table for the program

$$\begin{cases} \text{Minimize} \quad y_1 + y_2 + y_3 + y_4 \\ \text{subject to the conditions} \\ x_1 - x_2 - x_3 \qquad\qquad\qquad\qquad\qquad + y_1 \qquad\qquad\qquad = 3, \\ x_1 + x_2 \qquad\qquad\qquad\qquad\qquad\qquad\qquad + y_2 \qquad\quad = 4, \\ 2x_1 - x_2 \qquad - \mu_1 + \mu_2^+ - \mu_2^- - \nu_1 \qquad\qquad + y_3 \qquad = 1, \\ -x_1 + 4x_2 \qquad + \mu_1 + \mu_2^+ - \mu_2^- \qquad - \nu_2 \qquad\qquad + y_4 = 1. \end{cases}$$

Note that the pivot restrictions (III)′, (IV)′ imply that the pairs (x_1, ν_1), (x_2, ν_2), (μ_1, x_3) cannot be nonzero basic variables at the same time.

The following sequence of simplex tables then solves the problem:

		0	0	0	0	0	0	0	0	1	1	1	1
	b	x_1	x_2	x_3	μ_1	μ_2^+	μ_2^-	ν_1	ν_2	y_1	y_2	y_3	y_4
z − c	9	3	3	−1	0	2	−2	−1	−1	0	0	0	0
y_1	3	1	−1	−1	0	0	0	0	0	1	0	0	0
y_2	4	1	1	0	0	0	0	0	0	0	1	0	0
y_3	1	(2)	−1	0	−1	1	−1	−1	0	0	0	1	0
y_4	1	−1	4	0	1	1	1	0	−1	0	0	0	1
z − c	$\frac{15}{2}$	0	$\frac{9}{2}$	−1	$\frac{3}{2}$	$\frac{1}{2}$	$-\frac{1}{2}$	$\frac{1}{2}$	−1	0	0	$-\frac{3}{2}$	0
y_1	$\frac{5}{2}$	0	$-\frac{1}{2}$	−1	$\frac{1}{2}$	$-\frac{1}{2}$	$\frac{1}{2}$	$\frac{1}{2}$	0	1	0	$-\frac{1}{2}$	0
y_2	$\frac{7}{2}$	0	$\frac{3}{2}$	0	$\frac{1}{2}$	$-\frac{1}{2}$	$\frac{1}{2}$	$\frac{1}{2}$	0	0	1	$-\frac{1}{2}$	0
x_1	$\frac{1}{2}$	1	$-\frac{1}{2}$	0	$-\frac{1}{2}$	$\frac{1}{2}$	$-\frac{1}{2}$	$-\frac{1}{2}$	0	0	0	$\frac{1}{2}$	0
y_4	$\frac{3}{2}$	0	$\frac{7}{2}$	0	$\frac{1}{2}$	$\frac{3}{2}$	$-\frac{3}{2}$	$-\frac{1}{2}$	−1	0	0	$\frac{1}{2}$	1
z − c	3	0	−6	−1	0	−4	4	2	2	0	0	−3	−3
y_1	1	0	−4	−1	0	−2	2	1	1	1	0	−1	−1
y_2	2	0	−2	0	0	−2	2	1	1	0	1	−1	−1
x_1	2	1	3	0	0	2	−2	−1	−1	0	0	1	1
μ_1	3	0	7	0	1	3	−3	−1	−2	0	0	1	1
z − c	1	0	2	1	0	0	0	0	0	−2	0	−1	−1
μ_2^-	$\frac{1}{2}$	0	−2	$-\frac{1}{2}$	0	−1	1	$\frac{1}{2}$	$\frac{1}{2}$	$\frac{1}{2}$	0	$-\frac{1}{2}$	$-\frac{1}{2}$
y_2	1	0	2	1	0	0	0	0	0	−1	1	0	0
x_1	3	1	−1	−1	0	0	0	0	0	1	0	0	0
μ_1	$\frac{9}{2}$	0	1	$-\frac{3}{2}$	1	0	0	$\frac{1}{2}$	$-\frac{1}{2}$	$\frac{3}{2}$	0	$-\frac{1}{2}$	$-\frac{1}{2}$
z − c	0	0	0	0	0	0	0	0	0	−1	−1	−1	−1
μ_2^-	$\frac{3}{2}$	0	0	$\frac{1}{2}$	0	−1	1	$\frac{1}{2}$	$\frac{1}{2}$	$-\frac{1}{2}$	1	$-\frac{1}{2}$	$-\frac{1}{2}$
x_2	$\frac{1}{2}$	0	1	$\frac{1}{2}$	0	0	0	0	0	$-\frac{1}{2}$	$\frac{1}{2}$	0	0
x_1	$\frac{7}{2}$	1	0	$-\frac{1}{2}$	0	0	0	0	0	$\frac{1}{2}$	$\frac{1}{2}$	0	0
μ_1	4	0	0	−2	1	0	0	$\frac{1}{2}$	$-\frac{1}{2}$	2	$-\frac{1}{2}$	$-\frac{1}{2}$	$-\frac{1}{2}$

STOP: $x_1^* = \frac{7}{2}$, $x_2^* = \frac{1}{2}$, $\mu_1^* = 4$, $(\mu_2^-)^* = \frac{3}{2}$.

Minimum value $= +7 \leftarrow$ (Substitute x_1^*, x_2^* into original objective function).

Exercises

1. A cylindrical tin can with a bottom and a lid is required to have a volume of 100 cubic inches. Find the dimensions of the can that will require the least metal by:
   (a) expressing the surface area of the can in terms of its radius and height, eliminating one of these variables and minimizing the resulting function of one variable;
   (b) using Lagrange multipliers;
   (c) using geometric programming.

2. Consider the problem of finding the point(s) on the parabola $x^2 - 4y = 0$ that is nearest the point $(0, 1)$.
   (a) Use Lagrange multipliers to solve this problem.
   (b) Attempt to solve the problem by using the equation $x^2 - 4y = 0$ to eliminate x from
   $$f(x, y) = x^2 + (y - 1)^2,$$
   and then minimizing the resulting function of y. What happens?
   (c) Show that the problem can be solved if we eliminate y instead of x in the function $f(x, y)$ of part (b).

3. Maximize
$$f(x, y, z) = (x - 1)^2 + (y - 2)^2 + (z - 3)^2$$
subject to
$$g(x, y, z) = x^2 + y^2 + z^2 - 1 = 0$$
by using Lagrange multipliers.

4. Determine all maxima and minima of
$$f(x_1, x_2, x_3) = \frac{1}{x_1^2 + x_2^2 + x_3^2}$$
subject to
$$h_2(x_1, x_2, x_3) = 1 - x_1^2 - 2x_2^2 - 3x_3^2 = 0,$$
$$h_2(x_1, x_2, x_3) = x_1 + x_2 + x_3 = 0.$$

5. Determine all maxima and minima of
$$f(x, y, z) = xz + y^2$$
on the sphere $x^2 + y^2 + z^2 = 4$.

6. Apply Lagrange multipliers to derive a formula for the distance from a point $P_0(x_0, y_0, z_0)$ to a plane
$$Ax + By + Cz + D = 0$$
not containing the point P_0.

7. Show that the problem
$$\begin{aligned} &\text{Minimize} \quad f(x, y) = x^2 + y^2 \\ &\text{subject to} \quad (x - 2)^3 - y^2 = 0 \end{aligned}$$
admits no Lagrange multipliers and explain why. Solve this problem graphically.

8. Let $f(x, y, z)$ be a coercive function with continuous first partial derivatives on R^3. Suppose $f(x, y, z) = 1$ defines a surface in R^3 and suppose $f(0, 0, 0) \neq 1$. Show that the vector from the origin to (x^*, y^*, z^*) of minimum norm on the surface $f(x, y, z) = 1$ is perpendicular to the surface at the point (x^*, y^*, z^*).

9. Let $f(\mathbf{x})$ and $g(\mathbf{x})$ be coercive functions with continuous first partial derivatives on R^n. Suppose the equations $f(\mathbf{x}) = 1$ and $g(\mathbf{x}) = 1$ define nonintersecting surfaces in

R^n. Show that the vector $\mathbf{x}^* - \mathbf{y}^*$ of minimum norm satisfying $f(\mathbf{x}^*) = g(\mathbf{y}^*) = 1$ is perpendicular to both surfaces.

10. Find the point on the ellipse

$$5x^2 - 6xy + 5y^2 = 4$$

for which the tangent line is at a maximum distance from the origin.

11. Prove that of all triangles with a fixed perimeter P, the equilateral triangle has the largest area. (Hint: The area of a triangle with sides a, b, and c is given by

$$A = \sqrt{s(s-a)(s-b)(s-c)},$$

where $(a + b + c)/2 = s = P/2$.)

12. Let A be a positive definite 3×3-matrix. Let $\mathbf{x}^*$ be the vector of largest norm on the ellipsoid

$$\mathbf{x} \cdot A\mathbf{x} = 1.$$

Show that $\mathbf{x}^*$ is an eigenvector of A and relate $\|\mathbf{x}^*\|$ to the corresponding eigenvalue of A.

13. Use Lagrange multipliers to prove the Cauchy–Schwarz Inequality.

14. Use Lagrange multipliers to prove that if $x_1, x_2, \ldots, x_n$ are positive real numbers, then

$$\prod_{i=1}^{n} x_i \leq \frac{1}{n} \sum_{i=1}^{n} x_i^n.$$

15. A water main consists of two sections of pipe of fixed lengths L_1 and L_2 carrying fixed amounts of Q_1 and Q_2 gallons per second each. For a given total loss of head h the diameters d_1 and d_2 of the pipe sections result in a cost

$$L_1(a + bd_1) + L_2(a + bd_2)$$

if

$$h = C\frac{L_1 Q_1^2}{d_1^5} + C\frac{L_2 Q_2^2}{d_2^5},$$

where C is a positive constant. Find the ratio d_2/d_1 for the diameters that minimizes the cost for a given total loss of head h.

(Ans.: $d_2/d_1 = (Q_2/Q_1)^{1/3}$.)

16. The output of a manufacturing operation is a quantity Q which is a function $Q = Q(x, y)$ where $x =$ capital equipment and $y =$ hours of labor. Suppose the price of labor is p and the price of investment is q in dollars and the operation is to spend b dollars. For optimum production, we want to minimize Q subject to $qx + py = b$.

Show that at the optimum, we have

$$\frac{\partial Q}{\partial K} = \frac{\partial Q}{\partial l},$$

where $K = qx$ and $l = py$. Thus, at the optimum, the marginal change in output per dollar's worth of additional capital equipment is equal to the marginal change in output per dollar's worth of additional labor.

17. The following exercises are related to multistage rocket design as discussed in Example (7.2.5):
    (a) At an altitúde of 100 miles above the earth's surface, the escape velocity (that is, the minimum speed required for an object moving directly away from the earth to escape the gravitational attraction of the earth) is approximately 24,700 miles per hour. A three-stage rocket vehicle is to be used to launch a deep-space probe of mass P by achieving the required escape velocity at an altitude of 100 miles. Compute the minimum total mass M_0 of a rocket vehicle capable of this mission and the corresponding total masses of the three rocket stages, given that all three stages have structural factors of $S = 0.2$ and constant engine exhaust speeds of $c = 6000$ miles per hour.
    (b) In (7.2.5), we computed the minimum total mass of an n-stage rocket vehicle capable of inserting a payload of mass P in a circular orbit 100 miles above the surface of the earth for $n = 2, 3, 4, 5$. Discuss what happens to the value of this minimum total mass as the number of stages approaches infinity.

18. Consider the quadratic program

$$\begin{aligned} &\text{Minimize} \quad f(x_1, x_2) = 5x_1^2 - 2x_1x_2 + 5x_2^2 \\ &\text{subject to} \quad x_1 + x_2 = 1; \qquad x_1 \geq 0, \quad x_2 \geq 0. \end{aligned}$$

Compare the solutions of this problem obtained by:
    (a) using the equality constraint to eliminate x_2 in $f(x_1, x_2)$ and minimizing the resulting function of x_1 on the interval $0 \leq x_1 \leq 1$;
    (b) viewing the program as a minimum Q-norm problem as in Example (4.4.4);
    (c) using Wolfe's Algorithm.

19. Use Wolfe's Algorithm to solve the following quadratic program:

$$\begin{aligned} &\text{Minimize} \quad f(x_1, x_2) = 2x_1^2 + x_2^2 - 2x_1x_2 - 5x_1 - 2x_2 \\ &\text{subject to} \quad x_1 \geq 0, \quad x_2 \geq 0, \\ &\qquad\qquad 3x_1 + 2x_2 \leq 20, \\ &\qquad\qquad 5x_1 - 3x_2 \geq -4. \end{aligned}$$

20. Suppose $X = (x_{ij})$ is an $n \times n$-matrix and suppose that $\mathbf{x}^{(i)}$ is the ith row vector of X. Prove Hadamard's Inequality:

$$\det(X) \leq \prod_{i=1}^{n} \|\mathbf{x}^{(i)}\|$$

by using the Lagrange Multiplier Method to show that if $\det(X)$, regarded as a function of its n^2 entries x_{ij}, has a maximum value subject to the constraints

$$\|\mathbf{x}^{(i)}\|^2 = x_{i1}^2 + \cdots + x_{in}^2 = d_i^2; \qquad i = 1, 2, \ldots, n,$$

then

$$\mathbf{x}^{(i)} = d_i\mathbf{e}^{(i)},$$

where $\mathbf{e}^{(i)}$ is the ith-unit vector in R^n.

Index

Undergraduate Texts in Mathematics

(continued from page ii)

Halmos: Naive Set Theory.
Hämmerlin/Hoffmann: Numerical Mathematics.
Readings in Mathematics.
Harris/Hirst/Mossinghoff: Combinatorics and Graph Theory.
Hartshorne: Geometry: Euclid and Beyond.
Hijab: Introduction to Calculus and Classical Analysis.
Hilton/Holton/Pedersen: Mathematical Reflections: In a Room with Many Mirrors.
Iooss/Joseph: Elementary Stability and Bifurcation Theory. Second edition.
Isaac: The Pleasures of Probability.
Readings in Mathematics.
James: Topological and Uniform Spaces.
Jänich: Linear Algebra.
Jänich: Topology.
Jänich: Vector Analysis.
Kemeny/Snell: Finite Markov Chains.
Kinsey: Topology of Surfaces.
Klambauer: Aspects of Calculus.
Lang: A First Course in Calculus. Fifth edition.
Lang: Calculus of Several Variables. Third edition.
Lang: Introduction to Linear Algebra. Second edition.
Lang: Linear Algebra. Third edition.
Lang: Undergraduate Algebra. Second edition.
Lang: Undergraduate Analysis.
Lax/Burstein/Lax: Calculus with Applications and Computing. Volume 1.
LeCuyer: College Mathematics with APL.
Lidl/Pilz: Applied Abstract Algebra. Second edition.
Logan: Applied Partial Differential Equations.
Macki-Strauss: Introduction to Optimal Control Theory.
Malitz: Introduction to Mathematical Logic.
Marsden/Weinstein: Calculus I, II, III. Second edition.
Martin: The Foundations of Geometry and the Non-Euclidean Plane.
Martin: Geometric Constructions.
Martin: Transformation Geometry: An Introduction to Symmetry.
Millman/Parker: Geometry: A Metric Approach with Models. Second edition.
Moschovakis: Notes on Set Theory.
Owen: A First Course in the Mathematical Foundations of Thermodynamics.
Palka: An Introduction to Complex Function Theory.
Pedrick: A First Course in Analysis.
Peressini/Sullivan/Uhl: The Mathematics of Nonlinear Programming.
Prenowitz/Jantosciak: Join Geometries.
Priestley: Calculus: A Liberal Art. Second edition.
Protter/Morrey: A First Course in Real Analysis. Second edition.
Protter/Morrey: Intermediate Calculus. Second edition.
Roman: An Introduction to Coding and Information Theory.
Ross: Elementary Analysis: The Theory of Calculus.
Samuel: Projective Geometry.
Readings in Mathematics.
Scharlau/Opolka: From Fermat to Minkowski.
Schiff: The Laplace Transform: Theory and Applications.
Sethuraman: Rings, Fields, and Vector Spaces: An Approach to Geometric Constructability.
Sigler: Algebra.
Silverman/Tate: Rational Points on Elliptic Curves.
Simmonds: A Brief on Tensor Analysis. Second edition.

Singer: Geometry: Plane and Fancy.
Singer/Thorpe: Lecture Notes on Elementary Topology and Geometry.
Smith: Linear Algebra. Third edition.
Smith: Primer of Modern Analysis. Second edition.
Stanton/White: Constructive Combinatorics.
Stillwell: Elements of Algebra: Geometry, Numbers, Equations.
Stillwell: Mathematics and Its History.
Stillwell: Numbers and Geometry. *Readings in Mathematics.*
Strayer: Linear Programming and Its Applications.
Toth: Glimpses of Algebra and Geometry. *Readings in Mathematics.*
Troutman: Variational Calculus and Optimal Control. Second edition.
Valenza: Linear Algebra: An Introduction to Abstract Mathematics.
Whyburn/Duda: Dynamic Topology.
Wilson: Much Ado About Calculus.

CPSIA information can be obtained at www.ICGtesting.com
Printed in the USA
LVOW090515100112

263154LV00002B/8/A

9 780387 966144